Grid-scale Energy Storage Systems and Applications

Grid-scale Energy Storage Systems and Applications

Edited by

Fu-Bao Wu
Bo Yang
Ji-Lei Ye

ELSEVIER

ACADEMIC PRESS
An imprint of Elsevier

Academic Press is an imprint of Elsevier
125 London Wall, London EC2Y 5AS, United Kingdom
525 B Street, Suite 1650, San Diego, CA 92101, United States
50 Hampshire Street, 5th Floor, Cambridge, MA 02139, United States
The Boulevard, Langford Lane, Kidlington, Oxford OX5 1GB, United Kingdom

Library of Congress Cataloging-in-Publication Data
A catalog record for this book is available from the Library of Congress

British Library Cataloguing-in-Publication Data
A catalogue record for this book is available from the British Library

ISBN: 978-0-12-815292-8

For information on all Academic Press publications visit our website at https://www.elsevier.com/books-and-journals

Publisher: Joe Hayton
Acquisition Editor: Lisa Reading
Editorial Project Manager: Naomi Robertson
Production Project Manager: Sruthi Satheesh
Cover Designer: Miles Hitchen

Typeset by TNQ Technologies

Contents

Foreword

Energy storage, a key supportive technology for the development of smart power grids, will play an important role in the grid access of large-scale renewable energy, distributed power generation, microgrids, electric vehicles, and other application fields. With continuous and rapid economic growth, China's power system faces the problems of continuous increase of power consumption loads, gradual greater peak–valley difference, insufficient peak modulation capacity, and unreasonable power source structure. The combination of energy storage with traditional power generation facilities and new energy power generation is an important means for effectively solving the problems of energy and the environment.

In the Energy Development Planning during the 12th Five-Year Plan Period issued in 2013, the National Energy Administration said, "striving to break through key technologies in energy saving, low-carbon, energy storage and smart control fields; coordinating the supportive grid and wind power development and building reasonable supportive energy storage facilities in terms of renewable energy; and focusing on breakthroughs in large–capacity storage in terms of energy industry's sci-tech equipment are the key targets for the next five years." In the 2013 Directive Catalog of Strategic Emerging Industries' Key Products and Services, the National Development and Reform Commission also pointed to the development of energy storage technologies. China has made significant achievements in key technology research, establishment of standard systems and executed demonstration projects in the fields of energy storage materials, energy conversion facilities and energy storage integration, and boosted the development of the domestic energy storage industry.

This book, focusing on the rapid development of energy storage technology at home and abroad and combining research and application achievements in energy storage and new energy fields, systematically introduces the development of energy storage technology, technologies for energy storage battery management, technologies for energy storage systems' operation and control, integrated application and economic analysis of energy storage systems, and key technologies and demonstration projects for application of energy storage technologies in new energy power generation connection with grid and microgrids.

The book consists of seven chapters. Chapter 1 was written by Ye Jilei, Chapter 2 by Xue Jinhua, Chapter 3 by Xue Jinhua and Ye Jilie, Chapter 4 by Sang Bingyu, Chapters 5 and 6 by Ye Jilei, and Chapter 7 by Hu Jinhang and Tao Qiong. The book received great support from Wang Wei, Tao Yibin,

Li Guanjun, Feng Xinzhen, Xu Xiaohui, Wang Deshun, Ji Liantao, Yu Bin, Zhao Shanglin, Liu Huan, and other colleagues. The book was completed under the guidance of Wu Fubao and Yang Bo.

The research achievements of many precursors were consulted, and operation data of numerous battery manufacturers and demonstration projects were quoted. I would like to extend my special gratitude to the China Energy Storage Alliance, Shanghai Municipal Electric Power Company, Shanghai Jiao Tong University, CEPRI Power Engineering Department, and Zhejiang Provincial Power Company, etc. DingJie and Zhu Lingzhi, leaders and experts from the CEPRI New Energy Department, also paid great attention to the book. I also extend my sincere gratitude to both of them.

Policies, laws, regulations, and marketing mechanisms related with the energy storage industry are being continuously improved. There is a long way ahead for the development of energy storage technology and its commercial and large-scale application. The book only makes a systematic elaboration on current energy storage technology and key issues of system integration and application. With the rapid development of energy storage materials, manufacturing technology, operation control, integrated application, and other technologies, a lot of new technologies will surely appear and require our close attention and in-depth research.

However, due to the limitation of the writer, shortcomings and imperfections may occur in the book. Your criticism and corrections will be highly appreciated.

The authors

Preface

With voices getting louder for dealing with global climate change and the increasingly severe situation in energy shortage and energy supply security, the role of new energy, such as wind energy, solar energy, biomass energy, and ocean energy used by various countries, has been continuously improved, thanks to the characteristics of their being clean, safe, and renewable. Among them, wind energy and solar energy, featuring relatively lower costs, proven technology, and higher reliability, have seen robust development and have been playing an important role in energy supply in recent years. China promulgated the Renewable Energy Law of the People's Republic of China in 2006. By means of franchise bidding, region-based on-grid power prices for wind power and PV power, Photo voltaic power price subsidies, and other measures, the authorities have gradually set up a subsidy and incentive system to support the development and exploitation of new energy. So far, China's wind power has entered a rapid development stage, achieved above 100% growth for five consecutive years, overtaken the United States in installed capacity, and become the world's number one wind power country since June 2012. By the end of 2013, China's installed capacity for PV power generation reached 19.42 million kW, to become the number two PV country, after Germany.

According to state plans, China's wind power installed capacity will reach 200 million kW by 2020. The wind power in the Three North Region, i.e., North China, Northeast China, and Northwest China, and the coastal regions of Jiangsu and Shandong will focus on large-scale centralized exploitation and set up nine 10,000,000-kW-level wind power bases with an installed capacity of 70% of the national total. In the central region, scattered exploitation will be dominant. PV power generation's installed capacity will achieve 100 million kW by 2020. Unlike wind power exploitation, China's PV power generation applies the modes of "large-scale exploitation and centralization for remote transmission" and "scattered development for local consumption." Large-size ground PV power generation stations are suitable for Northwest China, North China, and other regions with abundant solar energy, while PV distribution facilities are mainly set up in developed places in central and eastern China. In the upcoming years, China's new energy industry will still maintain rapid growth.

However, along with rapid growth, China's new energy industry also faces a range of urgent problems. Among them, grid integration of new energy power has become a focus. It includes the problem of new energy's grid access, the safety and stability of power systems containing large-scale new energy integration, the accommodation of new energy power, and the power distribution networks' technical and management problems arising from the distributed grid integration of new energy.

New energy power grid integration technology has been of wide concern to central and local governments, industry, enterprises, and the public. Since the 11th Five-Year Plan period, the Ministry of Science and Technology has initiated numerous sci-tech projects in the 973 Program, 863 Program, and other sci-tech supportive plans. Numerous enterprises in the industry have carried out extensive research and practices and made significant achievements in new energy grid integration technology and boosted the development of the new energy industry.

As a scientific research institute directly under the State Grid Corporation of China (SGCC), China Electric Power Research Institute (CEPRI) has chaired and participated in numerous sci-tech projects in the 973 Program, 863 Program, other sci-tech supportive plans and SGCC projects; carried out extensive research related to production practices; and conducted technical research and standard development in the modeling, simulation, analysis, planning, and other basic theories and methods, as well as experiments, tests, evaluation, verification, and equipment R&D for new energy grid integration. CEPRI is also engaged in the fields of forecast of wind power and PV power generation, resource evaluation, meteorological technology R&D and application, R&D and application of smart controls, dispatch and operation of new energy grid integration, and R&D and application of distributed power supply, microgrids, and energy storage's integration and operation technologies. Scientific achievements and field application of these R&D efforts have been playing an important role in the rapid development of China's new energy power generation industry.

The *Series on New Energy Power Generation and Grid Connection Technologies,* containing simulation, analysis, and planning technologies for new energy grid integration, forecast technologies for new energy power, smart monitoring technologies for new energy power generation, technologies for distributed power sources and microgrids, power storage technologies, and application, etc., is a summary of CEPRI's explorations and practices in new energy power generation and grid connection as well as considerable field application. It is the first set of works that gives a systematic elaboration on the grid access technical research and practices of large-sized and distributed new energy power from multiple view angles. We hope the publication of the series will attract more experts and scholars at home and abroad and professionals intending to be engaged in the new energy industry to further deepen and carry forward research and application on new energy grid connection technologies and play a more important role in boosting the technical progress of China's new energy power generation industry.

Xiaoxin Zhou
Academician of Chinese Academy of Sciences and honorary president of the China Electric Power Research Institute

Introduction

This book, based on the demands of sustainable development of electric systems and the development trends of energy storage technology, discusses some important research fields that have developed rapidly and attracted wide attention from scientific researchers and technicians. It strives to highlight the academic significance and pragmatic value of this research. The book introduces the development of energy storage technology, technologies for energy storage battery management, technologies for energy storage systems' operation and control, integrated application and economic analysis of an energy storage system, application of energy storage technologies in new energy grid access and power generation, and application of energy storage technologies in microgrids. We hope that the publication of the book will boost the research and application of energy storage technologies, give full play to energy storage systems' important role in smart grids, and fuel the rapid development of the energy storage industry in China.

The book is of great reference value to researchers in relevant fields, technicians of electric power companies, R&D staff of energy storage systems, and technicians in new energy.

Chapter 1

Development of energy storage technology

Chapter outline

Grid-scale Energy Storage Systems and Applications. https://doi.org/10.1016/B978-0-12-815292-8.00001-0

Power production is the basic pillar for the functioning of modern society and one of the main energy sources for the development of all industries. The three major elements for a power system are power generation facilities, power transmission and distribution, and power consumption equipment. Power generators convert mechanical energy into electrical energy that is transmitted and distributed to users via transformers, converters, and electric wires. At the user's end, electric energy is converted to mechanical energy, heat energy, and light energy by means of electric motors, electric ovens, and electric lamps. These power generators, transformers, converters, electric wires, and electric consumers for the production, conversion, transmission, distribution, and consumption of power are connected to form a whole that is called a power system.

With the continuous development of society and the economy, the operation of the power system is changing dramatically. Currently, China's power grid operation confronts the following problems and challenges:

1. Unreasonable electric power production structure. At the end of 2013, China's coal-fired power accounted for 69.13% of the country's total installed capacity, while the installed capacity of well-modulated hydropower accounted for 22.45%, a low proportion in the total. The power source structure with coal-fired power as the main source leads to the insufficient peak modulation capability of the power system.
2. Robust development of intermittent energy development. Intermittent energy is gradually increasing penetration in the power grid, which brings enormous challenges to grid peak modulation, safe and stable operation, and quality of electricity supply.
3. Users' demand for power varies greatly at different times and in different seasons and regions. With the gradual increase in the power consumption gap between the peak time and valley time, the peak modulation problem is more obvious, which often leads to the low voltage at peak load time and operation under low voltage. At the valley load time, power equipment has a shorter operation time and excess capacity.

To solve these problems, energy storage technology can penetrate each link of the power system and play different roles in generation, transmission, transformation, distribution, and consumption. As a flexible part of a smart grid, an energy storage system can effectively realize demand-side management, eliminate peak–valley gaps, improve the operational efficiency of electric equipment, reduce power supply costs, enhance the capability of connecting large-scale renewable energy into the power grid, remove the bottlenecks of energy structure adjustment, play a role as a spare power source, improve the quality of power, and meet different demands of the increasingly developing modern power system.

This chapter introduces the definition of energy storage and the development process of energy storage at home and abroad. It also analyzes the demand for energy storage in consideration of likely problems in the future development of power systems. Energy storage technology's role in various parts of the power system is also summarized in this chapter. In addition, the prospects for application and challenges of energy storage technology in power systems are analyzed to offer reference methods for realizing sustainable development of power grids, solving the contradiction of imbalance between power supply and demand, and improving reliability of power supply.

1.1 Basic concept

Generally speaking, energy storage refers to a range of technologies and measures that convert an energy form into another energy form via certain media or devices, and release energy in a special form when necessary.

In a narrow sense, the storage of electric energy refers to technologies and measures that store electric energy with chemical or physical means and release the energy when necessary. Energy storage in this book refers only to the storage of electrical energy.

Traditionally speaking, the production, transmission, distribution, and consumption of electric energy are simultaneous. In other words, the power produced by power generation plants at any moment must equal the sum of the power used by consumers and grid loss. The application of energy storage technology in power systems may change this mode and solve the problem of the time and space mismatch between electrical energy production and consumption to achieve the objectives of optimizing power resource distribution, improving the quality of electric power, promoting utilization of renewable energy, saving energy, and reducing emissions.

1.2 The development history of energy storage technology

Electric energy storage is not a new technology. As far back as 1786, Italian physicists discovered the existence of bioelectricity. In 1799, Italian scientist Alessandro Giuseppe Antonio Anastasio Volta invented modern batteries. In 1836, batteries were used in communication networks. In the 1880s, New York used lead–acid batteries for power supply to road lamps in its DC power supply system to shut down power generators at night.

With the development of power technology, pumped hydro storage power stations will be gradually used in grid peak modulation. The world's earliest pumped hydro storage power station was the Netala Power Station set up in 1882 in Zurich, Switzerland. It was a seasonal pumped hydro storage power station with a lift of 153 m and power of 515 kW. In 1908, Italy built a pumped hydro storage power station on the Ubyangni Mountain. In 1912, Italy set up Veroni Pumped Hydro Storage Power Station that utilized the 156 m–high fall between two natural lakes and had an installed capacity of 7600 kW. In the 1950s, more than 50 pumped hydro storage power stations were put into operation across the world. From the 1960s, pumped hydro storage power stations had entered a robust development period. The United States, Japan, and Western Europe became the pioneers in the large-scale development of pumped hydro storage power stations. After the 1990s, developed countries slowed the development of pumped hydro storage power stations, while developing countries, with China as a key representative, started the large-scale construction of pumped hydro storage power stations.

Pumped hydro storage (PHS), an energy storage technology most extensively applied in the power system, is mainly used to balance peak and valley loads, regulate frequency and phase, back up in case of emergencies, make a black start, and offer spare energy for the system. By the end of 2011, PHS power stations with installed capacity of above 123,400 MW were put into operation. By the end of 2012, China's PHS units of above 20 GW were in operation. According to statistics, the world's total installed capacity of PHS power stations was 1600 MW in 1950, 3500 MW in 1960, 16,000 MW in 1970, 46,000 MW in 1980, 79,000 MW in 1988, 98,273 MW in 1998, and 127,000 MW in 2010. The growth trend is illustrated in Fig. 1.1. The Bath County Pumped Storage Power Station in the United States has the largest scale with the installed capacity of 2100 MW. The Power Station of San Fiorano in Italy has the highest water head of 1417 m. Japan's Kannagawa PHS Power Station, boasting the highest unit capacity, has an installed capacity of 2820 MW, unit capacity of 470 MW, and the water head of 695 m. PHS power stations can be built according to discretionary capacity and classified into daily, weekly, and seasonal based on regulation periods. The major restricting factors for the wider application of PHS power stations are great difficulty in site selection, long construction period, and high investment.

Compressed air energy storage (CAES) refers to a gas turbine generation plant for peak load regulation. To achieve the same power output, a CAES

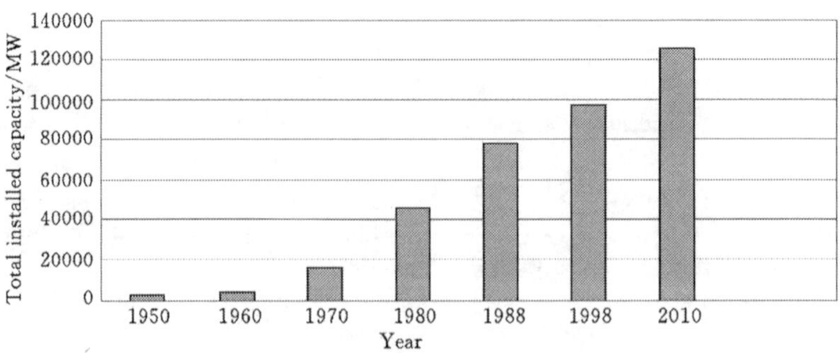

FIGURE 1.1 Total installed capacity of global PHS power stations.

plant's gas consumption is 40% lower than that of conventional gas turbine generators. Conventional gas turbine generators need to consume two-thirds of the input fuel for air compression when generating power, while a CAES plant can compress air in advance with cheap power from the valley load period of the grid, and then release the stored energy and some gas for power generation according to the actual demand. The compressed air is often stored in proper underground mines or caves under lava. To build such caves through lava takes 1.5−2 years.

The first CAES plant was a 290-MW power generation unit built in Hundorf, Germany, in 1978. In 1991, the second commercial CAES plant of 110 MW was set up in McIntosh, Alabama. A total of 30 months and $65 million were spent for the construction of the plant, which could be connected to the grid within 14 min. The third commercial CAES plant has the highest capacity of its kind in the world. Built in Norton, Ohio, it has an installed capacity of 2700 MW and consists of nine power generation units. It can store compressed air in a 670 m−deep limestone mine.

In terms of battery energy storage, the lead−acid battery is the oldest and most mature storage battery technology. It is a low-cost general technology for energy storage and can be used in areas such as electric energy quality modulation and UPS. But, the short service life and environmental pollution during manufacturing restrict its large-scale application in power systems. The zinc−bromine battery was developed by Exxon in the early 1970s, and after years of research and development, quite a few sets of zinc−bromine battery energy storage systems with capacity of dozens of kW were set up and demonstrated a net efficiency of up to 75%. In the early 1980s, the University of New South Wales in Australia developed the vanadium redox flow battery (VRB), and further research was done in Australia, Japan, and Canada. Currently, the VRB technology of Canada's VRB Power Systems and Japan's Sumitomo Electric have entered the application stage. In the 1970s, M. S. Whittingham of Exxon created the first lithium battery with titanium sulfide as the positive pole material and lithium as the negative pole material. In 1982, Illinois Institute of Technology discovered that lithium ions can be embedded into graphite, and it is a rapid and reversible process. The first applicable lithium-ion graphite battery was successfully created by Bell Laboratories. In 1991, Sony Corporation launched the first commercial lithium-ion battery. In 1996, it was discovered that the olivine-structured phosphate, for example $LiFePO_4$, has more advantages than traditional positive pole materials, so it has become the dominant positive pole material.

The research and development of electric storage technology has received great attention from the energy, transport, power, and communication industries of all countries, which quickly raised the technical and economic level of the technology. During the past decade, the governments of the United States, Japan, Australia, and European countries have included energy storage technology in their national development plans and successfully demonstration the application of numerous energy storage systems that showed considerable application potential.

By the end of 2010, the total global installed capacity of electric energy storage was 128 GW, accounting for 3.0% of the world's total installed capacity of power generation. According to project statistics of China Energy Storage Alliance, between 2000 and 2013, the world's accumulated installed capacity of energy storage reached 739 MW (excluding PHS, CAES, and heat storage projects). In comparison with 2012, the total installed capacity of global energy storage demonstration projects increased 104 MW, an annual growth rate of 14%. Currently, the international energy storage industry is growing at an annual average growth rate of about 9.0%, far higher than the world's power industry's growth rate of 2.5%. It means that energy storage has become an emerging industry in numerous countries. China has included large-scale energy storage technology in the National Energy Plan during the 12th Five-Year Plan Period and has been actively guiding and promoting the development of the energy storage industry.

1.3 Demands and functions of energy storage technology in power systems

1.3.1 Demand analysis of grid development in energy storage technology

1.3.1.1 Peak–valley gap intensifies demand for energy storage technology

Currently, China is undergoing a rapid industrialization process with robust power demand. In recent years, newly built installed capacity and power generation have kept increasing. Statistics indicate that China's installed capacity for power generation, at the end of 2013, reached 1.247 billion kW, up 9.25% year on year, and replaced the United States to become the world's number one; and it will achieve an estimated 1.5 billion kW by 2020. However, it also brings numerous problems and challenges:

1. The increasing peak–valley power consumption gap in a grid brings more outstanding peak modulation contradiction. In 2012, in the *Rolling Research Report on the 12th Five-Year Plan for the Power Industry,* the China Electricity Council forecast that the whole society's maximum load and power consumption will be approximately 966 million to 1.064 billion kW and 602 million to 661 million kWh at an average growth of 8.6% and 8.9%, respectively. The growth rate of the maximum load will be higher than that of power consumption, which will continuously expand the peak–valley gap.
2. The continuous expansion of grid scale increases the risk of a grid's safe and stable operation. The application of new technologies in power

generation, transmission, and control increasingly improves the grid's complexity and raises higher requirements on the safety and stability of large-scale grids.

3. The development of the economy and society raises higher requirements on the grid's electric energy quality and power supply reliability.

1.3.1.2 Demand for energy storage technology due to the large-scale development of renewable energy

The rapid development of renewable energy has become a common choice of countries to secure energy supplies, protect the environment, and deal with climate change, and an important measure to improve their competitiveness. In January 2006, the *Renewable Energy Law of the People's Republic of China* was formally implemented, giving priority to the development and application of renewable energy. In December 2009, *Amendment of Renewable Energy Law of the People's Republic of China* required the formulation of gross objectives and application of relevant measures to boost the development of renewable energy.

According to a report by World Wind Energy Association, by the end of 2013, the world's accumulated wind power installed capacity reached 318.12 GW, of which China's share was 91.41 GW, accounting for 28.7% of global total. According to the 12th Five-Year Plan for Wind Power Industry, by 2015, China's wind power was to have an installed capacity of 150 million kW and power generation of 190 billion kWh. *China's Wind Power Development Roadmap 2050* issued by the Energy Research Institute of the National Development and Research Commission indicated that China's wind power installed capacity will be 200 million kW by 2020, 400 million kW by 2030, and 1 billion kW by 2050.

According to the statistics of the China Photovoltaic Industry Alliance, by the end of 2013, the world's new photo voltaic installed capacity reached 39 GW, bringing the cumulative installed capacity to 139 GW; and China's new PV installed capacity reached 12.9 GW, bringing the cumulative installed capacity to nearly 20 GW. *Some Opinions of the State Council on Promoting the Healthy Development of PV Industry* issued in July 2013 said that the average annual increase of PV installed capacity would be about 10 GW between 2013 and 2015 and the total above 35 GW by 2015.

The power generation quantity of wind power and PV power varies greatly between day and night due to the characteristics of randomness, intermittence, and fluctuation. With the gradual increase of intermittent energy's proportion in the grid, the peak modulation, operation control, and power supply quality of grid confront enormous challenge. In July 2009, the *Research Report on China's Wind Power Development by National Energy Administration* pointed out that wind power's grid connection was having an influence on the safe operation of the power system, and wind power units mass connected to grids in some regions had an influence on the voltage, frequency, and stability of power due to the rapid increase of the installed capacity of wind power. Therefore, with the increase of renewable energy's installed capacity, it is necessary to equip large-scale energy storage systems to greatly improve the grid's capacity to accommodate renewable energy, increase the operationality and dispatchability of the grid access of renewable energy, and realize the safe and stable grid connection of intermittent energy.

1.3.1.3 Demands of distributed power generation and smart grid construction on energy storage

China's power system has taken the path of greater power grids and bigger units and followed the operation of centralized power transmission and distribution. The long distance between China's energy centers and power load centers determines the basic concept of the country's smart grid, which is "to build a firm smart grid with the UHV grid as the backbone network, the coordinative development of grids at various levels and the characteristics of informatization, digitalization, automation and interaction."

In terms of the key aspects in occidental smart grid development, the rapid rise of distributed power generation's proportion and the full competition power market led to the priority of smart development on the user's side and the power distribution network, and the main investment to the energy storage field. The US government has set large-scale energy saving technology as a strategic technology that supports the development of new energy. According to the smart grid aid plan issued by the US Department of Energy, the number and amount of energy storage projects exceeded those of all the other projects.

Although smart grid construction varies based on the actual situation in different countries, a high-quality, self-correcting, safe, clean, economical, and interactive smart grid is the common objective of all. Traditional grids mainly comprise power generation, power transmission, power transformation, power distribution, and power consumption, while energy storage is the key for building a smart grid and realizing the objective. With the advancement of smart grid construction in various countries, the application of energy storage technology needs rapid and widespread development.

1.3.2 Role of energy storage technology in power systems

1.3.2.1 Peak load shifting

With the continuous acceleration of urbanization and increase of electricity load, the load's peak—valley gap will be continuously widened. It is more and more difficult to meet the load growth demand by installing additional power generation, transmission, and distribution equipment. For power enterprises, on the one hand, plenty of investment is needed each year; on the other hand, the short peak load time leads to the decrease of asset utilization ratio. The construction of energy storage will effectively solve the problem.

Energy storage power stations of various forms can acquire electric energy to charge themselves from the grid during its low load period, and then switch to the power generation mode during the grid's peak load period to transmit electric energy to the grid. So, this contributes to reducing the loss of the system's power transmission network, creates economic benefits through reduction or replacement of new power generation plants, and improves the utilization ratio of power transmission and distribution equipment. On the other hand, due to the peak–valley load pricing policy, the operation of charging at the valley load time while consuming at the peak load time will save electricity cost for users.

Large-scale energy storage systems can realize the decoupling and load adjustment between power generation and power consumption and narrow the peak–valley load gap to some degree. Once energy storage systems reach a certain size, the construction of power sources and grids may be effectively delayed or reduced. Thus energy utilization efficiency and overall asset utilization ratio of grids will be improved. The current construction mode of power systems will be altered to promote the power systems' conversion from extensive expansion to intensive efficiency enhancement.

1.3.2.2 Improving grids' accommodation to new energy

New energy sources including wind energy, solar energy, ocean energy, and biomass energy are rapidly developing. Orderly integration is an important part of the construction of smart grids. Different from thermal power, hydropower, and other regular power sources, these new energy sources, due to their characteristics of intermittence and fluctuation, will influence the stable operation of the grid in case of large-scale integration. When they have a relatively low proportion in the grid's capacity, they have a small impact on the grid, and the grid's safe and stable operation can be secured by means of grid control and power distribution technologies. Once the proportion of the new energy source's installed capacity exceeds a certain value, there will be an obvious impact on regional grids. Especially in regions with a low proportion of hydropower, oil, and steam power sources, adjustment relying on slow power-adjusting thermal power units alone cannot completely respond to the rapid load changes, and it may even cause severe accidents.

Various measures can be applied to avoid or reduce the influence of new energy's mass integration on the stable operation of grids. One measure is to enhance the short-, medium-, and long-term forecasts on new energy power generation to better work on planned power balance. Another measure is to build energy storage systems, such as the establishment of light energy storage, wind energy storage, and light–wind combined energy storage systems to buffer and reduce the impacts on grids. The national wind, light, storage, and transmission demonstration project proves the feasibility of combined control of large-scale wind—light energy storage for the first time in the world.

So, with the continuous increase of installed capacity and scale of fluctuating renewable energy, the installation of additional energy storage equipment can rapidly provide active power support and strengthen the frequency modulation capacity of grids. The installation of large-scale energy storage equipment with good dynamic response, long service life, and high reliability at the power source side may effectively solve the problems of intermittence and uncertainties of large-scale integration of wind energy, solar energy, and other new energy sources, greatly improve the grid's capacity to accommodate renewable energy, and promote the intensive exploitation and utilization of renewable energy. It is also one of the important channels to relieve environmental pressure and cater to the low-carbon society requirements in the future.

1.3.2.3 Spare power sources

Securing power supply requires sufficient spare capacity by the power system. Power grids will confront high risks if a spare power source is insufficient. In case of grid failure and disconnection of main power sources, chain reactions may take place and lead to accidents. In case of a major disturbances in the power system, such as a short circuit, energy storage equipment may absorb or release energy instantaneously to give the adjustment equipment time for adjustment, avoid the instability in the system, and recover to normal operation. As for important electricity load departments, such as hospitals and governmental and military industrial departments, it is of critical importance to build spare power sources for accidents or emergencies. It is known that more than 200 industrial and commercial users have set up sodium—sulfur battery storage stations to offer spare power sources in case of accidents and emergencies and improved the reliability of power supply for users. In summary, in case of grid failures and power supply abnormality of the distributed power generation system, energy storage systems may provide stable electric energy for users.

1.3.2.4 Improving quality of electric energy

As for power loads that are especially sensitive to transient electric quality problems such as transient voltage drop and short-time power cuts, the momentary voltage fluctuation of power supply can possibly lead to flawed products and severe economic loss. These users may be equipped with power-type energy storage technology with supercapacitors, superconductors, and flywheels as typical facilities to realize rapid active power or reactive power conversion between energy storage equipment and the power system, reduce the power system's harmonic distortion, voltage fluctuation and flickering, remove transient drop and rise of voltage, and secure high-quality power supply.

1.4 Application outlook and challenges of energy storage technology in power systems

Electricity storage technology, breaking the limits on generation and consumption, is applicable to multiple application fields to solve problems difficult to handle with traditional methods. As a key support technology, energy storage technology has been preliminarily applied in power generation with new energy and distributed power sources, smart grids, electric vehicles, industrial and household users, and other fields. Many countries have set up several demonstration projects, launched relevant supportive policies, and boosted the rapid development of energy storage technology.

1.4.1 Application outlook

1.4.1.1 New energy power generation side

Renewable energy power generation–based distributed energy supply technology has become the development focus in the energy field. However, the fluctuations and intermittence of wind energy, solar energy, and other renewable energy sources increasingly sharpen the contradiction between new energy and the grid. Meanwhile, big grid failures also stress the urgent requirement of power supply to an efficient and large-scale energy storage technology. Relying on energy storage technology to store and stably transmit the power generated with wind and solar energy can provide a rapid active power support, enhance the grid's frequency modulation capacity, and enable large-scale wind and solar power to be conveniently and reliably integrated into regular grids. Therefore, energy storage technology has become one of the core technologies for the large-scale application of new energy power generation.

China's power generation technology with wind energy, PV, and other new energy has taken the lead in the world. However, wind energy and solar energy power generation is not stable due to the limitations of weather, season, and time. According to the *Wind Power and PV Power Generation Regulatory Report* released by the State Electricity Regulatory Commission of the People's Republic of China in January 2011, unpurchased wind power reached 2.776 billion kWh, and there is no such data for PV power for the period from January to June 2010. With the gradual increase of wind power in the grid, its large-scale power fluctuations have a negative influence on the safe and economic operation of grids. To ensure the safe operation, the grid's accommodation scale for wind power is restricted, resulting in lots of curtailment of wind electricity or refusal of new wind power units into the grid. Large-scale energy storage systems can realize the time and space shift of energy and efficiently restrict the fluctuations of wind power and will have huge application potential in new energy grid integration as an effective means for improving the existing grids' wind power accommodation capacity.

1.4.1.2 Urban power distribution network side

The supply and demand contradiction of electricity and energy has become a bottleneck that limits the stable and sustainable development of China's economy. Especially in large and medium-sized cities, with the gradual increase of electricity demand, the load gap between power consumption peak and valley keeps rising. In daytime, power consumption is pretty high, and power supply is even limited in some places. But power consumption is rather low in the dead of night, and much power generation capacity is wasted. And under low load, power generation efficiency obviously decreases. However, due to the continuously expanding peak—valley gap, it is difficult for installed capacity to meet the demand of peak load. Building new power plants to meet peak load demand is also not a good idea. The capacity of new power plants should match peak consumption. With the increasing peak—valley gap, the load rate will quickly drop, resulting in a low facility utilization rate at nonpeak time. Taking into consideration construction cost and resource conservation, it becomes more and more difficult to meet the increasing peak load by installing new equipment for power generation, transmission, and distribution.

Energy storage power stations of various forms can acquire electric energy to charge themselves from the grid during the valley load period and transmit electric energy to the grid during the peak load period. So, this contributes to reducing the loss of the system's power transmission network, postponing or eliminating the need for new power generation plants, and meeting the demand of increasing peak load. Moreover, energy storage stations can play the role of a spare power source, increase the system's spare capacity, and secure urban power supply.

1.4.1.3 Distributed power supply and microgrid side

Based on the demands of environmental production, energy savings, emission reductions, and sustainable development, the application of distributed energy on the power consumption side of users close to solar energy, wind energy, biomass energy, and other renewable energy can both effectively save remote transmission via great grids and resource utilization. Distributed power supply such as wind power and PV power is intermittent and fluctuant, and it cannot achieve off-grid operations. The increase of penetration rate of distributed energy will influence grid operation and control. To improve the utilization rate of distributed energy and release the distributed energy's impact on the grid, microgrids have appeared. A microgrid is a kind of mini-type grid, and users may be either power supply departments or large- and medium-sized enterprises and institutions, power generation departments, small users with high environmental protection requirements, or users in remote regions.

To ensure voltage and frequency stability of microgrids under different operation modes, microgrids should apply reasonable distributed power source and energy storage design to realize various operation modes and transition between these operation modes. During the connection and disconnection of distributed power sources, the short power imbalance can be controlled with energy storage equipment for rapid energy storage/release, thus achieving emergency balance control to active power and ensure the voltage stability of key load. In addition, due to the uncertainty of PV and wind power generation in the microgrid, the change of load is random to some degree. By installing energy storage equipment and having reasonable control, the system's energy balance can be maintained.

1.4.1.4 End user side

Large-size users are from electrolysis, electroplating, metallurgy, and other industrial sectors as well as electric buses, light rail, subways, and other transport sectors with high, intensive power consumption. To reduce power consumption costs, the grid peak—valley gap can be fully utilized. Through energy storage technology, valley electricity can be used to charge the energy storage system and then used for production and operation in peak time. This can not only reduce investment in power supply equipment to improve electric energy's utilization efficiency but also dramatically reduce operation costs for enterprises.

Regarding household users, the market demand for household emergency power management systems is gradually expanding due to energy supply shortage and frequent natural calamities. Energy storage technology can be used for a household emergency power management system or combined with PV power generation to adjust output power during the periods of high electricity charge and high power consumption, secure emergency power and reduce consumption at peak time, and provide all necessary energy for households.

Electric vehicles are being developed mainly for energy saving, environmental protection, and safety. Governments of various countries have issued numerous supportive and preferential policies for new energy vehicles. The development of electric vehicles combines energy storage and power systems. In addition to the transport function, the vehicles are also a possible spare power source. By means of the vehicle to grid (V2G) interaction, the vehicles can play the role in the grid's peak and frequency modulation. The development of electric vehicles will promote the application and spread of energy storage technology and generate more development potential for the energy storage industry.

1.4.2 Challenges

Energy storage technology has received policy and industrial support in occidental countries that have a huge energy storage market. However, the development of energy storage industry still confronts severe challenges from many aspects.

1.4.2.1 Technical challenges

Apart from the large-scale application of PHS, the maturity, reliability, and economy of other energy storage technologies still needs further verification, and users' selection of energy storage technologies requires further tests in the market. Greater input is in urgent need for large-scale R&D, demonstration, and industrialization of energy storage technologies. Especially for energy storage with large-capacity batteries and supercapacitors, breakthroughs are required to be made in basic technology, equipment R&D, operation control, and system integration, etc. By means of the construction and operation of demonstrative projects like smart grid systems, large-scale intermittent power source access systems wind energy/light energy/storage energy complementary power generation systems and distributed combined cooling, heating, and power production, we can accumulate technical data and operation experience for China's energy storage industry, improve the localization level of energy storage equipment production, and lay a solid foundation for the industrialization of energy storage.

How to integrate energy storage technology with electric power systems and realize large-scale application for the best effect still needs the test of practice. The electric industry has high requirements on products' reliability. Conventionally, the minimum standard for electric products requires the reliability test and trial use of 5 years at least. Testing and proven use for a certain time are required for the mass production and application of energy storage products. Thus, industrialization and large-scale application of energy storage technology require a relatively long period.

1.4.2.2 Challenges of policy and mechanisms

The launch and effective implementation of industrial support policies have boosted the operation of various energy storage demonstration projects and the spread of commercialization. The vagueness of a profit-making mechanism has led to the absence of development plans, supportive standards and criteria, operation supervision, and auditing systems. There has been an urgent need to establish supportive policies and marketing mechanisms that adapt to the development of China's electric power market and energy storage industry, improve the enthusiasm of industrial investment, realize the diversification of investment subjects, encourage power generation companies, grid companies, users, independent third-party energy storage enterprises, and other investors to invest in this industry.

1.4.2.3 Economic challenges

Since the key materials, production process, system integration, and other technologies of energy storage systems are in the initial stage, energy storage systems face the problems of high cost and short service life. In addition, the application of energy storage technology in the electric power market still needs in-depth exploration to enable it to play more roles, create multiple economic benefits, and maximize the value of the technology. Meanwhile, the economy of energy storage technology will greatly influence the market's choice in technical routes.

Therefore, during the large-scale application of energy storage, a reasonable commercial mode with multiple benefits will be explored in addition to cost reduction. Cost can be controlled by reducing the initial investment in energy storage systems or extending the energy storage system's service life with certain investment. Also, improving the charging and discharging efficiency of energy storage systems may reduce the power consumption cost of energy storage systems.

Further reading

[1] Zhenxiang H. Analysis on electric power system. Hangzhou: Zhejiang University Press; 1997.

[2] Zhenya L. Electric power and energy in China. Beijing: China Electric Power Press; 2012. p. 155.

[3] National Natural Science Foundation of China, Chinese Academy of Sciences, 2011–2020 Strategic development report on China's energy science. http://wenku.baidu.com/view/d33c655–fbe23482fb4da4cec.html.

[4] EPRI. Electricity energy storage technology options—a white paper primer on applications, costs, and benefits. http://www.doc88.com/p–86915917253.html.

[5] Wenliang Z, Ming Q, Xiaokang L. Application of energy storage technologies in power grids. Power System Technology 2008;32(7):1–9.

[6] Jianxing Z, Zhang Y, Cao Z. Development opportunities and challenges of large – scale battery energy storage technology in power grids. Power and Energy 2013;34(2):182–6.

[7] Hong J, Jin Y. Market and economic analysis on the energy storage industry. Energy Storage Science and Technology 2012;1(2):103–11.

Chapter 2

Technologies of energy storage systems

Chapter outline

Grid-scale Energy Storage Systems and Applications. https://doi.org/10.1016/B978-0-12-815292-8.00002-2

So far, a variety of methods for energy storage have been explored and developed, among which, electrochemical, physical, and electromagnetic methods are the main forms of energy storage applied to new energy power generation. The electrochemical way refers to battery energy storage, such as lead–acid, lithium-ion, vanadium redox battery (VRB), etc. The physical way includes pumped hydro storage (PHS), compressed air energy storage (CAES), and flywheel energy storage; the electromagnetic way includes supercapacitor energy storage and superconducting magnetic energy storage (SMES). There are obvious differences in energy density, power density, and application properties of different energy storage technology (EST), which give them both strengths and weaknesses in power system applications. Thus, some new types of EST, such as lead–carbon battery, Li/S battery, heat pump energy storage (ES), and gravity ES, have been studied to improve the technical and economic characteristics of the original ES, which actively promotes the innovation and development of ES industry.

This chapter introduces the working principles and characteristics, key technologies, and application status of electrochemical energy storage (ECES), physical energy storage (phES), and electromagnetic energy storage (EMES), respectively, and briefly several new types of EST. Finally, the ESTs suitable for new energy generation are proposed in this chapter based on the multiangle comparison and analysis made from aspects of technical maturity and performance parameters of different ESTs.

2.1 Electrochemical energy storage

2.1.1 Lead–acid battery

2.1.1.1 Working principle and characteristics

A lead–acid battery is composed of a positive plate, negative plate, electrolyte, separator, slot, and cover, etc. It takes PbO_2 as the positive active material, Pb as the negative active material, sulfuric acid as an electrolyte, and its open circuit voltage is 2v. The positive and negative active materials are converted into $PbSO_4$ after the discharge of the battery, and the electrochemical reaction is as follows:

$$\text{Anode reaction:}\ Pb + HSO_4^- - 2e \leftrightarrow PbSO_4 + H^+ \tag{2.1}$$

$$\text{Cathode reaction: } PbO_2 + 3H^+ + HSO_4^- + 2e \leftrightarrow PbSO_4 + 2H_2O \quad (2.2)$$

$$\text{Total battery reaction: } PbO_2 + Pb + 2H^+ + 2HSO_4^- \leftrightarrow PbSO_4 + 2H_2O \quad (2.3)$$

Water begins to break down when the charge state of the positive plate reaches about 70% in the process of battery charging:

$$2H_2O \rightarrow O_2 + 4H^+ + 4e \quad (2.4)$$

A lead–acid battery is divided into ordinary nonsealed, flooded lead–acid batteries and valve-regulated lead–acid batteries, according to the structure and working principle. The electrode reaction mechanism of the valve-regulated lead–acid battery during charge and discharge is the same as that of an ordinary lead–acid battery while the battery structure and working principle changes a lot for its adopting oxygen composite technology and lean liquid technology. Oxygen composite technology makes the hydrogen and oxygen produced in the charging process recombine into water and return to the electrolyte; lean liquid technology ensures that the oxygen can move rapidly and massively to the negative electrode to increase charging current. Oxygen composite technology and lean liquid technology can not only improve the specific efficiency, specific power, and specific energy of a lead–acid battery, but they can also lengthen its cycle life and reduce the maintenance cost.

The strengths of lead–acid battery are as follows: (1) low investment cost; (2) it is easy for charge and discharge control for the open circuit and discharge depth are basically linear; (3) themonomer capacity ranges from tens to thousands of ampere hours, so it is safe and reliable for MW-level energy storage station after series-parallel connection; and (4) mature recovery technology and high utilization rate.

The weaknesses of lead–acid battery are these: (1) the specific energy is low, generally 30–50 WH/kg; (2) short cycle life, usually 500 times; (3) heavy metal waste water containing lead is produced in the production process, which is acidic and easily causes pollution.

2.1.1.2 Key technologies

The key technologies of lead–acid battery include the preparation technology of grid alloy, positive and negative electrode and separator material, battery sealing, and maintenance-free technology, etc. The main trend of lead–acid battery will be the optimization of the preparation technology of key raw materials for batteries, improvement of the battery structure design, manufacturing process, and the applicable range of battery working conditions; besides, advanced lead–acid batteries will be developed, such as lead–carbon batteries. These new technologies are expected to lead to breakthroughs in energy density and cycle life of lead–acid batteries.

2.1.1.3 Application status

The merging and integration of the lead–acid battery industry in Europe and America have been extremely rapid. The annual output of lead–acid batteries in the United States amounted to $10.5 billion, accounting for one-third of the global output. Japan is leading the world in lead–acid batteries for motorcycle seals. The lead–acid battery has been widely used in power plants, substations, and other fields abroad. It plays an important role in maintaining the safety performance, stability, and reliable operation of power systems. As a result, it ranks first in various types of chemical power sources in terms of yield and output value. Table 2.1 lists some MW-level lead–acid battery energy storage systems and their roles.

2.1.2 Lithium-ion battery

2.1.2.1 Working principle and characteristics

A lithium-ion battery takes imbedded and extracted metal oxide or sulfide embedded as the positive electrode, organic solvent inorganic salt system as the electrolyte, and carbon material as the negative electrode. Li^+ comes out of the positive electrode into the negative electrode lattice, and the positive electrode is in a Li^+ lean state in charging process; Li^+ comes out from the

TABLE 2.1 Some large lead–acid battery energy storage systems.

Name	Location	Rated power/ capacity (MW/ MWh)	Functions	Installation time
BEWAG	Berlin	8.5/8.5	Hot standby, frequency control	1986
Crescent	North Carolina	0.5/0.5	Peak value adjustment	1987
Chino	California	10/40	Hot standby, load balancing	1988
PREPA	Puerto Rico	20/14	Hot standby, frequency control	1994
Vernon	California	3/4.5	Improve power quality	1995
Metlakatla	Alaska	1/1.4	Improve the stability of isolated power grid	1996
ESCAR	Madrid	1/4	Load balancing	Late 1990s

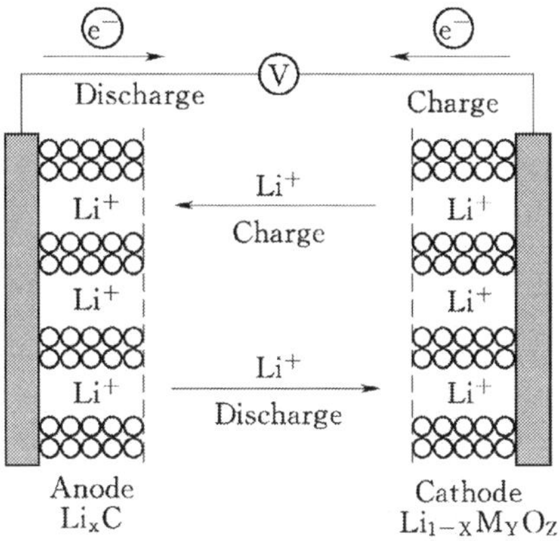

FIGURE 2.1 Working principle of a lithium-ion battery.

negative electrode and inserts into the positive electrode, and the positive electrode is rich in Li^+ in the discharging process. To keep the charge balance, the same number of electrons should be transferred through the external circuit during charging and discharging and move with Li^+ between positive and negative electrodes at the same time. Thus the redox reaction on the negative electrode is induced to maintain a certain potential, as shown in Fig. 2.1.

A lithium-ion battery is divided into lithium cobalt acid, lithium nickel acid, lithium manganese acid, lithium iron phosphate, etc., according to different positive electrode materials. Table 2.2 is a comparison of the characteristics of different lithium-ion batteries.

As the first commercial lithium-ion battery, the lithium cobalt oxide battery ($LiCoO_2$) has mature technology and a high market share. The theoretical capacity is 274 mAh/g, the practical capacity is greater than 140 mAh/g, and the open circuit voltage is 3.7 V. The main Strengths of $LiCoO_2$ are stable voltage in charging and discharging process and good circulating performance. The main disadvantages are these: the raw materials are more expensive, the safety performance of anti-over-charging is poor, and it is not suitable for a large-scale power battery.

The theoretical capacity of lithium nickel oxide ($LiNiO_2$) is 274 mAh/g, the practical capacity is 190–210 mAh/g, and the open circuit voltage is 2.5–4.2 V. The main advantage of $LiNiO_2$ is its low cost for the nickel resources, which are relatively abundant. The main disadvantages are the synthesis conditions are harsh, the circulation stability is poor, and the safety performance needs to be improved.

The theoretical capacity of lithium manganese oxide ($LiMn_2O_4$) is 148 mAh/g, the practical capacity is 90~M20 mAh/g, and the open circuit voltage is 3–4 V. The main advantages are low cost and higher safety performance than $LiCoO_2$, so it occupies an important position in the global power battery field. The main disadvantages are low theoretical capacity and poor

TABLE 2.2 Comparison on characteristics of different lithium-ion batteries.

Anode material	Theoretical capacity (mAh/g)	Practical capacity (mAh/g)	Open circuit voltage (V)	Cost	Safety performance	Cycle index (100% DOD)
$LiCoO_2$	274	140–160	3.7	H	Fair	300–500
$LiNiO_2$	274	190–210	2.5–4.2	M	Poor	>300
$LiMn_2O_4$	148	90–120	3–4	L	Good	100–200
$LiFePO_4$	170	110–165	3.4	L	Excellent	>2000

circulation performance. Newly developed $LiMnO_2$ cathode material in both theoretical capacity and the practical capacity is greatly improved, the theoretical capacity is 286 mAh/g, the practical capacity is about 200 mAh/g, and open circuit voltage ranges from 3 to 4.5 V, but there still exists the problem that the structure is unstable in the process of charging and discharging.

The theoretical capacity of the lithium iron phosphate ($LiFePO_4$) is 170 mAh/g, the practical capacity is greater than 110 mAh/g, and the open circuit voltage is 3.4 V. The main advantages are these: good cycle performance, more than 80% of the capacity is maintained after 2000 monomer 100% DOD cycles, high safety performance, stable voltage in discharging with continuous charge and discharge under condition of 1–3°C, instantaneous discharge ratio is up to 30°C. The main disadvantages are the low theoretical capacity, temperature performance is poor, and the discharge capacity decrease to 70%–80% under 0°C.

Lithium titanate ($Li_4Ti_5O_{12}$) can greatly improve the circulation performance of the system when it takes the role of negative electrode material, and the capacity retention rate of the single battery is more than 80% after 3000 cycles at 100% DOD; it has good safety performance, high magnification (40–50°C) discharge, quick charge (10°C), and a wide temperature range. The main disadvantages are the low working power (2.5 V), poor electronic conductivity, easy gas in the production process, and high price for its small-scale production at present.

2.1.2.2 Key technologies

The research trends of positive electrode materials include nanocrystalline, nanocrystalline coating carbon, suitable doping, particle surface coating, and stability of materials.

Carbon materials and lithium titanate are the main negative electrode materials. The price of carbon materials is relatively low and has wide application, but the safety performance is poor. The most suitable anode material is lithium titanate, which has been used by Toshiba in Japan to produce new lithium-ion batteries.

The quality of diaphragm material directly determines the safety and cycle life of lithium-ion batteries, and the research focuses on safety, yield, high performance, and low cost. High-end membrane materials, especially for power lithium-ion batteries, have higher demands for products' consistency. The stability of mass production batch of diaphragms produced with China's own intellectual property rights is relatively poor.

The lithium-ion battery model is used to solve the problems in design, production, and application. It includes a three-dimensional thermal analysis model, life prediction model, ion diffusion model, thermal safety model, and battery current distribution model. The lithium-ion battery involves knowledge of material, catalyst, reaction dynamics, heat transfer, mass transfer, flow, and other fields. As a result, many models are not yet mature, and the accuracy and dynamic response characteristics need to be further improved for its complicated reaction mechanism.

2.1.2.3 Application status

The lithium-ion battery energy storage system mainly plays the role of smoothing output of hybrid wind/PV generation, load shifting, and frequency regulation of the power network in many countries, among which the United States leads the world. The battery system was tested by Electric Power Research Institute in 2008, and it carried out research on the application of lithium-ion batteries distributed energy storage in an energy storage project in 2009. In addition, a demonstration application of the MW-grade lithium-ion battery energy storage system was also carried out to smooth wind power, adjust the frequency and voltage of the power system, etc.

State Grid corporation of China, China southern power grid company successively built MW-grade lithium-ion battery energy storage power station demonstration project; some energy storage manufacturers also established the kW- to MW-grade lithium-ion battery energy storage system and carried out PV battery energy storage hybrid power system and load shifting tests. The 20-MW lithium-ion battery energy storage system established at the first phase of the national wind/PV energy storage and transmission demonstration project has played a role of smoothing power output of wind/PV, tracking the operation of plan generation, load shifting, and system frequency regulation.

2.1.3 Vanadium redox battery

2.1.3.1 Working principle and characteristics

VRB takes vanadium-ion solution with different valences as its positive and negative active materials, which are stored in their respective electrolyte storage tanks, as shown in Fig. 2.2.

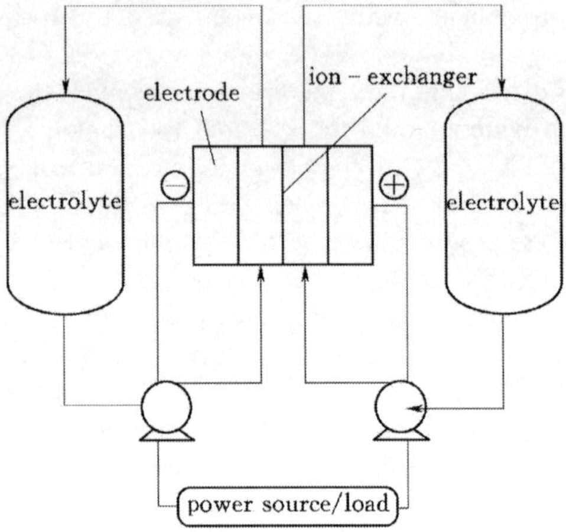

FIGURE 2.2 Schematic diagram of VRB.

The positive counter electrodes of VRB are VO^{2+}/VO^{2+}, while the negative one is V^{2+}/V^{3+}. VRB is the only battery system that uses the same element among all chemical powers, which avoids the cross-contamination of different kinds of active materials from principle in one battery. The standard electric potential difference of VRB is 1.26 V, the total battery efficiency is 75%−80%.

The electrolyte is operated by the pump and circulates through the positive and negative chambers of the battery leading to the oxidation and reduction reactions on the electrode surface in the charging and discharging process. Both reactions are reversible on the carbon felt electrode with fast reaction kinetics and high current efficiency and voltage efficiency, making it the most successful liquid-flow cell to date:

$$\text{Cathode reaction: } VO^{2+} + H_2O - e \leftrightarrow VO_2^+ + 2H^+ \tag{2.5}$$

$$\text{Anode reaction: } V^{3+} + e \leftrightarrow V^{2+} \tag{2.6}$$

$$\text{Total battery reaction: } VO^{2+} + H_2O + V^{3+} \leftrightarrow VO_2^+ + V^{2+} + 2H^+ \tag{2.7}$$

These are the strengths of VRB: (1) battery power output depends on the size of the battery stack, while the energy storage capacity depends on the reserves and concentration of the electrolyte, so the energy storage capacity and power are independent of each other, which makes the battery system design flexible and modular combination easy; (2) all reactions are completed in the liquid phase; vanadium ions are the only electrolytical ions to change status in the charging and discharging process, so VRB can realize ultradeep discharge (100%) without causing irreversible damage to the battery, resulting in long service life; and (3) there is low cost for the repeated use and regeneration of vanadium electrolyte.

There are also disadvantages of VRB: (1) the circulating pump is necessary to maintain the flow of electrolyte, which reduces the overall energy efficiency; (2) battery working temperature should be controlled in 5−45°C for vanadium with low valence is not easy to dissolve and separate crystals out when temperature is low, while service life reduces when temperature is high for the five-valence vanadium is easily decomposed into vanadium oxide; (3) the capacity unit price is relatively high, even with the price level of the sodium−sulfur battery, which is not superior to the price of the equivalent iron−lithium battery.

2.1.3.2 Key technologies

Electrolyte is the most important factor that affects the battery, followed by the diaphragm and electrode material.

1. Electrolyte. It is special that the electrochemical active material is dissolved in electrolyte. The voltage and volumetric energy of the battery increase when the concentration of electrolyte increases, but the resistance and viscosity, etc., are the same. In addition, due to the low solubility of pentavalent vanadium, the positive electrode solution will precipitate red polyvanadate and block the porous electrode surface when the battery is close to being fully charged, which leads the battery to gradually lapse. Therefore, high concentration and stability of electrolyte should be developed.
2. Diaphragm. The diaphragm basically decides the battery life and conversion efficiency; therefore it is required to have high corrosion resistance and ion exchange capacity to isolate positive and negative electrolyte to prevent the mixing of vanadium ions in different valence states, but only to free hydrogen ions. The diaphragm material is mainly divided into cation exchange membrane and anion exchange membrane. If the cation exchange membrane is adopted, a small amount of V^{2+} or V^{3+} ions will be combined with water to pass through the membrane and enter the anode of the battery; if the anion exchange membrane is selected, the neutral $VOSO_4$ and the anion $VO_2SO_4^-$ will also cross into the negative pole of the battery. At present, the cation exchange membrane, which is resistant to corrosion, and only allow H^+ with low impedance, is an important research direction in this field.
3. Electrode material. There are certain requirements for the electrode material: the electrode material should have high activity to positive and negative electrochemical reactions and reduce the activation of the electrode reaction; have excellent conductivity to reduce the battery's polarization in charging and discharging process; have a better three-dimensional structure to facilitate the flow of electrolyte and reduce the loss of pump consumption when the battery is working; and have high electrochemical stability to prolong battery life.

2.1.3.3 Application status

Sumitomo Electric Industries (SEI) and Kansai Electric Power have jointly developed VRB since 1985 and focused on the research and development of a fixed-type VRB power storage system, which is mainly used for peak regulation of power plants and wind and solar power generation systems. SEI already has the full range of technologies for producing and assembling a VRB system, putting it into commercial operation. In addition, countries such as Germany, Austria, and Portugal are also carrying out studies on the energy storage system based on VRB, which are expected to be applied to the PV power generation and wind power generation systems.

China began to develop VRB in the 1990s and has made some breakthroughs in key materials and prototypes; some demonstration models of different sizes been successively developed. Prudent Energy Corp. has realized the asset acquisition of VRB Power energy systems in Canada, including all patents, trademarks, technical secrets, and equipment materials owned or controlled by VRB Power. Table 2.3 lists the typical demonstration projects and functions of VRB.

2.1.4 Zinc–bromine

2.1.4.1 Working principle and characteristics

The zinc–bromine battery was jointly proposed by Exxon and Gould in the 1970s. The negative active material of zinc–bromine battery is metallic zinc, the positive active material is bromide separated by a porous diaphragm, and the electrolyte is zinc bromide. During the charging process, the zinc is deposited on the surface of the negative pole in metallic form, and the bromine is formed and stored at the base of the positive electrolyte. The theoretical open circuit voltage of zinc–bromine battery is 1.82 V, and the total efficiency is about 70%. The electrochemical reactions on the electrode when charging and discharging are as follows:

$$\text{Cathode reaction: } Zn^{2+} + 2e \leftrightarrow Zn \tag{2.8}$$

$$\text{Anode reaction: } 2Br^{-} \leftrightarrow Br_2 + 2e \tag{2.9}$$

$$\text{Battery reaction: } Zn + Br_2 \leftrightarrow ZnBr \tag{2.10}$$

The strengths of zinc–bromine battery are these: (1) the theoretical energy density is 430 Wh/kg and actually can reach 70 Wh/kg; (2) it has good circulation performance where the discharge depth will not damage the battery when it is at 100%; (3) it works under normal temperature, so no complex thermal control system is required; and (4) since both zinc and bromine in the electrolyte are common substances, they are cheaper than vanadium batteries.

There are also disadvantages of the zinc–bromine battery: (1) the aqueous solution of bromine is corrosive to the battery material; (2) during the charging

TABLE 2.3 Typical demonstration projects and function of vanadium flow battery.

Time/ year	Location	Capacity	Functions
2007	State of Florida	2 × 5 kW × 4 h	PV/energy storage generation
2007	Remote base station in Kenya	5 kW × 4 h	Telecommunication backup power supply
2006	Some Hill Wind Farm in Ireland	2 MW × 6 h	Wind power generation energy storage
2006	Riso National Library in Denmark	15 kW × 8 h	Wind power generation and energy storage
2004	Castle Valley project in Utah	250 kW × 8 h	Load shifting regulation
2003	King Island Wind Farm of Oceania	200 kW × 8 h	Wind power generation, energy storage, diesel generator
2001	Sapporo, Hokkaido Wind Farm in Japan	4 MW/6 MWh	Wind power generation and energy storage
1999	Kansai Electric Power	450 kW/ 1 MWh	Load shifting regulation
1997	Yokohama factory of Sumitomo Electric Industries Ltd.	200 kW/ 800 KWh	Smooth load fluctuation
2004	Columbia Air Force Base of United States	12 MW/ 120 MWh	Standby power

process, zinc tends to form the crystallized structure that causes the battery short circuit; and (3) bromine is easy to dissolve in the zinc bromide solution, which makes the bromine easy to diffuse to react directly with the zinc, causing an increase of the battery's self-discharge.

2.1.4.2 Key technologies

The battery material and structure need to be improved for the aqueous solution of bromine is corrosive to the battery material.

Deposition uniformity must be controlled for the deposition of zinc on the current collector may form a crystallized structure. The influencing factors include the flow velocity of the electrolyte, the length of the stretch direction, the thickness of the pore layer of the carbon electrode, and the addition of an organic inhibitor.

Battery self-discharging is influenced by the bromide diffusion coefficient in the diaphragm, the thickness of the diaphragm, the positive electrolyte liquid tank of oil phase volume and distribution coefficient of water phase and organic phase that liquid bromine has in the anode, and the electrolyte flow dynamics condition, etc. The degree of self-discharge can be reduced by adding complexant in the positive electrolyte.

2.1.4.3 Application status

ZBB is the leading international producer of the zinc–bromine battery and has successfully developed the 50–500 kWh battery pack. The Department of Energy, the Sandia National Laboratory, and the California Energy Commission have demonstrated the system performance, reliability, and technical competitiveness. In 2011, ZBB coproduced a zinc–bromine energy storage battery and its management system with a number of companies in China, which is currently at the stage of product development and technical verification.

Japan has been working on the development of zinc–bromine battery technology for electric utility since 1980. In 1990, 1 MW/4 MWh zinc–bromine battery was installed in Fukuoka Imajuku Kyushu Electric Power company substation by New Energy and Industrial Technology Development Organization, Kyushu and Meidensha, which is currently the world's largest zinc–bromine battery pack. The battery pack has completed more than 1300 cycles, with energy efficiency of 66%.

2.1.5 Sodium sulfur

2.1.5.1 Working principle and characteristics

Conventional secondary batteries such as lead–acid batteries and lithium-ion batteries are made of a solid electrode and liquid electrolyte, whereas the sodium–sulfur (NAS) battery, by contrast, is composed of a molten liquid electrode and solid electrolyte. The negative active material is molten metal sodium, the positive active material is sulfur and sodium polysulfide molten salt, and sulfur fills in conductive porous carbon or graphite felt. Also, the diaphragm is beta-Al_2O_3. The working principle of the NAS battery is shown in Fig. 2.3. The charge and discharge processes are reversible, and the whole process is controlled by the concentration diffusion.

The open circuit voltage of NAS battery will be reduced linearly to 1.74 V with the change of the sulfur Na_2S_x, as shown in Fig. 2.4. The NAS battery is ohmic in most of the discharge range. Therefore, the discharge characteristics of the battery are completely determined by the internal resistance, i.e., the battery discharge capacity is independent of the discharge ratio. This is a very striking feature of NAS batteries that differs from other batteries.

These are the strengths of NAS: (1) the theoretical specific energy is 760 Wh/kg, which is 3–4 times that of a lead–acid battery; (2) the discharge

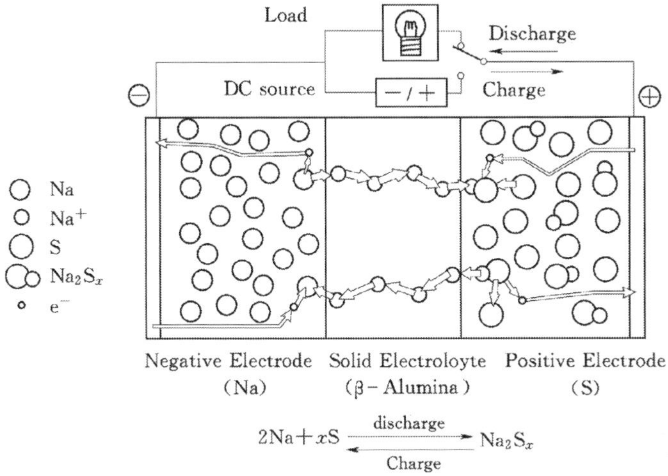

FIGURE 2.3 Working schematic diagram of NAS batteries.

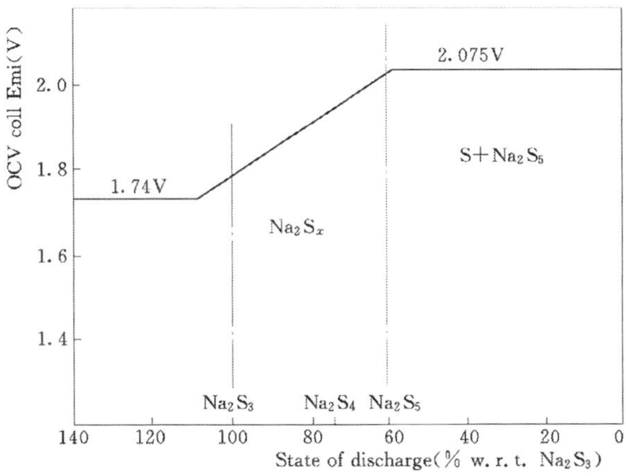

FIGURE 2.4 Changes of the open circuit voltage of sodium−sulfide batteries.

current density can be as high as 200−300 mA/cm^2, which can release three times the intrinsic energy of the electric current; and (3) it has a long cycle life, where it can discharge more than 2500 times under the condition of 100% DOD.

The disadvantages of NAS are as follows: (1) the working temperature of NAS is 300−350°C, when sodium and sulfur are in liquid under high temperature. If battery short circuit is caused by damaged ceramic electrolyte, sulfur and sodium will come into direct contact to cause an intense exothermic reaction that produces high temperature up to 2000°C followed by serious safety performance problems; (2) the working rate of the battery is low, which is not suitable for high power and fast charging and discharging; and (3) due to the corrosive nature of sulfur, the battery protector needs to be treated with strict corrosion resistance.

2.1.5.2 Key technologies

The materials and preparation process of NAS batteries hopes to improve and perfect the beta-Al_2O_3 solid electrolyte sintering technology of ceramic tubes and study on the beta-Al_2O_3 ceramic tube nondestructive testing technology. A goal is to design a solid polymer electrolyte battery production system and realize the automatic continuous production process for creating a high-capacity monomer NAS battery with excellent performance and good consistency.

Study on the interface characteristics of NAS batteries is important. In the battery exists an electrolyte/molten electrode and electrolyte insulating ring, and the electrolyte/sealants interface for beta-Al_2O_3 is solid electrolyte and diaphragm material. So it is needed for the corresponding interface characteristics research to ensure the mechanical properties and thermal matching properties of various combinations in the battery.

Study on the performance stability, reliability, and degradation mechanism of NAS batteries is key. The performance degradation of the battery is caused by a variety of mechanisms, and the root causes and their impacts on performance degradation need to be studied.

Regarding the manufacture and equipment technology of NAS battery modules, the key technologies to standardize the NAS battery module include connection technology, internal structure optimization, and a thermal management system.

2.1.5.3 Application status

At present, NGK and Tokyo Electric Power subsidiaries are still the sole producers of NAS batteries in the world, which provide more than 200 sodium–sulfur battery storage systems (>300 MW). In recent years, the NAS battery energy storage system has been widely used in such fields as load leveling, standby power supply, smoothing instantaneous fluctuation of power output, and ensuring stable output for a long time. In 2008, NGK installed a 34-MW NAS battery energy storage system at the 51-MW wind power station in six villages in Tsing Sen county, which stabilized the output power of the day by storing the electricity generated by night fans. The instantaneous power fluctuation of the wind power can be smoothed and the planned curve can be operated in a constant power mode for 2–4 h after the configuration of the NAS battery energy storage system.

In China, Shanghai Electric Power Company and the Shanghai Silicate Institute have made a breakthrough in the key technology of a NAS battery with large capacity and small batch preparation (2 per annum): a 100 kW/800 kWH NAS battery energy storage power station has had a successful demonstration run. Though the production process, major equipment, cost control, and market demand are still in need of development, and there is no application example of using a NAS battery energy storage system in China. Therefore, more investment and breakthrough in the commercial-scale production and marketization of NAS batteries in China is needed.

2.2 Physical energy storage

2.2.1 Pump hydro storage

2.2.1.1 Working principle and characteristics

PHS is a more widely applied energy storage technology, and its basic principle is converting electrical energy into potential energy that is stored in an upper reservoir through pumping water from a lower reservoir when the power system is at a low load; on the contrary, the water in upper reservoir will be operated under water turbine to generate electricity during the peak load of the power grid. The working principle of PHS is shown in Fig. 2.5.

Usually, the PHS power station is divided into two types according to the condition of natural runoff:

1. Pumped storage power station with no function of generating electricity. There is no or only limited natural runoff, and water recycles between the upper and lower reservoir. All the installations in the plant are pumped storage units, which are mainly responsible for tasks such as load shifting regulation, emergency standby, and not taking on conventional power generation tasks.
2. Mixed type pumped storage power station. There is natural runoff in the upper reservoir, and the water flow has reached the capacity to install conventional hydro generator sets to bear the load of the system. Part of the unit that is installed in the plant is the conventional water turbine generator set, and the other part is the pumped storage unit, which can take advantage

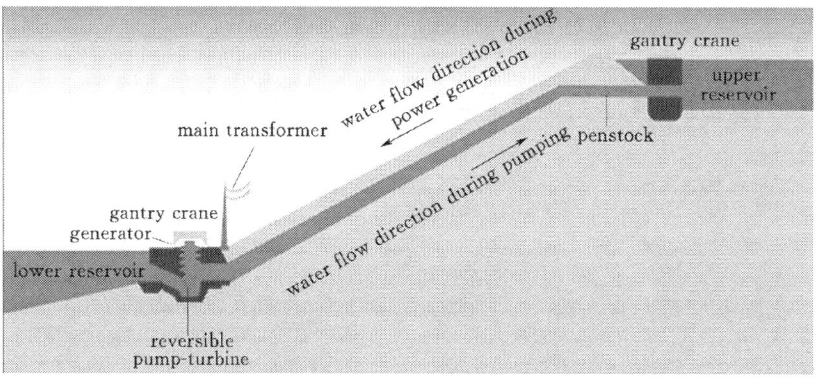

FIGURE 2.5 Working schematic diagram of PHS.

of the natural runoff to undertake tasks such as conventional power generation and water energy comprehensive utilization, as well as tasks such as load shifting regulation and emergency standby.

The pumped storage power station is the most mature and widely used large-scale energy storage technology. It has the strengths of large capacity (1 million kW), long life, and low operating cost. However, the construction of a pumped storage power station is constrained by geographic conditions, and it needs suitable upper and lower reservoirs. There is a certain amount of energy loss during energy conversion between pumping and generating, and the storage efficiency is about 75%–80%.

The pumped storage power station is generally built in the mountains far away from the power load. It is necessary to build a long-distance transmission system with a long construction period and high investment.

2.2.1.2 Key technologies

The key technologies of PHS include the following:

1. determination of reservoir capacity of upper and lower reservoirs, which includes the required storage capacity of the power generation, emergency storage capacity, and death storage capacity;
2. optimization of inlet/outlet port design, which includes the flow velocity distribution of the inlet/outlet, flow allocation problem, and inflow vortex problem;
3. high-pressure water pipeline design, which includes the selection of the lean pipe type and the hydraulic characteristics of the pressure pipeline;
4. starting mode of motor, which involves not only the structure of the unit, the electrical main wiring, and plant layout of the power station, but also the engineering investment of the electromechanical equipment. At present, the pump storage power station generally adopts the mode that frequency converter starts as the main and back-to-back start is the secondary.

2.2.1.3 Application status

The construction of pumped storage power stations is 130 years old. The development of PHS technology in many countries is relatively mature. For example, the installed capacity of Japan pumped storage power station has exceeded the installed capacity of conventional hydro power. Pumped storage power stations are mainly used for power system load in valley, frequency modulation, phase modulation, emergency standby, and black start, and provide spare capacity in the system. They can also improve the efficiency of the operation of thermal and nuclear power plants, as well as improving wind energy efficiency and quality of power supply. Until the end of 2010, the proportion of pumped storage power of different countries was as follows: France 13%, Japan 9.8%, Germany 11.2%, Britain 5.6%, the United States 2.2%, and China was 1.8%, with the highest proportion of 2.5% in east China.

Compared with the major countries such as Europe and the United States, the construction of pumped storage power station in China occurred relatively late. In the 1990s, a large number of large-scale pumped storage power stations were built in Guangzhou, Beijing's Ming tombs, and Tianshiping in Zhejiang province. By the end of 2013, pumped storage installed capacity of China reached 215.45 million kW, with a capacity under construction of 14.240,000 kW, and it is maintaining steady growth. Technologically speaking, the degree of localization of the monitoring, excitation, speed adjustment, protection, and automation system of pumped storage units under 250 MW has reached a certain level in China, but it needs to be improved. The equipment with capacity of 250 MW and above is gradually localized, among which the domestic excitation system of 300 MW has been successfully put into operation.

2.2.2 Compressed air energy storage

2.2.2.1 Working principle and characteristics

Based on the gas turbine technology, CAES consists of two cycle processes: the inflating compression cycle and the exhaust expansion cycle. During the process of compression, the surplus power of the load valley will drive the compressor and press the high-pressure air into the underground gas storage hole; while in peak load, the compressed air is preheated, and the fuel is burned inside the combustion chamber to generate power.

The principle of CAES is shown in Fig. 2.6.

The construction and power generation costs of CAES system are lower than that of a pumped storage power station. The probability of air leakage cracking is minimal, and it has both cold start and black start, so it is safe with

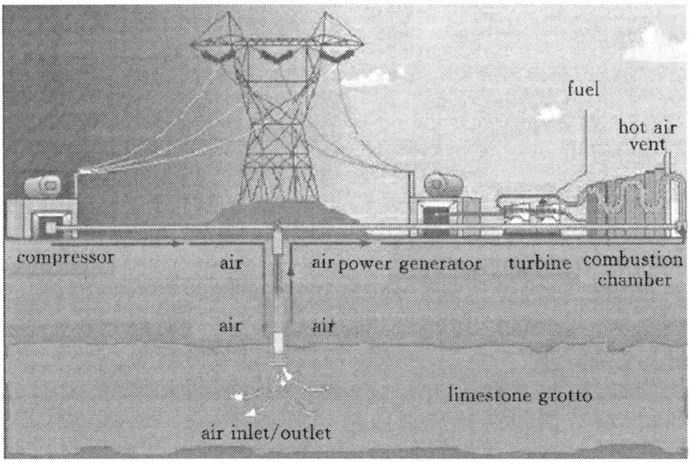

FIGURE 2.6 CAES working principle.

a long life. But its energy density is low, and it must be used in gas turbine power plants, not suitable for other types of power stations, so it is constrained by rising fossil fuel price and pollution control issues. In addition, similar to pumped storage power stations, special geographic conditions are also required to build large gas storage chambers for compressed air storage, such as sealed caves or abandoned mines.

To solve the main problems of a traditional CAES system, there have been some improved technologies in the world, including advanced adiabatic CAES system, small CAES system, and miniature CAES system. In the advanced adiabatic CAES system, the compressed heat is stored in the heat storage device, which will recover in the release process, and the theoretical storage efficiency of the system can reach more than 70%. The small CAES system is generally about 10 MW, and it is more flexible for using high-pressure tanks on the ground to store compressed air, which breaks the restrictions of dependence of a large traditional compressed air power station on gas storage caves. The scale of the micro-CAES system is generally several kilowatts to dozens of kilowatts. It also uses high-pressure vessels on the ground to store compressed air, but it is usually backup power for special areas, remote, isolated areas of the microgrid, and compressed air vehicle power.

2.2.2.2 Key technologies

The key technologies of CAES system include high-efficiency compressor technology, expansion machine technology, combustion chamber technology, heat storage technology, gas storage technology, etc.

1. The compressor and expansion machine are the core components of a CAES system, which play the decisive role for the performance of the whole system. Although CAES systems are similar to gas turbines, they have much higher air pressure than gas turbines. Therefore, compressors of large compressed air storage power stations usually use axial flow and centrifugal compressors to realize multistage compression and interstage and poststage cooling. As for the expander, the form of multistage expansion and intermediate reheat is always adopted.
2. Combustion chamber technology. Compared with a conventional gas turbine, the high-pressure combustion chamber of a CAES system has higher pressure. Therefore, more pollutants will be produced under higher temperature. As a result, the temperature will be controlled under 500°C.
3. Heat storage technology. Energy storage materials should have a large specific heat capacity, wide temperature range, and what is more, they should be free of pollution.
4. Gas storage technology. Large compressed air energy storage systems have large air capacity, so the gas is usually stored in underground salt mines, hard rock caves, or porous caverns. Though the micro-sized one can be free from the dependence on the gas storage cave.

2.2.2.3 Application status

The CAES is mainly used in peak valley electric energy recovery adjustment, balance load, frequency modulation, distributed energy storage, and power generation systems. At present, there are more compressed air storage power stations in the world, which are mainly in Germany, the United States, Japan, and Switzerland, as shown in Table 2.4. In addition, Italy, Israel, and South Korea are also actively developing compressed air storage power plants.

China's research on compressed air storage system is relatively late, but with the rapid increase of energy storage demand, relevant researches have gradually gained attention of some universities and research institutes. Engineering Thermal Physics of Institute of Chinese academy and other units have studied the thermal performance, economic performance and commercial applications of CAES station, but mostly concentrated in the theory and small experiment level. There is no commercially available compressed air storage power station.

2.2.3 Flywheel energy storage

2.2.3.1 Working principle and characteristics

The flywheel energy storage system mainly includes three parts: the rotor system for storing energy, the bearing system that supports the rotor, and the motor/generator system that realizes the energy conversion. The basic principle is using the electric motor to drive the flywheel to rotate at high speed, converting electric energy into mechanical energy to be stored; when necessary, the flywheel decelerates, and the motor runs as a generator, converting the kinetic energy of the flywheel into electricity, so the acceleration and deceleration of the flywheel realize the storage and release of electric energy. The basic structure of a flywheel energy storage system is shown in Fig. 2.7.

TABLE 2.4 Foreign compressed air storage demonstration power stations.

Name	Country	Total installed capacity (MW)	Generation date	Functions
Huntorf	Germany	290	1978	Hot standby and smooth load
McIntosh	USA	110	1991	Peak load regulation
Norton	USA	2700	2001	Peak load regulation
Sunagawacho	Japan	2	2001	Energy storage demonstration

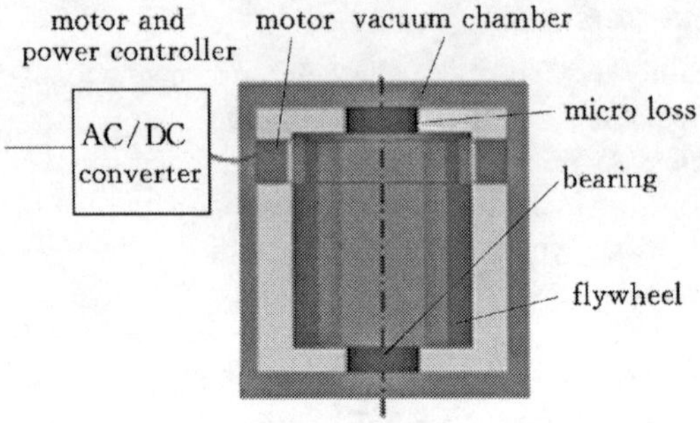

FIGURE 2.7 Working principle of flywheel energy storage.

Flywheel products can be divided into low-speed and high-speed flywheels.

The rotor of a low-speed flywheel product is mainly made of steel with high quality, and the rotor edge linear speed is generally not more than 100 m/s. This kind of product can adopt mechanical bearing, permanent magnetic bearing, or electromagnetic bearing. The whole system has low power density, so it improves the power and energy of the energy storage system mainly through increasing the quality of the flywheel. The rotor speed of a high-speed flywheel can reach up to 5000 r/min, and the rotor edge linear speed can reach up to 800 m/s. Therefore, high-strength materials such as glass fiber and carbon fiber are mainly used for manufacturing rotors. Such products can only use permanent magnet, electromagnetic, or superconducting bearings. At present, the research and application of permanent magnet and electromagnetic bearings in many countries are mature, and the latest research is high-speed flywheel products based on superconducting magnetic levitation.

A flywheel energy storage system has many advantages for it is running in a high-vacuum environment and has no friction loss, small wind resistance, a cycle efficiency of 85%–95%, long life, and is ecofriendly and free of maintenance. It also has disadvantages such as low energy density, high self-discharge rate, complex system, and high technical requirements for rotor and bearing.

2.2.3.2 Key technologies

Flywheel energy storage is now at the experimental stage, and there are still five main technical problems: the flywheel rotor, bearing, energy conversion system, motor/generator, and vacuum chamber.

1. Flywheel rotor. The flywheel rotor is the most important part of the flywheel energy storage system. The transformation of energy of the whole system depends on the rotation of the flywheel. It is necessary to develop the flywheel with high strength and suitable structure based on rotor dynamics design.

2. Support bearing. Bearing technology supporting a high-speed flywheel is one of the key factors restricting the efficiency and service life of the flywheel.
3. Energy conversion system. The core of the flywheel energy storage system is the conversion between power and mechanical energy, which adjusts energy input and output of the conversion process to coordinate the frequency and phase. The energy conversion unit determines the efficiency of the system and governs the operation of the flywheel system.
4. Generator/motor. The high speed of the flywheel energy storage rotor leads to the high speed of the flywheel motor, which requires high efficiency, low power consumption, and high reliability of the flywheel motor system. The current research on permanent magnet motors focuses on reducing loss and resolving the temperature sensitivity of a permanent magnet.
5. Vacuum chamber. The vacuum chamber is the auxiliary system of the flywheel energy storage system that makes the system independent from the outside environment.

2.2.3.3 Application status

The concept of flywheel energy storage in the modern sense was first proposed in the 1950s. In the 1970s, the US Energy Research Development Agency and the US Department of Energy (DOE) started funding the flywheel energy storage system application development. Japan and Europe carried out the relevant technology and product research and development successively. The flywheel energy storage has really entered the period of high-speed development due to the maturity of magnetic levitation, carbon fiber composite materials, and power electronics technology since the 1990s. Today, flywheel products based on permanent magnetic levitation and electromagnetic levitation bearings are mature, stable, and reliable. At present, the flywheel energy storage system is mainly used in the fields of uninterrupted power supply, renewable energy grid, and frequency modulation.

In the United States, Japan, and the United Kingdom, the development and application of flywheel energy storage technology has been developed, and a series of products have been formed to achieve megawatt power output. Meanwhile, China's research on flywheel energy storage system started late and mainly operated in colleges and universities. There is no mature product and demonstration application in China.

2.3 Electromagnetic energy storage

2.3.1 Supercapacitor energy storage

2.3.1.1 Working principle and characteristics

Based on the electrochemical electric double layer theory, a supercapacitor energy storage unit can provide powerful pulse power. The charge on the

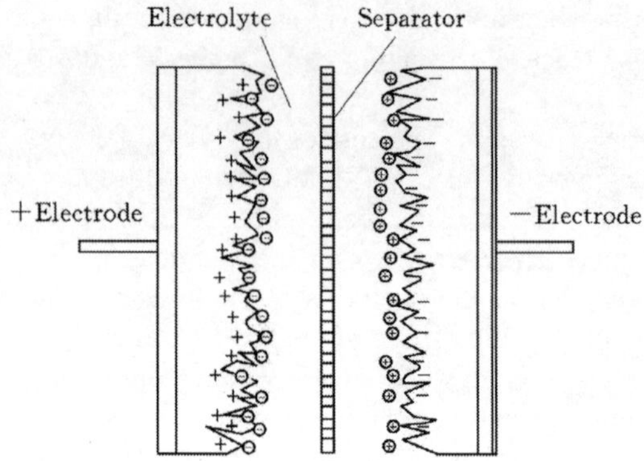

FIGURE 2.8 The schematic diagram of the supercapacitor.

surface of electrodes will attract the ions with opposite nature in the surrounding electrolyte solution and make them attach to the electrode surface to form a double charge layer, namely double layer capacitor. Due to the very small spacing between the charge layers (generally < 0.5 mm) and the special electrode structure, the surface area of the electrode is growing by a factor of 10,000, resulting in a great amount of capacity. Fig. 2.8 is the internal structure of the supercapacitor that includes the positive electrode, negative electrode, diaphragm, and electrolyte.

A supercapacitor can be divided into carbon-based supercapacitor, metal oxides supercapacitor, and polymer supercapacitor, etc., according to different electrode selection. The carbon-based supercapacitor is currently the most widely used one. The electrode materials of a carbon-based supercapacitor are made up of carbon materials. Organic electrolyte takes the role of the medium, and there is an ionic double electric layer between the activated carbon and the electrolyte. Energy is stored in the double electric layer and electrode through polarized electrolyte. There is no chemical reaction in the energy storage process of the double electric layer of supercapacitor, and the process is reversible. As a result, supercapacitors can repeatedly charge and discharge hundreds of thousands of times and not pollute the environment. Supercapacitors have several advantages: high power density and suitability for short-time high-power output; high charging speed and simple charging mode allow it to be charged with large current, and the charging process can be finished in a few seconds to minutes; long service life; the number of cycle usages of deep charge and discharge can reach 1−500,000 times; the circulation efficiency is 85%−98%; low temperature performance; most of the charge transfer is carried out on the surface of the electrode active material; and low attenuation of capacity with temperature decreases.

2.3.1.2 Key technologies

Supercapacitor body. Supercapacitor products have been industrialized in many countries, and the material purity, manufacturing process, and overall performance of products still need to be further improved in China.

Voltage balance. The working voltage of the supercapacitor has a great impact on its service life. The problem of capacitor voltage imbalance is the main factor limiting the use of series capacitor groups.

Control method. Advanced control methods can make output voltage of the inverter stable and reliable and realize fast dynamic response.

2.3.1.3 Application status

Supercapacitors have been widely used in electronics, electric toys, and other fields since Japan's NEC, Panasonic, and other companies launched industrial products in the 1880s. With the further reduction in the cost and the increase in the energy density, supercapacitors can be used in electric vehicles, rail transit energy recovery systems, and power systems, among which supercapacitors are used in power systems for short-time, high-power load smoothing and improving power quality. Especially in the distribution network, the role of maintaining voltage stability, suppressing voltage fluctuation and flashing, suppressing voltage drop, and instantaneous power supply is being gradually reflected.

2.3.2 Superconducting magnetic energy storage

2.3.2.1 Working principle and characteristics

The basic structure of SMES is mainly composed of superconducting coils, quench protection, cooling systems, converters, and controllers. As shown in Fig. 2.9, a superconducting coil can be used as an energy storage coil, which is powered by the power grid through the converter to generate a magnetic field in a coil for energy storage. The stored energy can be sent back to the grid or provided for other loads by inverters when needed.

A superconducting energy storage coil is almost free of loss, so the energy stored in the coil is almost undiminished. Compared to other energy storage systems, a superconducting magnetic storage has high conversion efficiency (about 95%) and quick reaction speed (up to a few milliseconds). The biggest drawback is the high cost and then the need for compressors and pumps to maintain low temperature of the liquefaction coolant, which makes the system more complicated and needs to be maintained regularly.

2.3.2.2 Key technologies

At present, SMES technology has developed greatly. But further rcsearch on the following technologies is needed to apply to power systems:

1. Explore and study new principles and new devices of superconducting electric power to maximize the superior performance of superconductors.

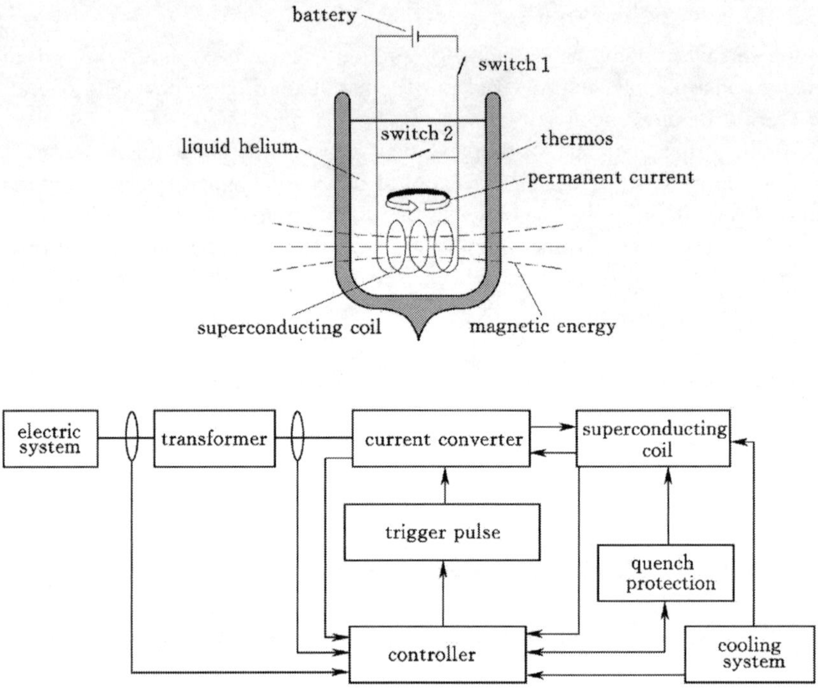

FIGURE 2.9 SMES energy storage schematic diagram and structure diagram.

2. Explore the superconducting materials with new high performance and high critical temperature. Develop new types of superconductors with low cost, higher critical temperature, and current density to further improve the superconducting wire/tape critical current density, mechanical characteristics, and thermodynamic properties.
3. Research cryogenic cooling technology and other related technologies, such as low-temperature systems with high reliability and conduction cooling technology, low-loss current leads, magnetic power supply, control and protection, etc. Combine superconducting technology with power electronics technology to integrate the function of superconducting transformers, superconducting energy storage, and active filtering, etc.

2.3.2.3 Application status

The research and development of SMES mainly began in developed countries such as the United States, Japan, and Europe in the 1970s. The SMES system is not only used to load peak, but it also can store the emergency standby power. For small- and medium-sized SMES, especially mini-SMES, its high-speed regulating characteristics of the active and reactive power can be used to improve the power factor, steady power grid frequency, control the instantaneous fluctuation of voltage, uninterruptible power supply to important users, and other functions. As a result, it greatly improves the quality of power supply, meeting military, industrial, and civilian power needs. In addition, an SMES system is commonly used in the power generation and wind power generation system and places that have strict requirements for the quality and reliability of power supply.

At present, only an American superconductor can provide the SMES products world-wide. Its products mainly include uninterruptible power supply and distributed power supply for low-temperature superconducting energy storage. While in China, there are laboratory products in Institute of Electrical Engineering of the Chinese Academy of Sciences, Tsinghua, etc. However, there is still a great distance from commercial application.

2.4 New type energy storage

2.4.1 Advanced lead–acid battery

Continuous development of an advanced lead–acid battery has made it still competitive. The UltraBattery and lead–carbon battery are the new types of lead–acid battery attracting much attention in recent years. Integrating characteristics of traditional lead–acid batteries and supercapacitors, these two types can significantly improve performance of the traditional lead–acid battery, especially the high current charge and discharge characteristics, and their service life can be three to four times that of the traditional lead–acid batteries.

2.4.1.1 UltraBattery

UltraBattery was jointly developed and produced by Australia-based CSIRO and The Furukawa Battery of Japan. At present the available reports on UltraBattery basically come from CSIRO and Furukawa. There are few research reports in China, and many studies are in the exploration stage.

UltraBattery is a hybrid energy storage device, composed of a lead–acid cell and asymmetric supercapacitor that are connected in parallel with an internal noncontrolled circuit. Fig. 2.10 is the schematic diagram of the UltraBattery structure. In the UltraBattery, the positive plate of the lead–acid cell is made of lead dioxide, and the negative plate is made of porous lead material. The UltraBattery integrates the lead–acid battery and supercapacitor in both the structure design and the use. When the UltraBattery charges or discharges at a very high rate, the supercapacitor can provide a high power to buffer the current and protect the lead–acid battery components and prolong service life of the battery.

The UltraBattery has the following advantages: energy density is similar to the lead–acid battery; high charge rate; less sulfation at the negative pole, and the cycle life is prolonged; and it is safer and less expensive than the lead–acid battery.

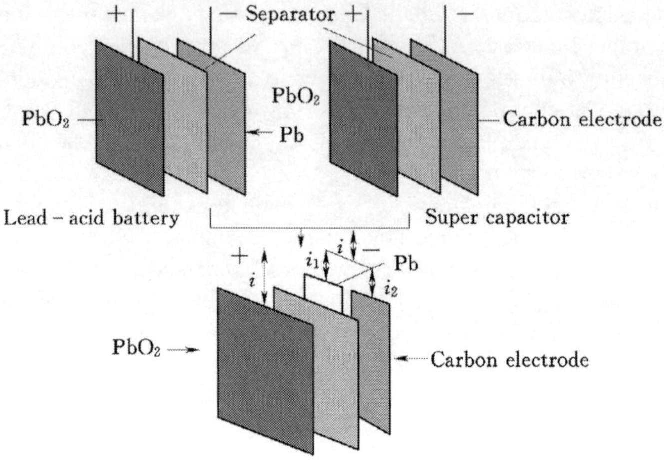

FIGURE 2.10 Schematic diagram of UltraBattery structure.

2.4.1.2 Lead–carbon battery

Developed by Axion in 2004, the lead–carbon battery is a mixture of lead–acid battery and supercapacitor. Fig. 2.11 shows the schematic diagram of the lead–carbon battery. The carbon materials with capacitance characteristics or high conduction characteristics (for example, activated carbon or carbon black) are directly injected in the negative pole in the process of creaming to improve utilization of lead active substance and constrain sulfation.

As the lead–carbon battery further simplifies the process on the basis of "combining" supercapacitor, and its fabrication is very close to the technique of the traditional lead–acid battery, it is easy to industrialize. Therefore the Advanced Lead–acid Battery Consortium has attached great importance to the development of the lead–carbon battery, and its members have carried out researches in the carbon material and lead–carbon negative pole. Today the research mainly focuses on the carbon material development, lead–carbon formula design, and battery structure optimization.

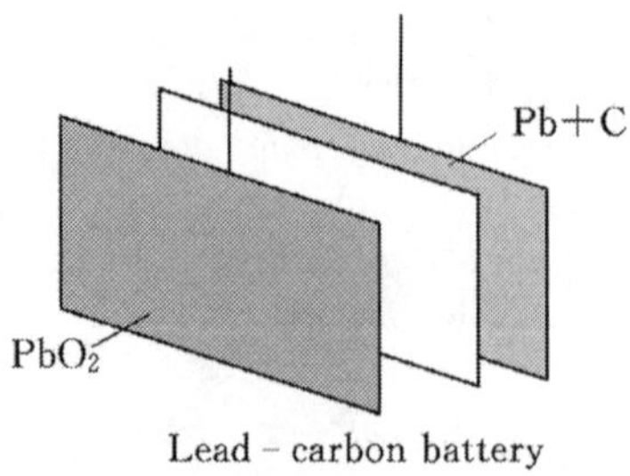

FIGURE 2.11 Schematic diagram of lead–carbon battery.

2.4.2 Lithium–sulfur battery

The lithium–sulfur battery is a member of the lithium-ion battery and is under development. Its advantage lies in the high energy density that is several times that of the traditional lithium-ion battery, theoretically 2600 Wh/kg, with open circuit voltage of 2 V. But the actual energy density is much lower than the theoretical value. For example, the actual value of lithium–sulfur battery developed by Sion Power is only 350 Wh/kg.

The lithium–sulfur battery is composed of the metal lithium negative pole and elemental sulfur positive pole. Its working principle is shown in Fig. 2.12. During discharge, the negative pole metal lithium dissolves in the electrolyte, and the lithium ion moves to the sulfur positive pole and reacts with sulfur to form polysulfide ion (Li_2S_X). During charge, the polysulfide ion decomposes, and the lithium ion returns to the negative pole. The whole chemical reaction process is as follows:

$$S_8 + 16e^- + 16Li^+ \rightarrow 8Li_2S \quad (2.11)$$

The following is the reaction equation of a series of polysulfide ion:

$$S_8 \rightarrow Li_2S_8 \rightarrow Li_2S_6 \rightarrow Li_2S_4 \rightarrow Li_2S_2 \rightarrow Li_2S \quad (2.12)$$

With a reaction that occurs from left to right, the solubility of polysulfide ion is lowered. When the reaction approaches the final stage, Li_2S_2 and Li_2S are separated out from the solution and form a solid precipitate that affects free movement of positive pole sulfur and consequently the battery capacity. Meanwhile, the dissoluble polysulfide ion moves to the lithium negative pole to cause serious self- discharge and shorten the service life of the battery. Currently the research focuses of lithium–sulfur battery are to improve sulfur content of the positive pole, design a stable conduction structure for the sulfur positive pole, develop a new type electrolyte that is compatible with both sulfur pole and lithium metal, etc.

2.4.3 Sodium-ion battery

The sodium-ion battery was developed by Aquion Energy of the United States in 2009. It is an asymmetric hybrid supercapacitor using low-cost activated

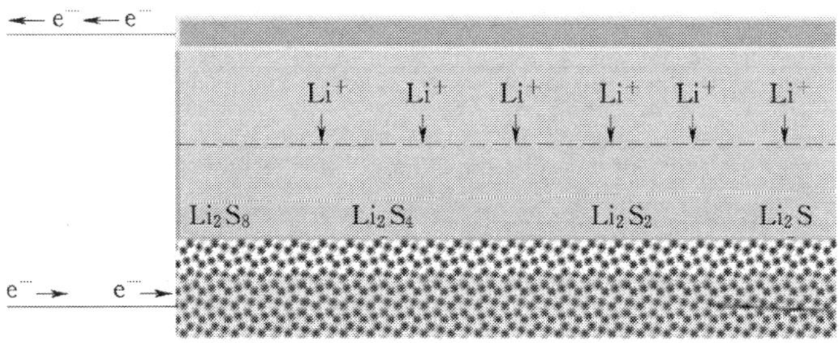

FIGURE 2.12 Working principle diagram of lithium–sulfur battery.

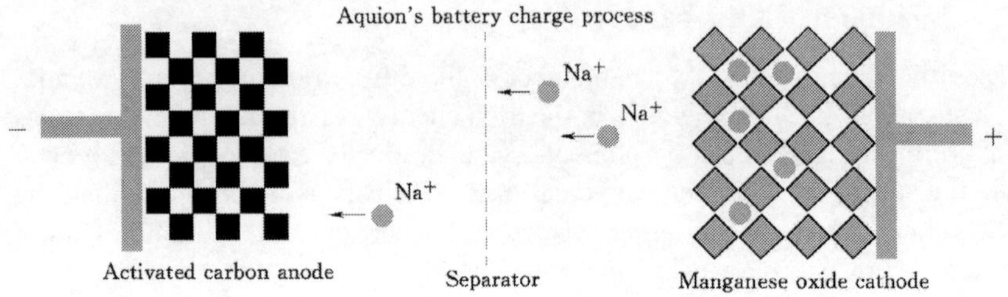

FIGURE 2.13 Working principle of asymmetric hybrid supercapacitor using water-based sodium ion electrolyte.

carbon anode, sodium manganese oxide cathode, and aqueous sodium ion electrolyte. Fig. 2.13 shows its working principle. During the battery charge, the cathode sodium ion is separated from the sodium manganese oxide and moves toward the activated carbon capacitor with negative charge. During the battery discharge, the sodium ion with positive charge returns to the cathode and combines with manganese to form manganese oxide and generate electric current.

The sodium-ion battery has the following advantages: low costs of raw materials; chemical reaction is free of corrosivity; long time charge will not cause battery damage, degradation, or self- discharge; and long cycle life.

At present the sodium-ion battery is at the technical verification stage. The DOE of the United States only has one sodium-ion battery project (14 kW/56 kWh sodium-ion energy storage testing system) in its global energy storage database. Aquion Energy started mass production in 2013 to explore the energy storage market of the high-capacity and low-cost battery.

2.4.4 Heat pump energy storage

Heat pump energy storage is a simple, low-cost energy storage technology. It generates hot air and cold air and stores them with mineral grains (or detritus). The heat pump compresses or inflates argon with the original Ericsson cycle system (or closed Brayton circulating system), and it conducts heat or cold air with the working fluid to assist the energy storage system. With innovation and development of modern technology and new materials, progresses have been made in the mechanical configuration, piston structure, valve design, and sealing design. The technical progresses enable the heat pump energy storage system to operate economically and maintain a high working efficiency at different temperatures.

Fig. 2.14 describes the charge and discharge process of the heat pump energy storage system. The charge process needs electricity to drive the heat

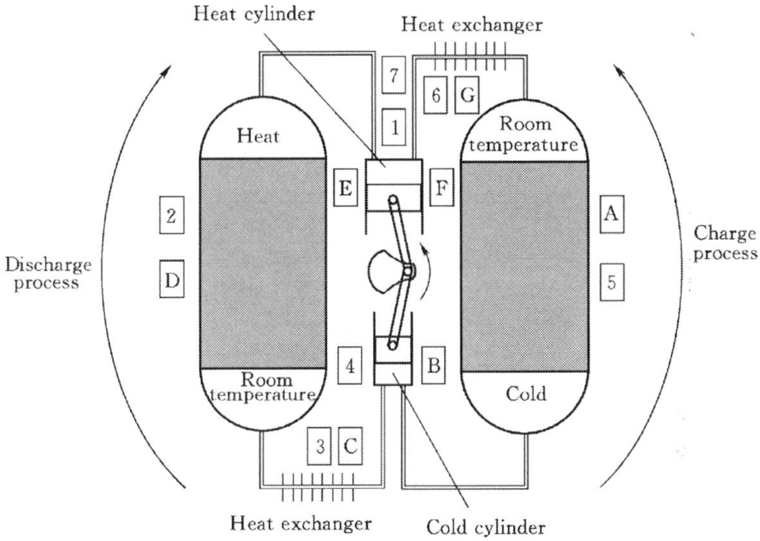

FIGURE 2.14 Schematic diagram of the charge–discharge cycle of the heat pump energy storage system.

pump to transmit heat and cold air and store them separately in the thermal storage tank and the cold storage tank till the hot frontal reaches the heat exchange critical value and the hot air and cold air pass the thermal storage tank and cold storage tank directly. At this moment the system is fully charged and stops charging. When in need of discharge, the heat pump serves as the generator to generate electricity.

2.4.4.1 System charge process

1. Drive the heat pump to inhale argon and compress the gas to 12 bar and heat to about 500°C.
2. Argon enters the heat exchange space that is filled with mineral grains, and heat is absorbed by the grains. At this moment the mineral grains are heated to 500°C, and argon is cooled to the ambient temperature (keeping at 12 bar).
3. Argon is discharged from bottom of the heat storage tank and passes through the heat exchanger to ensure the gas reaches the system ambient temperature.
4. Cooled argon enters a small expander to inflate gas to the system ambient pressure and cool to about −160°C.
5. The cooled argon enters the bottom of the cold storage tank and is cooled to −160°C when passing through the mineral grains. Argon restores to the system ambient temperature at the top of the cold storage tank.

6. Argon passes through another heat exchanger and is guaranteed to reach the system ambient temperature.
7. The returned argon reaches the system ambient pressure and temperature and reenters and flows through the compressor. The charge process starts again.

2.4.4.2 System discharge process

1. Argon flows through the top of the cold storage tank and is cooled by the mineral grains (keeping below the ambient pressure). The mineral grains are heated to the system working temperature.
2. Argon at a temperature of −160°C is discharged from the bottom of the cold storage tank and enters the small compressor at the bottom of the heat pump. Argon is compressed to 12 bar and the system ambient temperature.
3. Argon flows through the heat exchanger and is guaranteed to reach the system ambient temperature.
4. Argon enters the bottom of the heat storage tank and flows upward to reach the top of the tank. When passing through the mineral grains, the gas is heated to 500°C.
5. Heated argon leaves the heat storage tank to enter the expander at the top of the heat pump.
6. Gas is inflated to the system ambient pressure and temperature.
7. Argon flows through the heat exchanger again to ensure that it reaches the system ambient pressure and then enters the top of the cold storage tank. The discharge process starts again.

One of the biggest advantages of the heat pump energy storage system is the flexibility of site. Different from the pumped storage technology and compressed air energy storage technology, the heat pump energy storage technology does not depend on reservoir or underground cavern or other natural environments. It can be installed and constructed at any location when necessary and is not subject to the geologic constraints. Moreover, the materials used for the heat pump energy storage system are environmentally friendly, rich in reserve, and low cost. The floor area is smaller than the pumped storage power station.

2.4.5 Gravity energy storage

The gravity energy storage is developed from the principle of pumped storage, and its working principle is shown in Fig. 2.15. The gravity energy storage system consists of two underground silos (energy storage silo and backwater silo) with a diameter of 2−10 m and 500−2000 m depth. The energy storage silo is equipped with a series of gravity pistons, while the backwater silo is filled with free-flowing water. The two silos are located very close to each other and connected to form a closed loop by an interconnected channel.

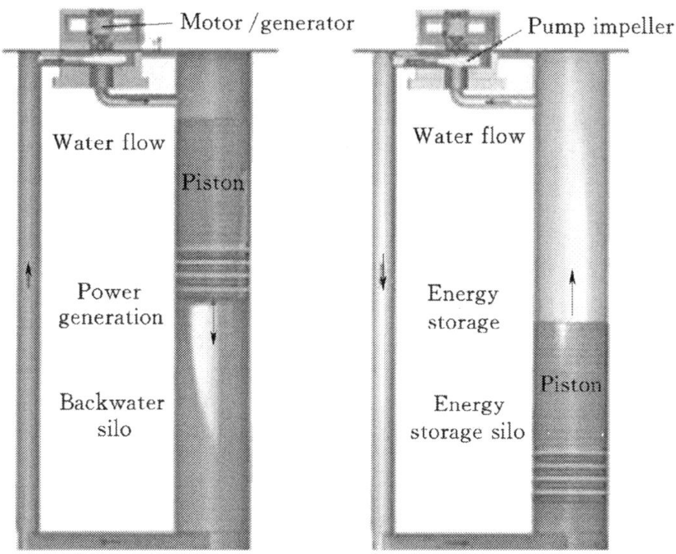

FIGURE 2.15 Structure of gravity energy storage system.

The working process of the system is simple. With upward and downward movement of the gravity pistons, it completes the energy storage and discharge.

In the process of energy storage, the water turbine of the gravity energy storage system serves as the water pump and runs reversely to use nonpeak energy or surplus energy generated by renewable energy to pump water above the piston of the energy storage silo to the backwater pipe, which flows to the silo below the piston through the interconnection pipe. Under the rising water pressure, the gravity piston is raised to the upper part of the silo, and electricity is transferred to the piston gravitational potential energy and stored.

In the process of power generation, the piston is released. With the piston position lowering, water below the piston is forced into the backwater silo through the interconnection channel and returns to the silo above the piston through the water turbine. In the process, the water turbine is driven by flowing water to run and drive the generator to generate power and transfer the stored gravitational potential energy into electricity.

Advantages of the gravity energy storage system are that efficiency is as high as 75%–80% and there are low production costs, about US$69–113/MWh. And the construction cost is US$1000/kW and US$1900/kW, respectively, for the peak load power station. Compared with the gas load shaving power stations, it is very competitive. Also, it has zero emissions, it does not need to continuously supplement work medium, it has a constant system pressure, rapid response, and high flexibility in site selection, and a short construction cycle.

Disadvantages of the gravity energy storage system are that at present there is no technical or commercial verification in practice. The main concerns are put on the silo wall leakage and the deep depth of the silo, high weight of the gravity piston, and maintenance difficulties, among others, that constrain development of the technology. Once any problem occurs, the system needs to shut down, and the repair is costly, especially when any fault occurs at underground 2000 m. Moreover, the sealing mechanism between the edge of the gravity piston and the silo wall is not well defined, which is the most important part concerning safe operation of the system and directly affects the energy storage and release effect of the gravity piston.

2.5 Comprehensive comparison of energy storage technologies

2.5.1 Technical maturity

See Fig. 2.16 for the technical maturity of various energy storage technologies. Based on the current situation of various energy storage technologies, they can be divided into the mature application, early stage of industrialization, and initial research states. The lead—acid battery is the most mature technology of chemical battery, and pumped storage is the most mature physical energy storage

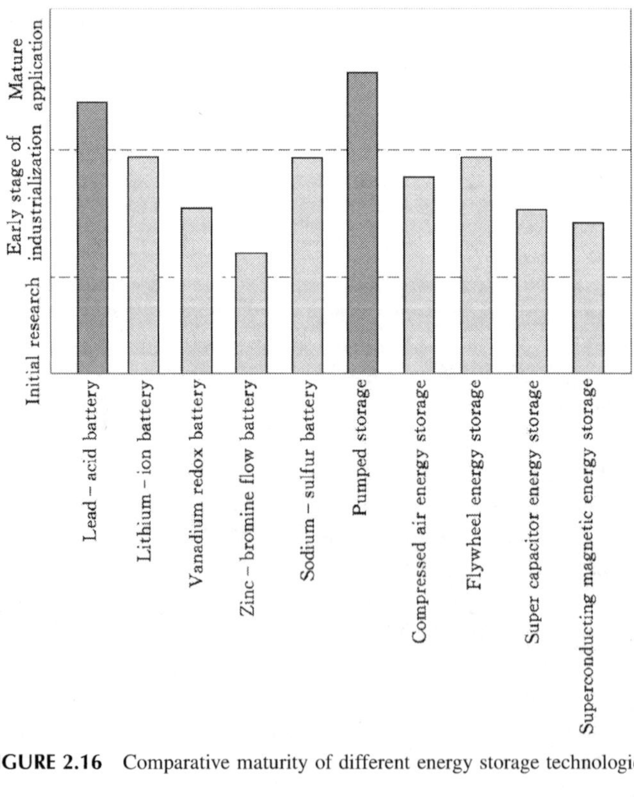

FIGURE 2.16 Comparative maturity of different energy storage technologies.

technology, which has been used for more than 100 years. Lithium-ion battery, vanadium redox battery, zinc–bromine flow battery, sodium–sulfur battery, CAES, flywheel energy storage, supercapacitor energy storage, and SMES are relatively mature technologies and commercially viable, but they are still at the early stage of industrialization and have some way to go to mass production.

2.5.2 Performance parameters

2.5.2.1 Power level and discharge time

See Fig. 2.17 for the power level and discharge time of different energy storage technologies. According to the energy storage applications, they can be divided into the following three categories:

1. Large-scale energy storage. The pumped storage and CAES are suitable for the applications of above 100 MW, and they provide hour-level electricity output in daytime and play an important role in load tracing and modulation and peak–load shifting, etc. The large-scale battery energy storage systems are suitable for medium-scale applications of 10–100 MW.
2. Energy quality adjustment. The battery energy storage, flywheel energy storage, supercapacitor, and SMES feature a rapid response, close to the ms level, and they are suitable for improving the energy quality, for example checking the voltage drop and voltage flicker. The typical power application is lower than 1 MW.
3. Backup power supply. Lead–acid batteries, lithium-ion batteries, and flow batteries not only have the characteristics of fast response (<1 s), but they also have a longer continuous discharge time. They are more suitable for the backup power supply of the system. The typical power application is 100 kW–10 MW.

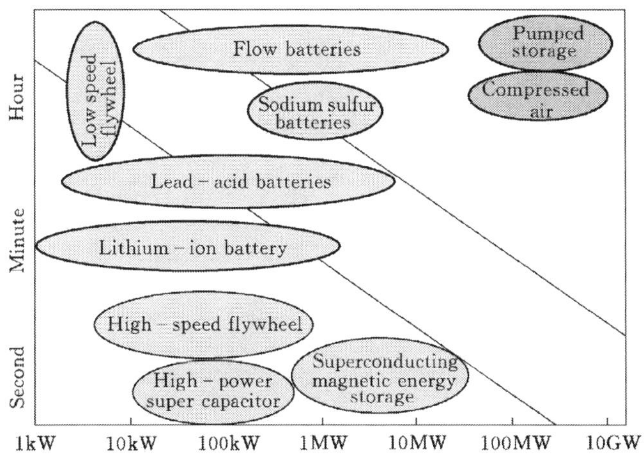

FIGURE 2.17 Comparison of power level and discharge time of various energy storage technologies.

2.5.2.2 Energy/power density

Energy density refers to the output energy of the energy storage system at unit volume or unit mass, while power density refers to the output power of the energy storage system at unit volume or unit mass. The energy storage system includes energy storage unit, auxiliary equipment, support structure, and electricity transform equipment. Similar products may have different energy density and power density because of differences of manufacturing process.

From Table 2.5, we can see the battery energy storage products have a relatively high energy density; the energy density of pumped storage, supercapacitor, and SMES is lower than 30 Wh/kg. However, flywheel energy storage, supercapacitor, and superconducting energy storage have a very high power density, and they can discharge in a large flow and response rapidly. They are suitable for coping with voltage sag and momentary power failure, improving users' electric power quality, constraining the low-frequency fluctuation of the power system, and improving the system stability. The energy density of lithium-ion battery and sodium—sulfur battery is 3—4 times that of the traditional lead—acid battery, while the energy density of vanadium redox battery and zinc—bromine battery is greater than that of conventional lead—acid batteries.

2.5.2.3 Self-discharge

See Table 2.6 for the comparison of characteristics of various energy storage technologies.

From Table 2.6, we can see that the self- discharge rate of lead—acid battery and lithium-ion battery is 0.1%—0.3% per day at the room temperature, and the resting period should not exceed dozens of days. The self- discharge of vanadium redox battery and zinc-bromine battery is relatively low and has a relatively long resting period. Because of high working temperature, sodium—sulfur battery needs a self- heating system to maintain operation, resulting in a self- discharge rate as high as 20% per day. The pumped storage and CAES have a very low self- discharge. The self- discharge rate of flywheel energy storage is 100% per day, and the best discharge time should be controlled within the minute level. The self- discharge of supercapacitor and SMES is 10%—40% per day, and the longest cycle is several hours.

2.5.2.4 Cycle efficiency

The single cycle efficiency is defined as the ratio of discharge capacity to charge capacity in a single cycle, regardless of the self- discharge loss. Fig. 2.18 compares the single cycle efficiency of various energy storage systems, which can be divided into two categories:

1. Lithium battery energy storage, flywheel energy storage, supercapacitors, and SMES systems have a high recycling efficiency, higher than 90%.
2. The other kinds of batteries and pumped storage, CAES systems have a relatively high cycle efficiency (60%—90%). CAES system includes the gas compression and inflation processes, and before rapidly compressing air, it needs preheating that consumes thermal energy; accordingly, its cycle efficiency is lower than that of the pumped storage system.

TABLE 2.5 Characteristics of various energy storage technologies.

Energy storage technology		Energy/power density		Typical rated power (MW)	Discharge time	Characteristics
		Wh/kg	W/kg			
Electrochemical energy storage	Lead–acid battery	30–50	75–300	<100	Minute to hour	Mature technology, relatively low cost, short cycle life, and environmental problems
	Lithium-ion battery	75–250	150–315	<100	Minute to hour	High specific energy, high cost, group application needs further improvement
	Vanadium redox battery	40–130	50–140	<100	Minute to hour	High battery cycle index, allowing deep charge and discharge, suitable for combination, but the energy storage density is low
	Zinc–bromine battery	60–80	50–150	<10	Minute to hour	
	Sodium–sulfur battery	150–240	90–230	0.1–100	1–10 h	Relatively high specific energy, high cost, operating safety needs to be improved
Mechanical energy storage	Pumped storage	0.5–1.5	–	100–5000	4–10 h	Applicable to large-scale application, mature technology, slow response, needs geographic resources
	Compressed air	30–60	–	10–300	1–20 h	Applicable to large-scale application, slow response; needs geographic resources
	Flywheel energy storage	5–130	400–1600	0.005–1.5	15 s–15 min	Relatively high specific power, high cost, high noise
Electromagnetic energy storage	Supercapacitor energy storage	0.1–15	500–5000	0.01–1	1–30 s	Quick response, high specific power, high cost, low energy storage
	Superconducting energy storage	0.5–5	500–2000	0.01–1	2 s–5 min	Quick response, high specific power, high cost, difficult in maintenance

TABLE 2.6 Comparison of characteristics of various energy storage technologies.

Energy storage technology		Self- discharge (per day) (%)	Cycle life		Investment cost	
			Service life (year)	Cycle index	US$ (kW)	US$ (kWh)
Electrochemical energy storage	Lead–acid battery	0.1–0.3	5–15	500–1000 (80% DOD)	280–910	70–420
	Lithium-ion battery	0.1–0.3	5–15	>1000 (90% DOD)	490–1400	280–1400
	Vanadium redox battery	Very small	5–10	>10,000 (100% DOD)	3500	140–1400
	Zinc–bromine battery	Small	5–10	>10,000 (100% DOD)	700–2520	140–980
	Sodium–sulfur battery	20	10–15	>4500 (90% DOD)	980–2800	140–280
Mechanical energy storage	Pumped storage	Very small	40–60	–	700–5040	84–210
	Compressed air	Small	20–40	–	560–1610	14–168
	Flywheel energy storage	100	>15	>20,000	140–420	1400–4900
Electromagnetic energy storage	Supercapacitor energy storage	20–40	>10	>100,000	140–560	420–5600
	Superconducting energy storage	10–15	>20	>100,000	140–560	980–9800

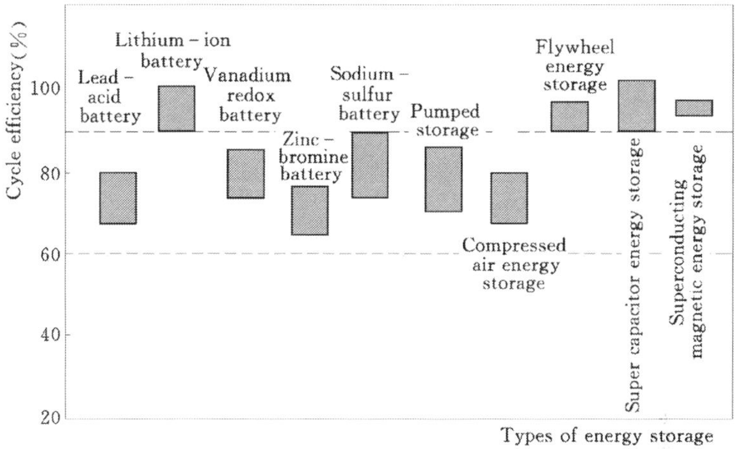

FIGURE 2.18 Comparison of single cycle efficiency of energy storage systems.

2.5.2.5 Cycle life

Table 2.6 compares the service life and cycle index of different energy storage technologies. Take the electrochemical energy storage, for example. Because of degradation degree of the performance of electrode materials, liquid flow batteries and sodium sulfur batteries have a much higher cycle index than lead—acid batteries and lithium-ion batteries. The service life of physical energy storage technologies such as pumped storage, CAES, and flywheel energy storage is mainly subject to the service life of mechanical parts of the systems and affected highly by the traditional mechanical engineering technology. The cycle index of an electromagnetic energy storage system can be as high as tens of thousand times.

2.5.2.6 Investment cost

Cost is an important element affecting whether the energy storage technology can be popularized. Table 2.5 lists the unit power cost and unit capacity cost, taking into account the energy conversion efficiency of energy storage systems.

The investment cost of a lead—acid battery is relatively low, but its cycle life is short and investment in its energy management applications is not the lowest. Considering differences of technology and applications, the cost does not include operation, maintenance, recycling, or other expenses.

Pumped storage and CAES have a relative low capacity cost; especially, CAES features a high economy performance among mature energy storage technologies. Cost of battery energy storage is lowered with technological improvement, but still higher than the pumped storage. Flywheel energy storage, supercapacitor, and SMES technologies feature a high power density and a low unit power cost, and they are applicable to the applications of high power and short time. Their unit capacity cost is very high.

2.5.3 Applications

Applications of different energy storage technologies can be summarized as follows:

1. For the applications of low power and long time, the lithium-ion battery is the best choice; the key technology is the battery grouping and lowering self- discharge rate of the system.
2. In remote areas with electricity supply from intermittent renewable energy, the key element of configuring an energy storage system with small capacity (dozens of kWh) is the price/performance ratio. At present, the lead–acid battery has a higher price/performance ratio, but its price is relatively high.
3. A high-capacity energy storage system is required in the large grid peak–load shaving (>100 MWh); pumped storage and CAES systems have obvious economic advantages; the capacity of the energy storage system used for load leveling of the distribution network is between 1 and 30 MW; the rapid response and configuration flexibility of the battery energy storage make it more applicable to such applications; however, the large-scale promotion of the battery energy storage needs further verification.
4. The key factor used for energy quality regulation lies in the instantaneous discharge capacity and cycle life. Flywheel energy storage and super-capacitor energy storage have more advantages than lithium-ion battery energy storage.

Of all energy storage technologies, the lead–acid battery technology is mature but constrained by the cycle life.

In the actual power system, not a single energy storage technology can fully meet all needs of various applications. In the field of new energy power generation applications, the DOE and the Electric Power Research Institute of the United States attach importance to development of mutual complementary integration and overall control technology of clean energy with diversified energy storage technologies such as compressed air and battery energy storage, and they have deployed a number of important demonstration projects. Europe tends to use water or underground compressed air to solve the large-scale PV and wind power access and utilization, and it uses battery energy storage to solve access and consumption of new energy power generation according to actual situation of resources. Japan has taken a lead in battery energy storage device manufacturing technology and attached importance to improving the large-scale new energy access capacity with wind power and solar power integration, and it has implemented many large-scale wind and solar power energy storage combined generation demonstration projects.

China mainly advocates pumped storage and battery storage. Pumped storage technology is mature, but it is constrained by water resources and geographic conditions, and the construction period is long and investment is high. Therefore, a large-capacity battery energy storage system has gradually become a research focus. At present, the initial formation of large-capacity lithium-ion battery industrialization has laid a foundation for large-scale energy storage applications, while breakthroughs have been made in research of flow batteries and sodium—sulfur batteries.

References

[1] Xiaoya Z, Yin Z, Sun Z. Application and prospect of advanced energy storage technology in smart grid. Electric Age 2011;(1):44—7.

[2] Li W, Hu Y. Present situation and research progresses in its service life of dynamic lead — acid battery. China Manufacturing Information 2011;40(4):70—2.

[3] Yang Y. Research status and development prospects of new lithium — ion battery cathode materials. Advanced Materials Industry 2010;(10):11—4.

[4] Zhao P, Zhang H, Wang H, et al. Research status and prospects of vanadium redox energy storage battery. Journal of Shenyang Ligong University 2009;28(2):1—6.

[5] Zhang H, Zhang Y, Liu Z, et al. Research progress in the technology of flow batteries. Progress in Chemistry 2009;21(11):2333—40.

[6] Xiao Y, Yan M. Zinc — bromine flow battery based energy storage technology. Oriental Motor 2012;(5):80—4.

[7] Sun B, Jiang J, Shi W, et al. Research on the status of sodium —sulfur battery energy storage applications. Modern Electric Power 2010;(6):62—5.

[8] Zhang X. Research on some problems of compressed air energy storage system. Beijing: Graduate University of Chinese Academy of Sciences; 2011.

[9] Zhang W, Zhu W. Key technology of flywheel energy storage and its development status. Transactions of China Electrotechnical Society 2011;26(7):141—6.

[10] Haigang W, Dai X, Zhang L, et al. New developments of flywheel energy storage technology. Acta Energiae Solaris Sinica 2002;23(6):748—53.

[11] Yuzhong H, Yan L, Jiang Y, et al. Research progress in superconducting magnetic energy storage system (I) — superconducting energy storage devices. Automation of Electric Power Systems 2001;25(12):63—8.

[12] Ge Z, Zhou L, Zhao W, et al. Recent development of ultrabattery. Chinese Journal of Power Sources 2012;36(10):1585—8.

[13] Rastler D. Electricity energy storage technology options. America: Electric Power Research Institute; 2010.

[14] Barnes FS, Levine JG. Large energy storage systems handbook. America: Taylor and Francis Group, LLC; 2011.

[15] Kintner — Meyer M, Elizondo M, Balducci P, et al. Energy storage for power systems applications; a regional assessment for the northwest power pool. America: Pacific Northwest National Laboratory; 2010.

[16] Schoenung S. Energy storage systems cost update. America: Sandia National Laboratories; 2011.

[17] China Energy Storage Alliance. Energy storage white paper of China 2011. China: China Energy Storage Alliance; 2011.
[18] China Energy Storage Alliance. Energy storage white paper of China 2012. China: China Energy Storage Alliance; 2012.
[19] China Energy Storage Alliance. Energy storage white paper of China 2013. China: China Energy Storage Alliance; 2013.

Chapter 3

Technologies for energy storage battery management

Chapter outline

Battery management is of particular importance given the great improvements in the manufacturing process of energy storage batteries. Battery management is not only important for normal and stable operation of batteries but also necessary for extending the battery's service life. The development of battery management systems is critical to the energy storage system made up of thousands of batteries. Through continuous technical upgrading, other

Grid-scale Energy Storage Systems and Applications. https://doi.org/10.1016/B978-0-12-815292-8.00003-4

countries have developed relatively mature battery management systems (BMSs), including representative Smart Guard, LGCPI Battery Packs, and BMS 4C. These BMSs not only boast accurate monitoring technologies and efficient balance modules but also have good methods for battery temperature control to ensure reliability and safety.

Key technologies for energy storage battery management mainly include SOC (state of charge) estimation, SOH (state of health) estimation, balance management, and protection. SOC is the key index that reflects the real-time residual capacity of energy storage batteries. SOH is the basis for judging whether the energy storage batteries have normal operation capacity. Balance management is necessary to secure energy storage batteries under good operational conditions. Protection technologies ensure that energy storage batteries operate properly. This chapter introduces the typical structures and main functions of BMSs and elaborates the key technologies for SOC estimation, SOH estimation, balance management, and protection.

3.1 Battery management systems

3.1.1 Typical structures

The main objective of a BMS is to ensure the safe and stable operation of batteries, improve the cycle efficiency, and extend the service life of batteries. Given the huge battery information and management demands in a large-scale energy storage system, the hierarchic management mode may be applied in a BMS. Based on the battery cluster modes in energy storage systems, a BMS has two kinds of typical structures, i.e., the two-tier topology with the application of battery modular management units (BMMUs) and battery cluster management units (BCMUs) and the three-tier topology which applies BMMUs, BCMUs, and battery array management units (BAMUs).

3.1.1.1 The structure of two-tier topology

The two-tier topology BMS as illustrated in Fig. 3.1 may be applied in the case of a small battery energy storage system and energy storage with a single cluster of batteries. The BMS, consisting of multiple BMMUs and one BCMU, applies a CAN bus for data transmission within the system to secure high reliability and efficiency of communications. Under such a structure, a BMMU can be flexibly adjusted based on the number of battery cubicles to facilitate information collection and management of energy storage batteries. The hierarchic relationship between a BMMU and BCMU is as follows.

A BMMU can have automatic online tests of the voltage and temperature of monomer batteries in the battery module, the terminal voltage of battery cubicles, charge/discharge current and temperature, sound an alarm on voltage, current and temperature values that are beyond limits, identify the maximum/minimum voltage and maximum/minimum temperature of monomer battery

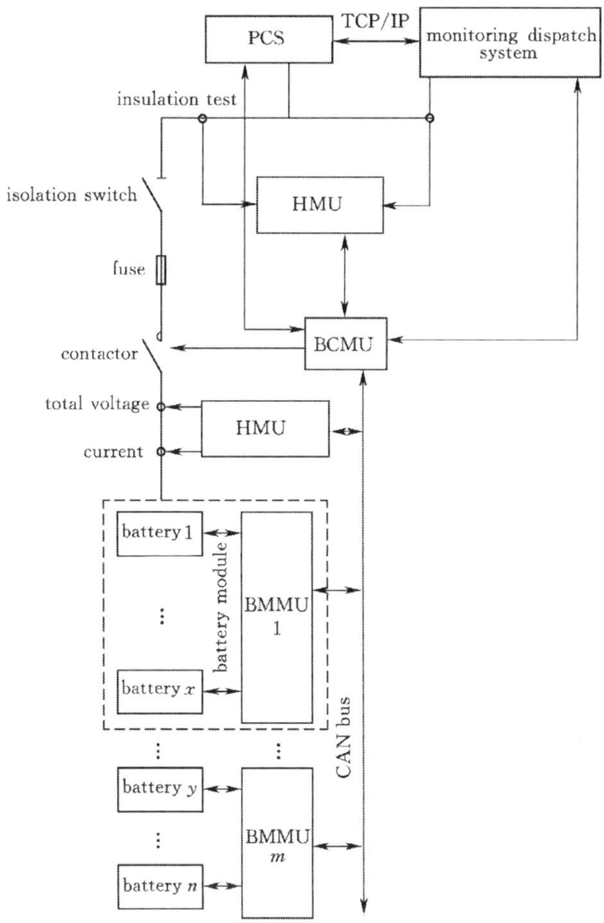

FIGURE 3.1 The two-tier topology in a BMS.

through analysis, work out the SOC of the battery module, and realize the balance among batteries inside the module. In general, a BMMU integrates the functions of battery operation information collection, charge balance management, and failure diagnosis and sends the information collected to the BCMU.

A BCMU and BMM have real-time communication with each other to obtain battery data (monomer battery's voltage and temperature, battery module's SOC) and obtain and identify failure information. In addition, a BCMU can carry out the POST (power-on-self-test) and the insulation test of the battery's positive and negative poles to the casing, control the contactor's closure, test the terminal voltage and current of battery clusters, calculate the SOC of battery clusters, and realize the balance between battery modules. After analyzing the comprehensive information of battery clusters, a BCMU can effect intelligent interaction with the PCS and monitoring dispatch system via an independent CAN bus or RS485.

3.1.1.2 The structure of three-tier topology

The three-tier topology BMS as illustrated in Fig. 3.2 may be applied in the case of a large battery energy storage system and energy storage with multiple clusters of batteries. The BMS, consisting of multiple BMMUs and BCMUs and one BAMU, applies CAN buses for data transmission within the system. Likewise, such topology is convenient for the replacement and upgrading of a single battery cluster, contributing to capacity expansion and maintenance of the whole battery storage system.

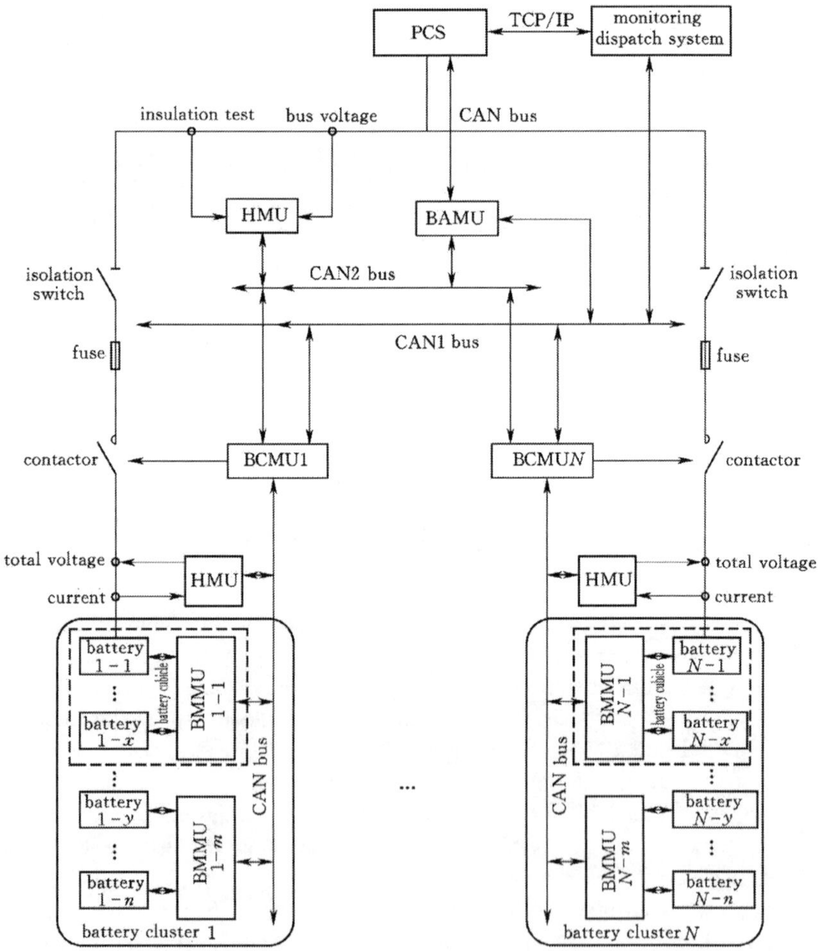

FIGURE 3.2 The three-tier topology in a BMS.

The hierarchic relationship among BMMU, BCMU, and BAMU is as follows:

The function of a BMMU in this three-tier topology is consistent with that of the BMMU in the previous two-tier topology.

The function of a BCMU in this three-tier topology is consistent with that of the BCMU in the previous two-tier topology.

The BAMU pools information of all BCMUs, monitors the operation status, and calculates the SOC and SOH of the whole battery energy storage system, and it communicates with the energy storage converter and higher level monitors. According to a control strategy, the BAMU issues control instructions to BCMUs to manage the access and disconnection of all battery clusters as a whole.

The topological structure of a BMS is not only restricted by the cluster mode of energy storage batteries but also related with the cost and spatial distribution of energy storage batteries, as well as the structure of battery cubicles. So, a BMS should be properly set up based on different management demands of various battery energy storage systems.

3.1.2 Main functions

To ensure normal energy exchange between the battery energy storage system with the grid, the BMS must have the real-time monitoring and uploading of relevant parameters of energy storage batteries, monitor the overall status of the batteries, and automatically repair in time. The main functions of the BMS should include the following aspects.

3.1.2.1 Battery parameter test and management

A battery parameter test is a basic function of a BMS. Battery parameters directly reflect the operation status of batteries and provide data support for other functions of a BMS. Battery parameters will be frequently used in follow-up calculations, and their reliability and accuracy will be carried forward. Therefore, the accuracy of battery parameters is of particular importance.

Parameters to be tested by a BMS include monomer battery voltage, total voltage, charge/discharge current, battery cubicle temperature, status of fling-cut switch, and insulation resistance. Since each battery cluster consists of hundreds of monomer batteries in series, the quality of each monomer battery will influence the performance of the whole battery cluster. Thus, the voltage of each monomer battery must be monitored and tested. Total voltage, as an important parameter of battery clusters, describes the external voltage characteristics of battery clusters. Charge/discharge current, the current flowing through battery clusters in case of energy exchange, is a key parameter for calculating the battery's SOC and an important factor that influences the battery's performance. The operation temperature and distribution of batteries are key parameters that influence the performance of the whole battery cluster.

To ensure the normal operation of batteries and the consistency of operation conditions, real-time monitoring and adjustment to temperatures must be carried out. Each battery cubicle consists of multiple battery modules. In case of failure of a certain battery cluster, the cluster should be disconnected via a diverter switch. The BMS should monitor whether the open/close position of the switches is properly placed. Insulation resistance, a parameter that reflects whether the battery energy storage system has electric leakage, concerns personnel safety.

3.1.2.2 Data communication management

The subsystems of a BMS often apply CAN bus or RS485 for data exchange and instruction control. Communications are required for the data transmission of BMS to energy storage and conversion devices and higher level monitors. So, it is important to strengthen the stability of communication links to ensure the normal operation of the whole battery energy storage system.

3.1.2.3 Online SOC diagnosis

SOC is the parameter for the residue capacity of batteries. To extend the battery's life and reduce damage to batteries, excessive charge and discharge should be avoided. SOC is one of the key indicators for monitoring charge and discharge. In addition, SOC is an important parameter under the control of a battery energy storage system. Accurate estimation of SOC is a necessary process in a BMS. There are many methods for SOC estimation that are related with charge/discharge current, temperature, and other factors.

3.1.2.4 SOH diagnosis

SOH, including the performance of capacity, power, and internal resistance, is a parameter for the life of battery clusters. The accuracy of SOH is influenced by multiple factors, such as monomer characteristics, discharge rate, temperature, and consistency of battery clusters.

3.1.2.5 Balance management

Balance management may avoid the inconsistency caused by production and operation of batteries in maximum and extend the battery's service life and energy utilization rate.

3.1.2.6 Failure diagnosis and protection

To ensure the battery's safety during the charge/discharge process, a BMS sets up safety ranges for the parameters of various batteries. In case of operations beyond the setup values of parameters, two levels for exceeding ranges are defined based on the severity. Alarms will work in case Level one is exceeded; and disconnections will work in case Level two is exceeded, which means a severe failure, and the BMS will directly disconnect the branch's contact and avoid the access of the battery cluster. Main limits in a BMS to be set up include charge/discharge overcurrent values, monomer overvoltage/undervoltage values, high/low temperature limit, SOC high/low values, and battery module overvoltage/undervoltage values.

The aforesaid functions are divided according to the application demands of the current BMS. The functions will be expanded to cater to the demands of various energy storage batteries and technical development in the future. SOC estimation, SOH estimation, balance management, and protection management technologies will be elaborated in the follow-up sections.

3.2 SOC estimation method

3.2.1 Definition

When the battery is fully charged, the state is called SOC = 100%, and when it is flat, the status is said to be SOC = 0%. Presently, from the perspective of the battery's capacity, a more unified definition of SOC is put forward by U.S. Advanced Battery Consortium (USABC) in the *Electric Vehicle Battery Test Manual:* under a certain discharge rate, SOC is the ratio of residual capacity to rated capacity.

In constant current discharge, the value of SOC is equal to the ratio of residual capacity to rated capacity under the same condition

$$\mathrm{SOC} = \frac{Q_C}{Q_t} \cdot 100\% = \left(1 - \frac{\Delta Q}{Q_t}\right) \cdot 100\% \tag{3.1}$$

Where Q_c is the residual capacity of the battery (Ah); Q_t is the capacity of the battery in constant current discharge (Ah); ΔQ is the discharge capacity of the battery.

As for electric vehicles by Honda, a Japanese brand, the definition of SOC is as follows:

$$\begin{aligned} &\mathrm{SOC} = \text{residual capacity}/\text{rated capacity} - \text{capacity attenuation factor} \\ &\text{Residual capacity} = \text{rated capacity} - \text{net discharge capacity} - \text{self}- \\ &\qquad \text{discharge capacity} - \text{temperature compensation capacity} \end{aligned} \tag{3.2}$$

The preceding equation takes into account the effects of an energy storage battery's self-discharge, temperature and aging factors. In theory, the method is ideal, but it involves large calculations and a complex relationship, lowering its credibility.

Many factors affect the accuracy of SOC measurement. Among them are open circuit voltage, temperature, charge/discharge current, and cycle times that are closely related to SOC. If any factor is ignored, the error of SOC estimation is increased.

3.2.2 The methods for SOC estimation

SOC can be calculated using the real-time test value for the external characteristics of the battery, such as battery voltage, charge/discharge current, battery resistance, and temperature. The common methods for estimation include the discharge experiment, time measurement, open circuit voltage, battery resistance, and the Kalman filter. Each method has its advantages and disadvantages, and the different methods for SOC estimation are shown in Table 3.1.

3.2.2.1 Discharge experiment

As a reliable method for SOC estimation, the discharge experiment uses constant current to continuously discharge. The product of current and time is the discharge capacity. The method is often used in the laboratory and is suitable for all batteries. But the significant disadvantage features long-term measurement, during which the battery stalls.

3.2.2.2 Time integration

Based on black box theory, the time integration method is frequently used for SOC estimation. When the black box exchanges energy with the external environment, time integration of current in and out the black box records the change of energy. Regardless of the battery's status change in the internal black box and the influence of other factors, the method is easy to be implemented. If the initial status of charge/discharge is SOC_0, the SOC can be gained by the following formula:

$$\mathrm{SOC} = \mathrm{SOC}_0 - \frac{1}{C_N}\int_0^t \eta I \mathrm{d}\tau \tag{3.3}$$

Where C_N is the rated capacity of the energy storage battery; η is the charge/discharge efficiency; I is the charge/discharge current.

However, the method faces some challenges: first, it is difficult to define the initial value of SOC; second, it is difficult to measure η, namely charge/discharge efficiency; third, at high temperature, the fluctuation of charge/discharge current is obvious, leading to a larger error. Although the current measurement can be solved by a high-performing sensor, it will push up the cost. The charge/discharge efficiency, η, can also be gained by a large number of preliminary experiments and an established empirical formula.

If the battery is in full charge/discharge mode and adopts constant current to charge itself, it will have a relatively stable initial value after the charge completes (SOC = 100%); At the same time, charge efficiency is also high (above 95%), which can be approximated as 11 or a constant value, so the error accumulated in each charge/discharge cycle is basically removed when the initial SOC value is recalibrated to complete the next charge. If the battery works in the float charge mode or the frequent handoff between charge and discharge, the initial value of the battery cluster is difficult to calibrate and the accumulated errors cannot be corrected, resulting in a larger error.

TABLE 3.1 Different methods for SOC estimation.

Methods	Application fields	Advantages	Disadvantages
Discharge experiment	All battery systems; judging the initial capacity of batteries	Easy operation; accurate data; no relation with SOH	No test support online; time-consuming; experiencing energy loss when changing the state of the battery
Time integration	All battery systems	Online test; reduced limitation by battery itself; giving full play to the advantages of microcomputer monitoring	Recording the battery's capacity externally only without solving the internal relationship of battery's capacity and SOC; high cost of measuring accurate current; sensitive to interference
Open circuit voltage	Lead–acid batteries; lithium batteries	Easy operation; low cost	No test support online; the battery is required not to be operational for a long time
Battery resistance	Lead–acid batteries; nickel–cadmium batteries	Online measurement providing information for SOH	The internal resistance with low value has complex origins; limited by battery's operation conditions, such as current and temperature; only applied into the low SOC
Kalman filter	All battery systems	Online measurement	Requiring appropriate battery model; having difficulties in determining parameters
Fuzzy inference and neural network	All battery systems	Online measurement	Requiring a number of training data on batteries

3.2.2.3 Open circuit voltage

The open circuit voltage (OCV) is easier. It is generally applied to the energy storage battery whose SOC changes significantly with OCV, especially in the beginning or ending of the discharge. The SOC and OCV for nickel metal hydride batteries have a certain linear relation (proportional relation). When the performances of lithium batteries and lead–acid batteries are totally stable, their OCV and SOC have a linear relation.

However, the method is influenced by the following factors:

The first is standing time. If the standing time is so brief that the battery voltage does not recover fully, the battery's current OCV will not be correctly reflected. If the standing time is long and self-discharge obvious, the actual value of SOC will be lower than expected, resulting in errors in measurement.

The second factor is previous charge/discharge status. Regardless of previous standing time charge/discharge status, SOC is unrelated to OCV. Under the same OCV, the errors in SOC between a resting battery that completes the charging process and a resting battery that completes discharging will be up to above 50%.

The third factor is temperature. In the case of experiencing a significant temperature change, even though the battery is in the same SOC, it will have a significant difference in OCV.

3.2.2.4 Battery resistance

This method uses alternating current on different frequencies to stimulate the battery and measure the AC resistance of the internal battery. SOC estimation is gained by an established calculation mode. The discharge capacity reaching 80% of the energy storage battery is the deciding point of the ampere-hour (Ah) method and the current-resistance (IR) method. To explain further, if the total capacity of the battery is C (Ah), then the discharge capacity that does not affect the battery's service life is 0.8 C (Ah). Within the discharge range of 0%–80%, the discharge current is taken as a sample every 1/8 of a second. Later, based on the integration of discharge current and time, the discharge capacity and the SOC of the battery can be calculated. It is namely Ah method. During the discharge range from 80% to 100%, the battery's internal resistance is measured first, and then the SOC of battery is calculated in accordance with the determined relationship between internal resistance and capacity. It is called resistance method, a supplement to the SOC calculated by ampere-hour method. In theory, the method is easy because it only considers discharge current and internal resistance of the battery. However, it ignores the influence of temperature, service life, and a monomer battery's unbalance in a cluster. Meanwhile, given that the internal resistance is complex in origin and easily influenced by a future discharge system, its calculation is not very accurate.

3.2.2.5 *Kalman filter*

The core idea of Kalman filter theory is to make optimal estimation for the status of the power system to realize minimum variance. The Kalman filter takes SOC as the internal status of the battery system and realizes minimum variance for SOC estimation by recursive method. This method is a set of recursive formulas consisting of filter calculation and filter gain calculation. The former is based on recurrence of the current, voltage, temperature, and other inputs to gain the SOC estimation; while the latter carries out recursive calculation based on a variable's statistical characteristic, and then it calculates the filter gain and estimation error.

The advantage of Kalman filter is to maintain high accuracy in the estimation process. On one hand, it has a strong modification effect on the initial value error. Even if the standing time of a battery cluster is not long enough and recursive initial value is inaccurate, their effects on SOC estimation are gradually weakened until they disappear. The estimation is therefore unbiased; On the other hand, it has a strong inhibitory effect on the noise, so it is especially suitable for a hybrid electric vehicle with rapid current change.

The disadvantages of the method mainly include that, first, its accuracy relies on the battery's electrical mode, which is the key of the algorithm, and second, it is a complex calculation. Nevertheless, these disadvantages can be overcome by using easy and reasonable battery models as well as processors with faster operating speed.

3.2.2.6 *Fuzzy inference and neural network*

These two methods establish the relationship between input and output via systematic input/output sample. The neural network uses the fuzzy logic inference and the neural network technology to estimate the residual capacity of the battery. The advantages of the method are that, first, it estimates the residual capacity of the battery using its OCV to avoid including the battery aging factor, and second, it rides on people's advanced experience in observing and researching objects, so it is simple and reliable; third, it makes full use of the strong fitting ability of neural network to curve, and the network structure required is very simple and easy to implement. The disadvantage of the neural network method is that the accuracy of the entire capacity estimation system depends not only on the accuracy of the neural network estimation, but also on the output of the fuzzy logic inference.

3.3 SOH estimation technology

3.3.1 Definition

The battery's SOH is used to reflect the battery's age. As the battery ages, its maximum discharge capacity decreases gradually. So SOH can be used as a parameter to judge the service life of the battery. The definition formula of SOH is as follows:

$$\mathrm{SOH} = \frac{Q_{\max}}{Q_{\mathrm{rated}}} \cdot 100\% \tag{3.4}$$

Where $Q_{\max}$ is the battery's maximum discharge capacity; Q_{rated} is the battery's rated capacity.

3.3.2 Methods for SOH estimation

The battery's methods for SOH estimation include full discharge test, internal resistance method, electrical conductance impedance method, electrochemical impedance spectrum method, Bayesian regression method, and fuzzy theory estimation method.

The full discharge test takes full discharge of the battery and measures its capacity. The method is time-consuming in practice and interrupts the battery's operation.

The internal resistance method means by putting load on the battery, its internal resistance can be calculated from the current and the voltage using Ohm's Law. Because the internal resistance will gradually increase as the battery ages with time, the aging of the battery can therefore be determined by measuring its internal resistance. But this method cannot accurately determine the maximum available capacity of the battery.

The electrical conductance impedance method measures the response of the voltage or the current through alternating signals at both ends of the battery. With the aging of the battery, the electrical conductance will decrease while impedance increases, thus determining the SOH of the battery. However, the method, like the internal resistance method, cannot accurately determine the maximum available capacity of the battery.

The electrochemical impedance spectrum method refers to that with the small amplitude sine wave as a disturbing signal; the impedance spectroscopy within the broad frequency range is measured to determine a battery's SOH.

Bayesian regression method is an algorithm based on relevant vector. It makes modifications on target parameters in accordance to some relative parameters of the battery, and then it estimates the battery's SOH. The method has a higher accurate inference and estimation for the unmeasurable state variable of the internal battery, not only getting the average estimated value of the battery system failure's time, but also concluding the probability distribution of failure in expected time.

On the basis of the principles of fuzzy mathematics and diagnosis, the theory estimation method is used to determine the symptoms of membership degree, and the fuzzy relationship matrix parameters and threshold, and then diagnose and estimate the battery's SOH. The method adapts well to complex and nonlinear systems, but the selection of parameters has a great influence on the SOH estimation.

3.4 Balance management technology

If the monomer batteries are inconsistent, the available capacity and cycle life of the battery cluster will consequently fall sharply. To prevent the escalation of this problem during the manufacturing process and usage, the battery cluster has to be balanced. Managing battery balance enables high-energy monomer batteries to charge at a slow speed while low-energy monomer batteries can charge faster. On the contrary, during the discharge process, it quickens the process in high-energy monomer batteries while slowing the speed at which low-energy monomer batteries discharge. This technology mainly puts the terminal voltage, the maximum available capacity, and the real-time SOC as its balance targets.

The control strategy with terminal voltage as a balance target aims at making voltages of the battery cluster gradually similar by first conducting real-time measurement on monomer batteries' voltages during the charge/discharge process and then discharging the batteries with high voltages and charging the batteries with low voltages. As the most widely used method, it is easy to operate and has a lower requirement on the algorithm. Its disadvantage is that by using monomer batteries' voltages as a balance target, it is difficult to ensure the precision and efficiency of the balance. In addition, the balance strategy cannot be applied to parallel monomer batteries.

With capacity and real-time SOC as balance targets, the control strategy aims at controlling the residual capacity of the batteries or making their SOC very close in the charge/discharge process. The capacity and SOC as batteries' parameters cannot be gained through direct measurement, but by first measuring factors such as voltage, current, and temperature among others followed by calculating these factors. At the same time, the accuracy of the calculation is influenced by the battery model, aging, self - discharge and temperature. Therefore, it is difficult to determine specific capacity and SOC of each monomer battery. Presently, this control strategy is rarely applied in practice.

Depending on the energy consumption of balance components during the process, the balance management technology can be divided into loss passive technology and nonloss active technology. The former, known as discharge balance or passive balance, refers to additional shunt resistance structure outside the battery. The technology with low efficiency achieves the current balance only in overcharge process and is unable to work in discharge process. The latter, also named energy transfer method or active balance method, adopts DC/DC circuit structure outside the battery. The technology with a high efficiency can balance the current in charge/discharge process. But it needs precise battery voltage acquisition as the basis of setting the balance, and its circuit structure is complex. Its reliability therefore still needs to be improved.

1. Loss passive technology. The basic structure of the loss passive balance circuit is shown in Fig. 3.3; the batteries E_1, E_2 … E_n are placed in paralleled with the shunt resistor R_1, R_2 … R_n, respectively. An upsurge in voltage of the

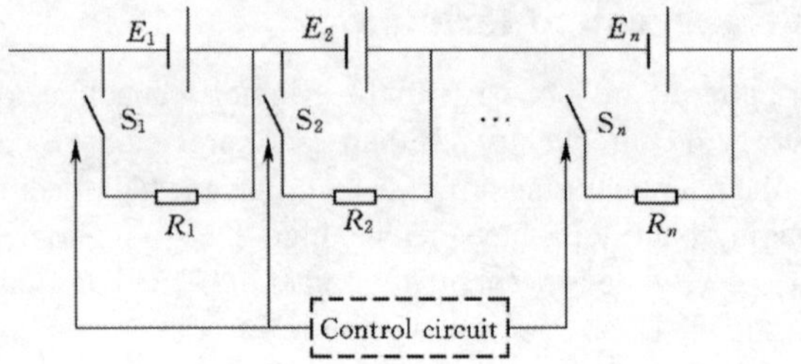

FIGURE 3.3 Loss passive balance circuit.

battery E_1 enables the control circuit to turn on the bypass control switch S_1, and the corresponding shunt resistor R_1 generates heat to prevent the E_1 voltage from exceeding the other monomer batteries' voltages. The voltage of each battery in the battery cluster is repeatedly detected by the control circuit and reaches the same value after many cycling rounds. The value of the shunt resistor R_1 is generally 10 times that of the battery resistance.

The technology that achieves balance target ensures each monomer battery in the battery cluster is parallel with a resistor to realize the discharge. The advantages are simple structure, high reliability, and low cost, while its disadvantages include larger energy consumption, low speed, and low efficiency. Furthermore, the heat produced by the resistance will affect the normal operation of the system. Therefore, the method is only suitable for a battery cluster with smaller capacity. For the large-capacity energy storage plants and electric vehicles with large output power, it is ideal to adopt nonloss active balance technology to reduce the energy loss.

2. Nonloss active technology. There are two structures of nonloss active technology: one is composed of energy storage components (inductors or capacitors) and control switches, and the other is the application of DC/DC converter technology to control the inductor, capacitor, and other energy storage components, thus achieving energy transfer and realizing the recharge or discharge of the monomer battery.

The switched capacitor topology is shown in Fig. 3.4. The capacitor C stores the monomer battery's energy with high voltage through the on/off of all

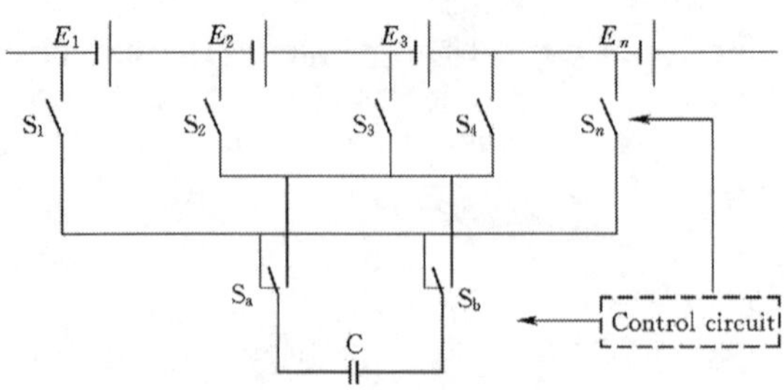

FIGURE 3.4 Switched capacitor topology.

switches, and then it releases the stored energy to the battery with a lower voltage. The energy storage components in this topology are capacitors or inductors because their principles are similar. The structure of the balance method with little energy loss is simple and easy to control. However, when the voltage difference between adjacent batteries is small, it takes a longer time to realize balance because the method is low in balancing speed and efficiency. Therefore, the method is not suitable for batteries that need to be quickly charged with large current.

The balanced circuit topology applying a DC/DC converter can be divided into centralized and distributed methods. In theory, the method has no energy loss and can reach balance quickly. It has become the preferred solution to realize balance of energy storage batteries.

The centralized DC/DC converter balance methods include forward and flyback structures, as shown in Fig. 3.5. Each monomer battery is connected parallel to secondary winding of the converter, with equal numbers of turns, so the monomer battery with a lower voltage obtains more energy and achieves the balance of the entire battery cluster.

The advantages of this topology are high speed, efficiency, and low energy loss, but the disadvantage is that when the voltages are relatively high and several batteries are in series, it is difficult to make the numbers of turns of the secondary windings consistent and compensate the voltage difference caused by the leakage inductance of the converter. In addition, it needs many of the components with large size, and is not easy to be modularized. The voltage stress of the switch device is also high.

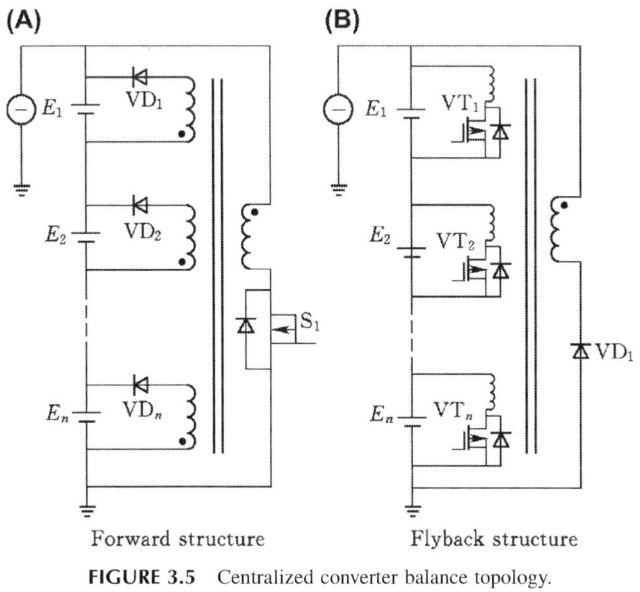

FIGURE 3.5 Centralized converter balance topology.

The structure of the decentralized balance method is to configure a parallel balanced circuit for each monomer battery. The method can be divided into isolated and nonisolated circuits with converters.

The nonisolated topology is based on the bidirectional balance of adjacent monomer batteries. The structure, without a converter, is relatively simple and suitable for occasions where the number of batteries in the series is few. The two common topologies are buck–boost circuit and Ćuk circuit, as shown in Fig. 3.6. The control strategy of the method is to balance the circuit when the voltage difference between the adjacent monomer batteries is within the allowable range.

The isolated topology is shown in Fig. 3.7. Each of the balanced circuits is a buck–boost circuit with isolated converters. The advantage is high balance

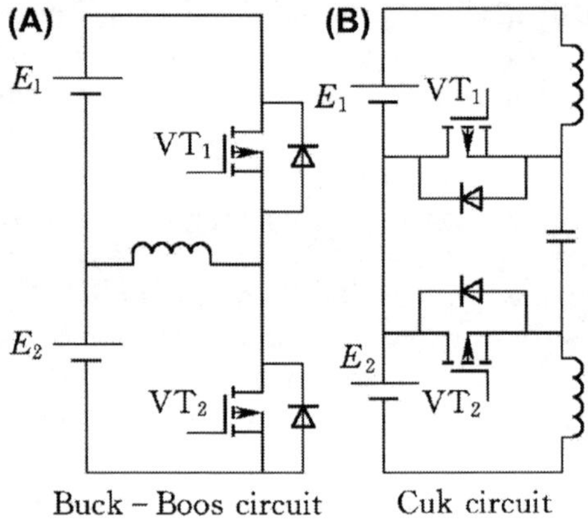

FIGURE 3.6 Nonisolated balance circuit topology.

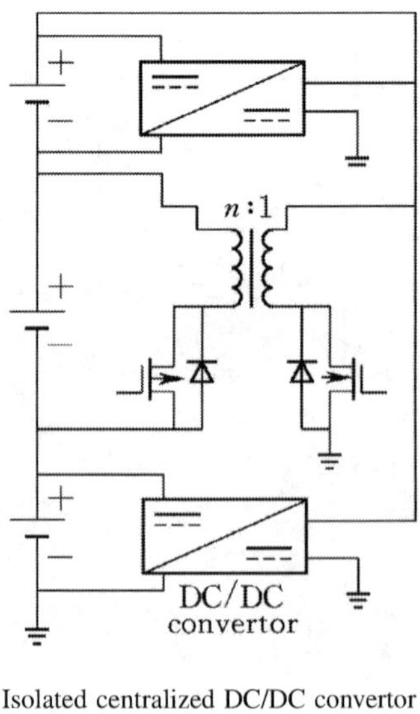

FIGURE 3.7 Isolated centralized DC/DC convertor balance circuit.

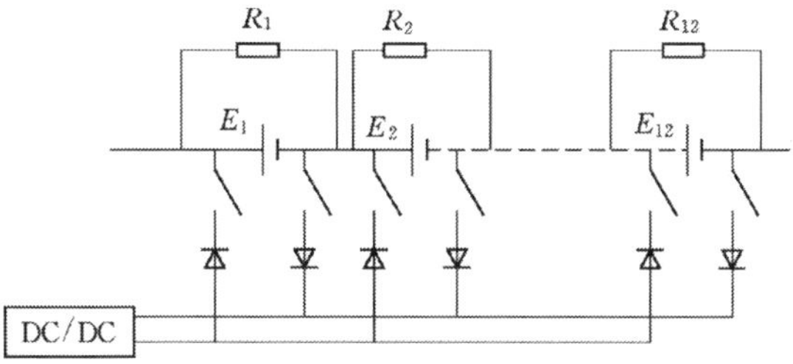

FIGURE 3.8 Resistance-dissipation discharge and DC/DC recharge balance circuit.

efficiency and the voltage on the switch devices is independent of the number of batteries in series. A balanced structure is ideal for a large number of batteries in series. The key disadvantage is that there are many magnetic components in the circuit that are large in volume and easily have mutual inductance. The leakage inductance exists in the convertor, making it difficult to keep the number of turns of the secondary windings consistent.

In practice, the energy storage system often uses the aforementioned technologies comprehensively; for example, the centralized DC/DC converter topology is used as a recharge circuit with a higher use frequency; a resistance-dissipation balance circuit is used as a discharge balance circuit, as shown in Fig. 3.8.

In the balance circuit, the current only flows in one direction, reducing the number and cost of switching devices. The control strategy is to primarily recharge energy and secondarily discharge energy to improve efficiency and lower the costs. In addition, the integrated use of switched capacitor andand distributed DC/DC converter method can circumvent the low efficiency caused by using many switches in the switched capacitor method, on one hand, while reducing the use of magnetic components in the distributed DC/DC converter and then reduce the volume of the circuit.

3.5 Protection technology

The battery protection technology refers to cases of failure, the alarm signal or trip instruction is sent out to isolate the failure and protect the battery. The protection for the BMS include overvoltage/undervoltage, overcurrent, short-circuit, and overtemperature protection. Fig. 3.9 is the circuit for battery protection based on R5421.

3.5.1 Overvoltage protection

During the charging process, when the control circuit detects voltage has reached the cutoff voltage, the "CO" pin will move from high voltage to zero voltage to turn off the field effect transistor V_2 and cut off the charging loop, thus realizing the overvoltage protection. At this moment, the V_2 has a diode,

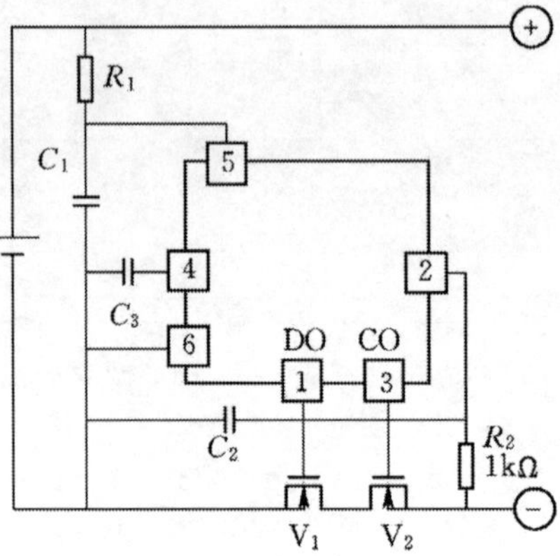

FIGURE 3.9 Circuit for battery protection based on R5421.

so the loads are discharged through them. The delay time from detecting the battery surpassing the cutoff voltage to signal cutting off of the V_2 is determined by C_3 and usually about 1 s, avoiding misjudgment caused by interference.

The cutoff charge voltage of different batteries can be determined by their charge characteristics. In general, batteries' charge characteristics can be divided into three types: the steady voltage, the negative voltage increment ($-\Delta V$), and the positive voltage increment (+AV).

The characteristic of the steady voltage is at the end of the charge, the battery's voltage can be automatically balanced, and typical examples are the traditional lead−acid batteries. Based on oxygen transfer rate in a large amount of liquid, when the voltage exceeds the end limitation of the charge voltage (2.35 V), the water will be electrolyzed and start to boil. The cutoff charge voltage can then be determined.

The charge characteristic of the negative voltage increment is as follows: in the initial charge process, the battery's voltage gradually increases. When the SOC reaches 100% and the battery continues to charge, the battery's voltage will drop rapidly, showing "$-\Delta V$" characteristic. If the system cannot reduce the charge current or stop charging on time, the charge current will drop rapidly with the increase of battery's voltage, resulting in a sharp increment in temperature, referred to as thermal runaway. The battery with this charge characteristic is highly sensitive to temperature, and the nickel−metal hydride batteries are taken as typical examples. The cutoff charge voltage can be determined by detecting the inflection point of the negative voltage increment ($-\Delta V$) during the charge process.

The charge characteristic of the positive voltage increment is in the initial charge process where the battery's voltage gradually increases. When the SOC reaches 100%, the battery's voltage will continue to rise if charging continues presenting +V characteristic. Exceeding the maximum allowable voltage will damage the battery, resulting in dangerous incidents such as bursting or explosion causing fire. The valve-regulated lead–acid batteries and lithium-ion batteries have the typical characteristics of the positive voltage increment. The cutoff charge voltage can be determined by detecting the inflection point of positive voltage increment ($+\Delta V$) during charging.

3.5.2 Undervoltage protection

The discharge curve shows that when the amount reaches some voltage value, it will drop sharply. In that case, if the battery continues to discharge, it can gain very little energy at the expense of its service life. Therefore, it is necessary for the battery to stop discharging when it attains a suitable voltage value. This is called the cutoff discharge voltage. The cutoff discharge voltage varies from the discharge rates, electrode plate types, and battery types, so its specific value should be determined by application requirements, battery's characteristics curve, and the data provided by the manufacturer. In general, the battery with a high-current discharge has a lower cutoff voltage; on the contrary, the battery with a small current discharge has a higher cutoff voltage.

During the discharge process when the control IC detects the battery voltage is lower than the cutoff discharge voltage, the "DO" pin will move from high voltage to zero voltage to turn off V_1 and cut off the discharge loop to achieve undervoltage protection. At this point, because V_1 has a body diode, the energy storage converter can charge the battery through the diode.

Since the battery's voltage cannot be reduced under the undervoltage protection, the current consumption used to protect the circuit is extremely small. At this time, the control IC enters into a low-energy state and protecting the circuit consumes current of less than 0.1 μA. The delay time between detecting declining battery voltage compared to the discharge cutoff voltage to sending a signal to cut off V_1 is determined by the C_3. It is usually about 100 ms to avoid misjudgment caused by interference.

As the battery stops operation due to undervoltage protection, the voltage will gradually increase. The battery then discharges under the vicinity of low voltage, making the circuit repeatedly turn on/off the power transistor. To avoid this, the lower limit self-locking circuit is used. Fig. 3.10 is the lower limit self-locking circuit based on CD4011.

Under the normal condition, $B_- = 0$ and $P_- = 0$, while under the under-voltage, $B_- = 0$ and $P_- = P_+ = B_+$. The waveform of each voltage point can be shown in Fig. 3.11. As long as the overdischarge signal indicates a low level, the gate signals of the field-affected transistor are all maintained at a low level to preferably control the switches of field-affect transistor.

3.5.3 Overcurrent protection

The charge/discharge current heavily influences the battery's service life and cycle performance. An upsurge in charge/discharge current results in

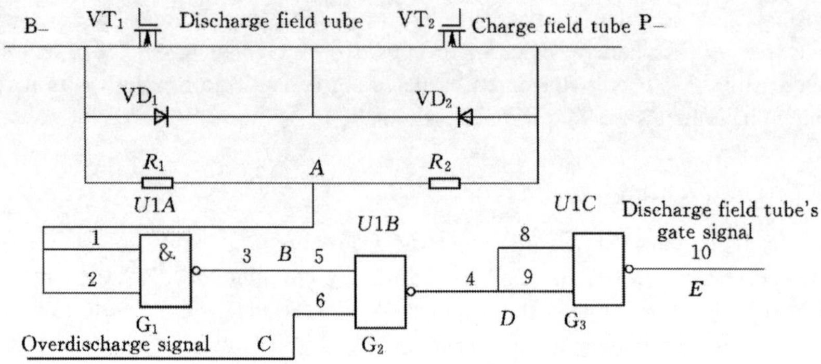

FIGURE 3.10 Lower limit self-locking circuit.

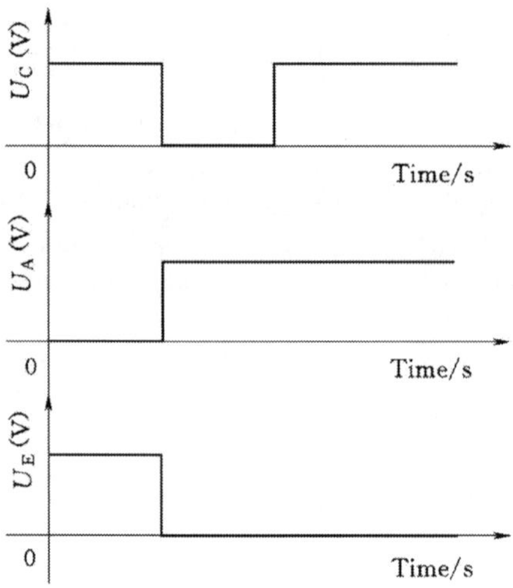

FIGURE 3.11 Waveform of each voltage point.

increasing the Ohmic drop and polarization effect, and dropping the discharge voltage in addition to shortening the battery's service life. Therefore, the overcurrent protection is required.

When the battery is in the normal discharge process, the current flowing through two MOSFET in series will produce a voltage U at both ends of MOSFET due to the on-resistance of MOSFET.

$$U = 2IR_{DS} \tag{3.5}$$

Where R_{DS} is the on-resistance of a single MOSFET.

In Fig 3.9, the "V" pin on the control IC tests this voltage. When the loop current is so large that U > 0, 1 V (this value is determined by the control IC), the "DO" pin will move from high voltage to zero voltage to turn off V_1 and then cut off the discharge loop. At that time, the discharge loop becomes zero, and then it achieves overvoltage protection. The delay time between detecting the overcurrent and sending a signal to cut off the V_1 is determined by C_3, and it is usually about 13 ms to avoid misjudgment caused by interference.

It can be seen from the preceding control process that the detection value of the overcurrent depends not only on the control value of IC, but also on the on-resistance of the MOSFET. When the on-resistance of the MOS- FET is large, in case of the same control value of the control IC, the value of the overcurrent protection is small.

3.5.4 Short circuit protection

The working principle of the short circuit protection is similar to that of the overcurrent protection, but they differ in judgment method and delay time. In the discharge process, if the current is so large that $U > 0.\ 9$ V (the value is determined by the control IC), the control IC determines that it belongs to the short circuit of the loads. The "DO" pin will quickly move from high voltage to zero voltage to turn off V_i and cut off the discharge loop, realizing the short-circuit protection. The delay time is usually very short, less than 7 μs.

3.5.5 Overtemperature protection

The overtemperature protection aims at keeping the battery in normal operating temperature range through cooling systems such as fan and thermal resistance heating devices. The key of the overtemperature protection is to analyze the relationship between the temperature displayed on the sensor and the heat source to determine the proper placement of the battery and then ensure cubicle thermal balance and rapid cooling. The temperature sensor measures both the natural and the battery temperature to determine the size of the damping vent inside the battery cubicle, thus minimizing power consumption.

3.6 Typical cases for battery management

3.6.1 Valve regulated lead–acid battery (VRLA battery)

1. The limitation on the charge/discharge voltage. The VRLA battery has a very strict requirement on the charge voltage (much higher than that of the lithium-ion battery). If the voltage is too high or too low, the service life of the battery will be seriously shortened. At room temperature (25 ± 5°C), the average charge voltage is 2.35 V. If the average charge voltage is adjusted according to the ambient temperature, the temperature compensation coefficient is −5 m V/(°C/per). The float charge voltage is generally 2.27 V, and the temperature compensation coefficient is −3 mV/(°C/per). Take the VRLA batteries provided by a manufacturer as an example: the relationships among the charge voltage, the float voltage, and the ambient temperature are shown in Table 3.2. In addition, if the batteries are different in grid and

TABLE 3.2 Average charge voltages and float charge voltages at different ambient temperature.

Ambient temperature (°C)	Float charge voltage (V/Per)	Average charge voltage (V/Per)
5	2.32	2.45
10	2.31	2.43
15	2.30	2.40
20	2.28	2.38
25	2.27	2.35
30	2.25	2.33
35	2.24	2.3
40	2.22	2.28

electrode material, their charge voltages also differ. For example, the charge voltages of batteries with low antimony alloy grid are lower than those of the batteries with lead–calcium alloy grid.

In the different discharge depths, the charge voltage has a great influence on the service life of the VRLA battery. When the discharge depth is 80%, and the charge voltage does not exceed 2.44 V, the battery's service life is at 100%; when the charge voltage is raised to 2.50 V, the service life of the battery drops to about 65%. When the charge voltage is lower than 2.35 V, the service life of the battery declines to about 10%. Therefore, the reasonable charge voltage should be based on specific circumstances.

For the VRLA battery, the cutoff discharge voltage is related to its discharge rate. The larger the discharge rate is, the smaller the cutoff voltage will be. Their relationships are shown in Table 3.3.

TABLE 3.3 Cutoff voltages in different discharge rates.

Discharge rate (A)	Cutoff discharge voltage (V/Per)	Discharge rate (A)	Cutoff discharge voltage (V/Per)
$I < 0.025C$	1.97	$0.1C \leq I < 0.2C$	1.83
$0.025C \leq I < 0.05C$	1.92	$0.2 \leq I < 0.5C$	1.75
$0.05C \leq I < 0.1C$	1.87		

Note: C refers to the ratio of charge/discharge current to rated capacity. For example, if the rated capacity is 1200 mAh, 0.2°C means that the charge/discharge current is 240 mA (0.2 times of 1200 mAh), and 1°C means the charge/discharge current is 1200 mA (one time of 1200 mA).

In the field of new energy power generation, the VRLA battery is in a charge/discharge mode most of the time. More specially, the battery operates in the power-stable mode until it reaches the cutoff charge/discharge condition. If the battery is often undercharged, it is necessary to recharge the battery every 2 months.

2. The limitation on the charge/discharge depth. The charge/discharge depth has a great impact on the service life of the VRLA battery. Assuming the discharge depth is at 80% and the service life is at 100%; when the discharge depth is at 100%, the service life is reduced to about 70%; when the discharge depth is reduced to 60%, the service life increases to about 170%; when the discharge depth is reduced to 25%, the service life increases to about 375%.

Therefore, it is necessary to strictly control the discharge depth of the battery. In general, the maximum allowable discharge depth is 80%, and when the discharge depth reaches 70%, an alarm is set off. With the increase of the service life and the change of the ambient temperature, the maximum allowable discharge depth is required to make adjustment in time.

3. The limitation on the charge/discharge current. Compared with other batteries, the VRLA battery has the strictest requirement on the charge current. This is mainly limited by the oxygen recombination. If the charge current exceeds the acceptable limit (the maximum oxygen recombination), the battery accelerates water loss. The acceptable charge current will constantly change during its recycle use, so the oxygen generated by the charge current must be completely recombined. In the charge process, if the voltage does not reach 2.35 V and the seal valve has opened frequently, it means the charge current is larger than the acceptable limit, and timely reduction of the charge current should be initiated. A low current may cause the passivation of the battery. Therefore, in the initial stage, the charge current should not be too small.

Taking the 116 Ah monomer battery as an example, the effects of the different discharge rates on the battery capacity are shown in Table 3.4. When the discharge current is $1I_{20}$, the rated capacity can be released; when the discharge current is $20I_{20}$, only 57% of the rated capacity is released. The impact of discharge current on the lea–acid battery is far greater than the lithium-ion battery.

4. The limitation on charge/discharge temperature. The best operation temperature range for the VRLA battery is 15 to 25°C. Below or above the best range, the battery's performance will be affected. Table 3.5 is the operation temperature range for the VRLA battery.

The ambient temperature affects the battery's capacity. As the temperature decreases, the battery' capacity also decreases. For example, if the temperature

TABLE 3.4 Effect of discharge current on capacity.

Discharge current rate	Capacity (Ah)	Percentage of the rated capacity (%)
$1I_{20}$	116	100
$2I_{20}$	106	96.2
$4I_{20}$	96.5	83
$7I_{20}$	87.5	75.4
$20I_{20}$	66.4	57.2

Note: I_{20} is the discharge current within 20 h, and its value can be accumulated by $C_{20}/20$ $(A)_o$. C_{20} is the rated capacity within 20 h.

TABLE 3.5 Operation temperature range for the VRLA battery.

Operation state	Operation temperature	Best operation temperature
Discharge	−40 to 50°C	15 to 25°C
Charge	−20 to 50°C	15 to 25°C
Storage	−20 to 40°C	15 to 25°C

decreases from 25 to 0°C, the discharge capacity will be reduced to about 80% of rated capacity. It shows that if the temperature is too low, the battery will be undercharged in the long term, resulting in sulfation of the battery's negative plate. Eventually, the battery is unable to discharge. As the ambient temperature increases, the battery's capacity increases within a certain range. For example, when the temperature rises from 25 to 35°C, the discharge capacity will rise to about 105% of the rated capacity. However, if the temperature continues to increase, the capacity will slowly increase and finally stop.

The ambient temperature is closely related to the service life of the VRLA battery. High temperature can shorten the battery's life and halve it with every increase of 10°C above 25°C.

3.6.2 Lithium iron phosphate battery

1. The limitation on the charge/discharge voltage. The requirement of the lithium iron phosphate battery for the charge voltage is relatively lower than that of the VRLA's. In fact, its charge voltage is feasible as long as it is within the maximum allowable charge voltage. If the charge voltage is lower than the specified value and/or is undercharged, it will reduce the battery's capacity, extending the battery's life without affecting the

basic performance of the battery. Within the operating temperature range, the lithium iron phosphate battery cannot adjust temperature in the charge/discharge process. In the case of the charge current being less than 3 C, the maximum allowable charge voltage is 3.65 V; in the case of discharge current being less than 2 C, the minimum allowable discharge voltage is 2.5 V.

However, the lithium iron phosphate battery is highly sensitive to the overcharge. If the charge voltage exceeds the allowable limit, it will damage the lithium iron phosphate battery more than the VRLA battery. For example, the lithium iron phosphate battery will decelerate performance, even causing dangerous incidents including burning or explosion. The serious overdischarge will cause irreversible capacity loss.

If the lithium iron phosphate battery is unused over a long time, it is necessary to charge it once every 3 months, so its SOC >80%.

2. The limitation on the charge/discharge depth. As long as the discharge voltage of the lithium iron phosphate battery is not below the allowable minimum, the requirement for the discharge depth is lower in the discharge process. The smaller discharge depth helps to prolong the battery's life. In the actual project, the maximum allowable discharge depth for the lithium iron phosphate battery is usually 10%, and an alarm is set off when the discharge depth reaches 15%.
3. The limitation on the charge/discharge current. For the lithium iron phosphate battery, its requirement for charge/discharge current is much stricter than that of the VRLA battery. However, the excessive charge/discharge current easily leads to the collapse of embedded Li^+ voids, resulting in irreversible capacity loss.

There is no convincing data to prove that the frequent and continuous large charge/discharge current influences the performance and service life of the lithium iron phosphate battery. The industry experts have reached a consensus that the charge/discharge current should be controlled within 50% of the rated capacity, and the larger charge/discharge current can occasionally be used for special needs. According to the statistics for the current demonstration projects, the standard charge rate of the lithium iron phosphate battery is 0.5°C, and its fast charge rate and instantaneous charge rate are 1 and 3°C (10 s), respectively. In addition, its standard discharge rate is 0.5°C, and its maximum continuous discharge rate is 2°C.

4. The limitation on charge/discharge temperature. The best operating temperature range for the lithium iron phosphate battery is within 15 to 25°C. In lower temperature, the battery's performance is poor. The operation temperature range for the lithium iron phosphate battery is shown in Table 3.6.

TABLE 3.6 Operation temperature range.

Operation state	Operation temperature	Best operation temperature
Discharge	−20 to 50°C	15 to 25°C
Charge	0 to 50°C	15 to 25°C
Storage	−30 to 50°C	15 to 25°C

The ambient temperature also affects the discharge capacity of the battery. The lower the temperature is, the smaller the battery capacity will be. The battery can discharge 100% of rated capacity at 10−60°C; the battery can discharge 90% of the rated capacity at −10°C; and the battery can discharge 70% of the rated capacity at −30°C.

References

[1] Wang Z. Research on battery management system for sodiun - sulfur battery. Wuhan: Wuhan University of Technology; 2012.

[2] Yuan Y. Research on battery management system of pure electric vehicle. Shanghai: Tongji University; 2009.

[3] Zhong W. Battery management system of electric vehicle. Nnanchang: Nanchang University; 2009.

[4] Zhu S. Energy storage battery handbook. Tianjin: Tianjin University Press; 1998. p. 73−6.

[5] Fu Z, Lin C, Chen Q. Key technology of thermal management system to EV battery packs. Journal of Highway and Transportation Research and Development 2005;(22):119−23.

[6] Qiao G. Research and design on battery management system of electric vehicle. Wuhan: Wuhan University of Technology; 2006.

[7] Jin X. Research on Smart management of power Li - ion battery. Shanghai: East China Normal University; 2008.

[8] Deng C. Research and design on $LiFePO_4$ battery management system for FCEV. Wuhan: Wuhan University of Technology; 2011.

[9] Li Z. Study on performance of lithium iron phosphate battery for pure electric vehicle. Beijing: Tsinghua University; 2011.

[10] Wang L. Research on $LiFePO_4$ battery system integration and management system of electric vehicle. Shanghai: Shanghai Jiao Tong University; 2010.

[11] Qiu B. Research on charge equalization and protection of power lithium iron phosphate battery pack. Chongqing: Chongqing University; 2013.

[12] Tong G. The design of $LiFePO_4$ storage battery management system. Nanning: Guangxi University; 2013.

[13] Zhang X. Research on lithium iron phosphate battery management system of electric vehicle. Chongqing: Chongqing University; 2008.

[14] Li N, Bai K, Chen H. Summary of equalization for $LiFePO_4$ Li - ion batteries. North China Electric Power 2012;(2):60−5.

Chapter 4

Operation control technology of energy storage systems

Chapter outline

Grid-scale Energy Storage Systems and Applications. https://doi.org/10.1016/B978-0-12-815292-8.00004-6

The operation control technology of energy storage systems (ESSs) defined in this chapter mainly centers on the operation control of the energy storage converter of the battery energy storage system (BESS). According to different operating modes and state switching process of the BESS, the ESS operation control mainly includes grid-connected operation control, off-grid operation control, and on-/off-grid switching control. Under each mode, the ESS should have the capability of rapid dynamic response, carrying out the active and reactive power modulation of the converter input and output voltage and current to respond to changes of the system. Under the sinusoidal steady state, the voltage and current have a coupling effect. To realize the decoupling control of active and reactive power, the coordinate axis transformation and decoupling of the voltage and current must be carried out. Section 4.1 introduces the related basic principles, including the coordinate transformation, pulse width modulation (PWM) modulation technology, bidirectional AC/DC converter theories and mathematical model, bidirectional DC/DC converter theories and mathematical model, and typical topological structure of the EES. Section 4.2 introduces the ESS on-grid operation control technology, including AC/DC converter control, DC/DC converter control, island detection, and low voltage ride through (LVRT) technology. Section 4.3 expounds on the ESS off-grid operation control technology, including V/f control, black start control, and multiple machines parallel coordinated control. Section 4.4 centers on the ESS dual-mode switching control system, including the smooth switching control from on-grid to off-grid and simultaneous control from off-grid to on-grid. Section 4.5 verifies the aforementioned control technology with tests in the practical verification system.

4.1 Basic principles

4.1.1 Coordinate transformation

The three-phase output capacitor on the AC side of the energy storage converter can be regarded as a spatial three-phase winding, as shown in Fig. 4.1. The physical quantity passing through the three-phase winding distributed in sinusoidal distribution is the spatial phasor f_S. Consider the three-phase cross-section as the spatial complex plane, and randomly choose a stationary co-ordinate (*Re*-Im) at the complex plane; if the real axis Re is the spatial reference coordinate, the phasor R can be written as $R = /R/e^{j\theta}$ where/R/is the phasor module (amplitude), θ is the spatial electrical angle between the phasor axis and reference axis Re, known as spatial phase.

4.1.1.1 Stationary coordinate transformation from ABC axis system to αβ axis system

The three-dimensional axes in Fig. 4.1 constitute the *ABC* spatial three-dimensional coordinate system, as shown in Fig. 4.2. The position of *ABC*

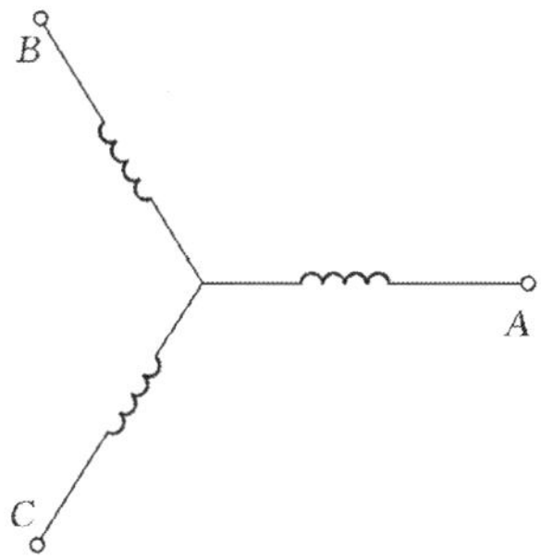

FIGURE 4.1 Converter three-phase output winding.

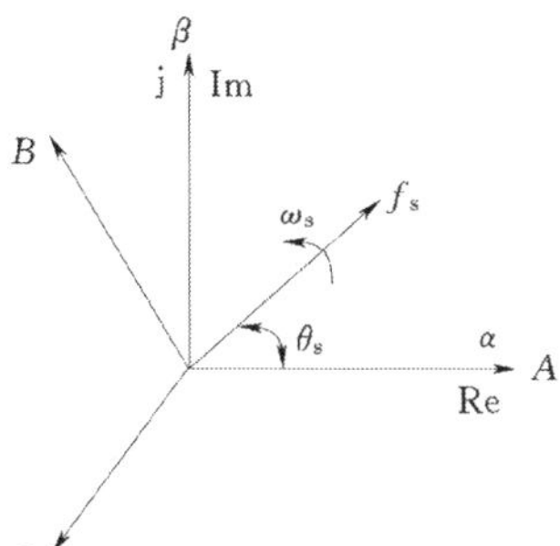

FIGURE 4.2 Coordinate transformation of $\alpha\beta$ coordinate system of the three-dimensional *ABC* coordinate system.

coordinate system in the spatial complex plane can be determined by the spatial electrical angle between *A* Phase and Re; if *A* axis coincides with Re axis, the spatial position angle of *B* axis is $a = e^{j120}$ degrees, *C* axis's position angle $a^2 = e^{j240}$ degrees, while a and a^2 are spatial operators.

Now take the real axis Re of the spatial complex plane as α axis, and imaginary axis Im as β axis, thus a spatial stationary $\alpha\beta$ two-dimensional coordinate system is formed, and the spatial phasor can be described with the stationary $\alpha\beta$ coordinate system as f_a^s. The spatial phasor of voltage and current at the three-phase output end of the converter u_s and i_s can be written as follows:

$$u_s^\alpha = u_\alpha + ju_\beta \tag{4.1}$$

$$i_s^\alpha = i_\alpha + ji_\beta \tag{4.2}$$

In Fig. 4.2, the spatial phasor of the original *ABC* coordinate system can be expressed with $\alpha\beta$ coordinate system. Here the coordinate transformation refers to the due relationship between the two coordinate system coordinate components while meeting the principle that the spatial phasor f_s remains unchanged.

To meet the constraint of unchanged power, the $\alpha\beta$ coordinate system coordinate components should be 3/2 times the coordinate components of the original *ABC* coordinate system. Because the two coordinate systems are stationary, and α axis coincides with *A* axis, in *ABC* coordinate system $u_s = |u_s|\, e^{j\theta s}$, while in $\alpha\beta$ coordinate system $u_s^\alpha = |u_s| e^{j\theta s}$, i.e., $u_s = u_s^\alpha$, so we have the following:

$$\sqrt{\frac{2}{3}}\left(u_A + u_B e^{j120^\circ} + u_C e^{j240^\circ}\right) = u_\alpha + ju_\beta \tag{4.3}$$

Now supposing the imaginary parts and the real parts of the two sides of the equation are equal respectively; we can get the following:

$$\begin{pmatrix} u_\alpha \\ u_\beta \end{pmatrix} = \sqrt{\frac{2}{3}} \begin{pmatrix} 1 & -\frac{1}{2} & -\frac{1}{2} \\ 0 & \frac{\sqrt{3}}{2} & -\frac{\sqrt{3}}{2} \end{pmatrix} \begin{pmatrix} u_A \\ u_B \\ u_C \end{pmatrix} \tag{4.4}$$

This transformation is called Clark transformation, applicable to the voltage and current space vectors, and the unified transformation matrix is this:

$$\text{Clark}_{ABC\rightarrow\alpha\beta} = \sqrt{\frac{2}{3}} \begin{pmatrix} 1 & -\frac{1}{2} & -\frac{1}{2} \\ 0 & \frac{\sqrt{3}}{2} & -\frac{\sqrt{3}}{2} \end{pmatrix} \tag{4.5}$$

The corresponding Clark inverse transformation matrix is this:

$$\mathrm{Clark}_{\alpha\beta\rightarrow \mathrm{ABC}} = \sqrt{\frac{2}{3}}\begin{pmatrix} 1 & 0 \\ -\frac{1}{2} & \frac{\sqrt{3}}{2} \\ -\frac{1}{2} & -\frac{\sqrt{3}}{2} \end{pmatrix} \tag{4.6}$$

4.1.1.2 *Rational coordinate transformation from αβ coordinate system to* dq *coordinate system*

The coordinate transformation from $\alpha\beta$ coordinate system to synchronous rational dq coordinate system is shown in Fig. 4.3. In Fig. 4.3, dq axes are mutually perpendicular axes rotating at the synchronous electrical angle speed ω_s in space. Θ_d is the electrical angle between d axis and α axis. The dq coordinate system is rotating at the synchronous speed. The following results:

$$\theta_d = \omega_s t + \theta_0 \tag{4.7}$$

Where θ_0 is the initial position angle between d axis and α axis and can take an arbitrary value.

Therefore dq coordinate can be any synchronous rational coordinate system.

Where u_s is a space vector expressed in the ABC coordinate system, given in the ABC coordinate system $u_s = u_s = |u_s|e^{j\theta s}$, if expressing the phasor in the dq coordinate, $u_s^d = |u_s|e^{j\theta_\delta}$, we will get this:

$$u_s^d = u_s e^{j-\theta_d} \tag{4.8}$$

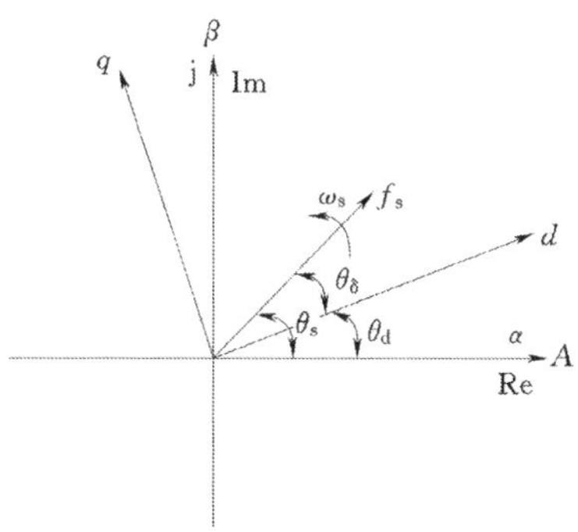

FIGURE 4.3 Vector transformation from stationary $\alpha\beta$ coordinate system to synchronous rational dq coordinate system.

As $u_s = u_s^\alpha$, we get the following:

$$u_s^d = u_s^\alpha e^{j-\theta_d} \tag{4.9}$$

Where $e^{j-\theta_d}$ is the transformation factor from $\alpha\beta$ coordinate system to dq coordinate system. On the contrary, $e^{j\theta_d}$ is the transformation factor from dq coordinate system to $\alpha\beta$ coordinate system:

$$u_s^\alpha = u_s^d e^{j\theta_d} \tag{4.10}$$

Thus u_s^α and u_s^d can be expressed as shown:

$$u_s^\alpha = u_\alpha + ju_\beta \tag{4.11}$$

$$u_s^d = u_d + ju_q \tag{4.12}$$

Based on Euler formula $e^{j\theta_d} = \cos\theta_d + j\sin\theta_d$, from Formulas (4.9), (4.11), and (4.12) we can get this:

$$\begin{pmatrix} u_d \\ u_q \end{pmatrix} = \begin{pmatrix} \cos\theta_d & \sin\theta_d \\ -\sin\theta_d & \cos\theta_d \end{pmatrix} \begin{pmatrix} u_\alpha \\ u_\beta \end{pmatrix} \tag{4.13}$$

From Formulas (4.10)–(4.12) we get this:

$$\begin{pmatrix} u_\alpha \\ u_\beta \end{pmatrix} = \begin{pmatrix} \cos\theta_d & -\sin\theta_d \\ \sin\theta_d & \cos\theta_d \end{pmatrix} \begin{pmatrix} u_d \\ u_q \end{pmatrix} \tag{4.14}$$

Formulas (4.13) and (4.14) are Park's transformation and inverse transformation formula, respectively.

It is clear that the rotational coordinate transformation and stationary coordinate transformation are essentially the same, the rotational coordinate transformation is the transformer factors, which reflect the relationship of the pole coordinate of the two complex planes, while the stationary coordinate transformation is the transformation matrix that reflects relationship of coordinate components of the two complex planes.

4.1.1.3 Physical significance of coordinate transformation

Under the sinusoidal steady state, the three-phase symmetrical AC voltage can be expressed as shown:

$$u_A = \sqrt{2}U_m \cos(\omega_s t + \varphi_0) \tag{4.15}$$

$$u_B = \sqrt{2}U_m \cos(\omega_s t + \varphi_0 - 120°) \tag{4.16}$$

$$u_C = \sqrt{2}U_m \cos(\omega_s t + \varphi_0 - 240°) \tag{4.17}$$

Where U_m is the voltage effective value; φ_0 is the A phase voltage phase initial angle.

First transfer the three-phase symmetrical sinusoidal voltage from stationary ABC coordinate system to stationary $\alpha\beta$ coordinate system:

$$\begin{pmatrix} u_\alpha \\ u_\beta \end{pmatrix} = \sqrt{\frac{2}{3}} \begin{pmatrix} 1 & -\frac{1}{2} & -\frac{1}{2} \\ 0 & \frac{\sqrt{3}}{2} & -\frac{\sqrt{3}}{2} \end{pmatrix} \begin{pmatrix} u_A \\ u_B \\ u_C \end{pmatrix} = \begin{pmatrix} \sqrt{3}U_m \cos(\omega_s t + \varphi_0) \\ \sqrt{3}U_m \sin(\omega_s t + \varphi_0) \end{pmatrix} \tag{4.18}$$

And then change the transferred $\alpha\beta$ coordinate components to dq coordinate system:

$$\begin{aligned} \begin{pmatrix} u_d \\ u_q \end{pmatrix} &= \begin{pmatrix} \cos\theta_d & \sin\theta_d \\ -\sin\theta_d & \cos\theta_d \end{pmatrix} \begin{pmatrix} u_\alpha \\ u_\beta \end{pmatrix} \\ &= \begin{pmatrix} \cos(\omega_s t + \theta_0) & \sin(\omega_s t + \theta_0) \\ -\sin(\omega_s t + \theta_0) & \cos(\omega_s t + \theta_0) \end{pmatrix} \begin{bmatrix} \sqrt{3}U_m \cos(\omega_s t + \varphi_0) \\ \sqrt{3}U_m \sin(\omega_s t + \varphi_0) \end{bmatrix} \\ &= \begin{bmatrix} \sqrt{3}U_m \cos(\varphi_0 - \theta_0) \\ \sqrt{3}U_m \sin(\varphi_0 - \theta_0) \end{bmatrix} \end{aligned} \tag{4.19}$$

From Formula (4.19), we can see that u_d and u_q have changed to DC amount, i.e., the two voltage phasors of dq coordinate do not change anymore but stay fixed, and amplitude and phase are subject to the phase voltage effective value, A phase voltage phase initial angle φ_0, and dq coordinate system initial position angle θ_0. From Fig. 4.3, we can see the angle θ_δ between the space vector and d coordinate should be this:

$$\theta_\delta = \varphi_0 - \theta_0 \tag{4.20}$$

As dq coordinate system is randomly selected, if d axis of the synchronous coordinate system is directed to the space vector f_s, i.e., $\theta_\delta = 0$, $\theta_s = \theta_d$, we get this:

$$\begin{pmatrix} u_d \\ u_q \end{pmatrix} = \begin{pmatrix} \sqrt{3}U_m \\ 0 \end{pmatrix} \tag{4.21}$$

Therefore, the coordinate transformation changes the sinusoidal steady-state variables of ABC coordinate system to the DC components of dq coordinate system, reduces the system's order, and realizes control of u_s through control of u_d and u_q. Under the dynamic state, the amplitude and phase of u_s change constantly, but we can realize control of u_d and u_q in the dynamic process as well, i.e., dynamic control of u_S, providing conditions for high-performance control.

4.1.2 PWM modulation technology

4.1.2.1 Basic principle of PWM control

The sampling control theory has an important conclusion: when applying the narrow pulse with the same impulse but in different shapes on inertial links, the effect is basically the same. Impulse is the area of the narrow pulse. Here the same effect means the waveform of the output response of the links are basically the same. If analyzing each output waveform with Fourier transform, their low frequency is very close; there are only slight differences in the high frequency. PWM is based on the conclusion to control switch-on and switch-off of the semiconductor switching device to produce a series of pulses with the same amplitude but different width at the output port, and to replace sinusoidal waveform or other needed waveform with these pulses. Modulating the pulse width according to a certain rule can change not only the output voltage of the inverter circuit output, but also the output frequency.

From the aforementioned principle, we can know that if the frequency, amplitude, and pulse count within a cycle of the output waveform of the inverter circuit are given, the width and interval of pulses of PWM waveform can be calculated precisely. Controlling the connection and disconnection of the switching devices of the inverter circuit, we can get the needed PWM waveform. This method is called the calculation method, and it is very complicated because when the frequency, amplitude, or phase of the sine wave is changed, the result will change accordingly.

There is another method called modulation method, which regards the desired output waveform (generally sine wave or trapezoidal wave or sine wave injected with zero sequence harmonics or square wave) as the modulation wave, and compares it with the triangle wave (or sawtooth wave), which is N times the modulation wave frequency, to produce a group of rectangular pulse sequences with the same amplitude and the width in proportion to the modulation wave to have equivalent effect with the modulation wave and switch on or switch off the switching tube. Controlling width of the pulses with a certain rule can realize modulation of the output voltage and frequency of converter.

4.1.2.2 PWM common methods

The basic principle of PWM control was advanced quite early, but it was not realized before the 1980s due to the constraints of the level of electrical appliance development. It was not until the 1980s that the PWM technology was applied with the rise and rapid development of full-control electronic devices. With the development of electric technology, microelectronic technology, and automatic control technology and application of new theories such as modern control theory and nonlinear system control ideas, PWM technology has experienced unprecedented development. So far, many PWM technologies have been developed, but the following two are widely used:

1. SPWM method

SPWM (sinusoidal PWM) is a PWM method that is relatively mature and widely used. An important conclusion of the sampling control theory should be mentioned: the effect is basically the same when the narrow pulse with the same impulse but in different shapes is applied to the inertial links. Based on this theory, SPWM method controls on/off of the inverter circuit switching devices with the SPWM waveform, which is the PWM waveform that changes the impulse width in line with the sinusoidal rule and has equal effect with the sine waveform to make the output pulse voltage area equal to the area of expected output sine wave in the corresponding zone, and adjusts the frequency and amplitude of the output voltage of inverter circuit by changing the frequency and amplitude of modulation wave.

According to polarity of the modulation pulse, SPWM technology can be divided into the monopolar modulation and bipolar modulation. Under the same conditions, the harmonic component of the monopolar SPWM modulation wave is smaller than that of the bipolar modulation wave.

The monopolar modulation compares a sine wave with an isosceles constant amplitude triangular wave whose polarity is positive at the positive semicycle and negative at the negative half-cycle. At the positive half-cycle of the sine wave, if the amplitude of the sine wave is higher than that of the triangular wave, the comparator outputs a positive level, otherwise the comparator outputs zero level. At the negative half-cycle of the sine wave, if the amplitude of the sine wave is smaller than that of the triangular wave, the comparator outputs a negative level, otherwise the comparator outputs zero level. The resulting PWM signals consist of positive, negative, and zero levels. Because the triangular wave only changes in a polarity in the half-cycle of the modulation wave, the resulting PWM wave is a control method changing in a polarity, named monopolar modulation; see Fig. 4.4A.

Bipolar modulation compares the bipolar triangular wave with positive and negative alternating with a sinusoidal modulation wave. When the amplitude of the sine wave is higher than the triangular wave, the comparator outputs a positive level; otherwise it outputs a negative level. Therefore PWM signals with negative and positive level are obtained. This method is called bipolar modulation; see Fig. 4.4B.

2. SVPWM method

SVPWM (space vector pulse width modulation) is also known as magnetic flux sine PWM method. With the effect of a three-phase waveform overall as the prerequisite, the ideal circular rotating magnetic field trajectory that approximates the motor air gap as the purpose, we approach the basic circle

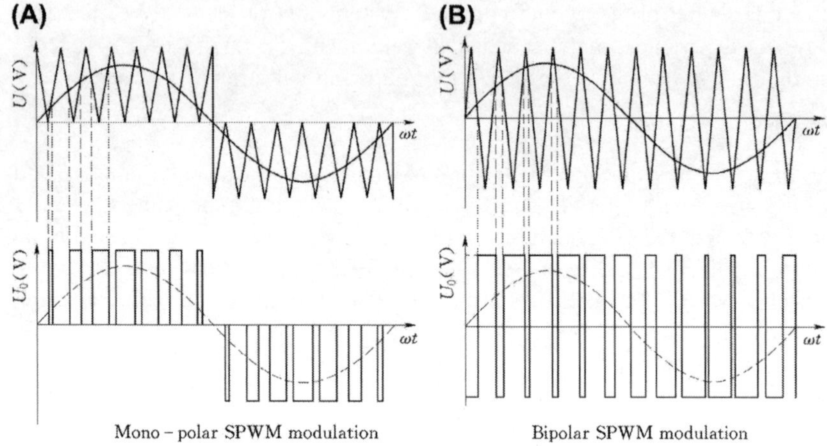

FIGURE 4.4 SPWM modulation schematic diagram.

magnetic flux with actual magnetic flux generated by the switching mode of inverter and decide on/off of the inverter according to the comparison result to form the PWM waveform. This method starts from the angle of the motor, regards the inverter and motor as a whole, and controls with the mode of an inscribed polygon approximating a circle to enable the motor to obtain the circular magnetic field with constant amplitude (sine magnetic flux).

The space vector of voltage does not represent an actual physical quantity, but a mathematical treatment for control and analysis. SVPWM is mainly used in a three-phase PWM rectifier and inverter. Its concept is to form a composite vector U, and it defines the three-phase PWM rectifier working status as six basic vectors and two zero vectors. These basic vectors divide the plane into six sectors. The composite vector U makes a circular motion with the origin of coordinates as the center of circle, the amplitude as radius, and grid frequency ω as the angular speed. When U falls on a sector, it can be composed with the basic vectors and zero vector of the sectors.

Based on the SVPWM philosophy, the three-phase rectifier bridge switching function is derived:

$$S_k = \begin{cases} 1, \text{Upper bridge arm switches on, lower bridge arm switches off} \\ 0, \text{Lower bridge arm switches on, upper bridge arm switches off} \end{cases} \tag{4.22}$$

Where k = a, b, c.

In normal operation, only one of the upper and lower bridge arms is switched on.

The switch functions S_a, S_b, S_c have eight statuses (000—111), including two zero vectors (U_0, U_7), and six nonzero vectors ($U_1 \sim U_6$); see Fig. 4.5.

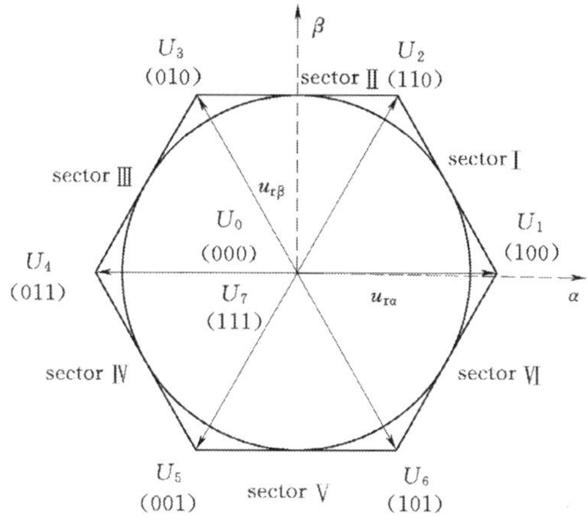

FIGURE 4.5 Diagram of basic voltage space vectors.

Using the linear combination of the eight basic voltage vectors, we can combine more new voltage space vectors with different phases from U_1-U_6, finally compose a group of voltage space vectors with the same amplitude but different phase, and approximate the magnetic linkage circular of a round rotating magnetic field. At each sector, two adjacent voltage vectors and a zero vector are chosen, forming the random voltage vector in line with the principle of volt-second balance, $U_{ref}T = U_xT_x + U_yT_y + U_0T_0$, where U_{ref} is the expected voltage vector; T is the sampling period; T_x, T_y, T_0 are the action time of the corresponding two nonzero voltage vectors U_x and U_y and zero voltage vector U_0 within a sampling period; U_0 includes the zero vectors U_0 and U_7. This expression means the integral effect value of vector U_{ref} generated in time T is the same as the sum of integral effect of U_x, U_y, and U_Q in time T_x, T_y, and T_0.

In other words, SVPWM control finally forms a PWM pulse wave with the same amplitude but different width by distributing the eight basic voltage vectors and their action time to realize the circular trajectory tracking magnetic flux. The higher the switching frequency of the primary circuit power devices is, the closer is the hexagon trajectory to the circular rotating magnetic field. The space voltage vector modulation can improve the bus DC voltage use rate by 15%.

4.1.3 Bidirectional AC/DC converter principle and mathematical model

4.1.3.1 Principle overview

From the development of electrical and electric technology, we know rectifier is an AC/DC converter used early. Rectifiers went through the development process from noncontrolled rectifiers (diode rectifiers) and phase-controlled rectifiers (thyristor rectifiers) to PWM rectifiers (switchable power switches). Though the traditional phase-controlled rectifier has been used for a long time, with relative mature technology and wide application, it still has the following problems:

1. The thyristor commutation causes the distortion of voltage waveform on the grid side.
2. The harmonic current on the grid side "pollutes" the grid with harmonic wave.
3. Under depth control, the power factor of the grid side is reduced.
4. Under closed-loop control, the dynamic response is relatively slow.

Though the diode rectifier has improved rectifiers' grid side power factor, it will produce harmonic current on the grid side to "pollute" the grid; moreover, the disadvantage of the diode rectifier is also the uncontrollability of the DC voltage. In light of the disadvantages mentioned, PWM rectifier has comprehensively improved the traditional phase control and diode rectifier. The key improvement is replacing the half-controlled power switch or diode with the full-controlled power switch and replacing the phase-controlled rectifier or uncontrolled rectifier with PWM chopping controlled rectifier. Therefore a PWM rectifier can achieve the following excellent performance:

1. the grid side current is a sine wave;
2. grid-side power factor control (e.g., unity power factor control);
3. bidirectional transmission of electric energy;
4. rapid dynamic control response.

Based on whether energy can flow bidirectionally, two different kinds of PWM rectifiers with different topological structures are derived, i.e., reversible PWM rectifier and irreversible PWM rectifier. The power conversion system (PCS) is a reversible PWM rectifier with energy flowing bidirectionally. The rectifiers with bidirectional energy flow embody not only the AC/DC rectification characteristics, but also the DC-AC inversion characteristics. Specifically, this type of PWM rectifier is actually a new type reversible PWM rectifier, abbreviated as a converter in the following.

With the bidirectional electric energy transmission, when the converter absorbs energy from the power grid, it is operating in the rectifying status; when the converter transmits electric energy to the grid, it is operating in the active inverter working status. The unity power factor is when the converter is in the rectifying status, and the grid side voltage and current are in the same phase (positive resistance characteristics); when PWM rectifier is operating in the active inverter status, its grid side voltage and current are in opposed phase (negative resistance characteristics). In summary, the bidirectional AC/DC converter is actually a controllable four-quadrant operation converter at the AC and DC sides. For a better understanding, the following paragraphs will first

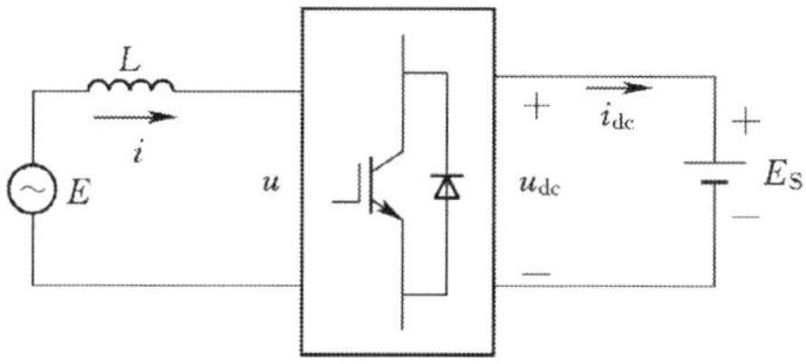

FIGURE 4.6 AC/DC converter single-phase equivalent circuit model.

expound on the principles of converter four-quadrant operation with the model circuit. Fig. 4.6 is an AC/DC converter single-phase equivalent circuit model. From Fig. 4.6, we can see the converter model circuit consists of an AC circuit and DC circuit. The AC circuit consists of AC electromotive force e and grid side inductance L, etc. The DC circuit is energy storage battery ES; the power switch bridge circuit consists of a voltage source or current source bridge circuit. Ignoring the losses of the power bridge circuit, we can get the following from the power balance relationship on AC and DC sides:

$$iv = i_{dc}v_{dc} \tag{4.23}$$

Where v, i is the AC side voltage and current of the model circuit; v_{dc}, i_{dc} is the DC side voltage and current of the model circuit.

From Eq. (4.22), it is not difficult to understand: with the AC side control of the model circuit, we can control the DC side, and vice versa. The following paragraphs will mainly analyze the converter operating status and control principles from the AC side of the model circuit.

Under the steady state, see Fig. 4.7 for the AC side vector relationship of the converter.

In Fig. 4.7, $\dot{E}$ is the vector of electromotive force of AC grid; $\dot{V}$ is the AC side voltage vector; V_L is the AC side inductance voltage vector; $\dot{I}$ is the AC side current vector.

To simplify the analysis, only the fundamental wave component is taken into account, and the harmonic component is neglected, as is the AC side resistance. In this way, we can analyze the following from Fig. 4.7: When using the grid electromotive force vector as reference, controlling AC voltage vector $\dot{V}$ can realize the four-quadrant operation of the converter. Suppose $\dot{I}$ remains unchanged, and $|\dot{V}_L| = wL|\dot{I}|$ remains unchanged, the movement trajectory of the converter AC voltage to the end point of vector $\dot{V}$ forms a circle with $|\dot{V}_L|$ as radius. When the end point of voltage vector $\dot{V}$ is at Point A of the circle trajectory, the current vector $\dot{I}$ is lagged behind by the electromotive force $\dot{E}$ by 90 degrees, and the converter grid side shows the pure inductance characteristics, as shown in Fig. 4.7A; when the voltage vector $\dot{V}$ end moves to Point B of the circle trajectory, the current vector $\dot{I}$ is parallel and

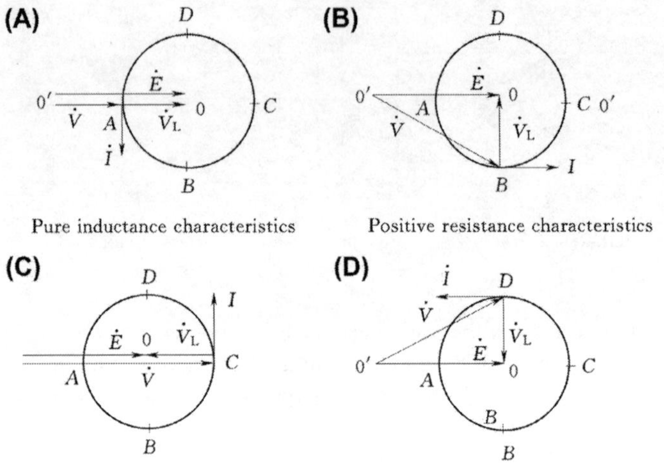

FIGURE 4.7 Relationship of vector at AC side steady state of the converter.

has the same direction with the electromotive force $\dot{E}$, and at this moment, the converter grid side shows positive resistance characteristics, as shown in Fig. 4.7B. When the end point of voltage vector $\dot{V}$ moves to Point C of the circle trajectory, the current vector $\dot{I}$ is 90 degrees ahead of the electromotive force $\dot{E}$, and at this moment, the converter grid side shows pure capacitive characteristics, as shown in Fig. 4.7C. When the end point of voltage vector $\dot{V}$ moves to Point D of the circle trajectory, the current vector $\dot{I}$ is parallel with the electromotive force $\dot{E}$ but in the opposite direction, and at this moment, the converter grid side shows negative resistance characteristics, as shown in Fig. 4.7D. Points A, B, C, and D are the four points of special working status of the converter four-quadrant operation. With further analysis, we can summarize the converter four-quadrant operation rules as follows:

1. When the end point of voltage vector $\dot{V}$ runs at AB trajectory, the converter is operating at the rectification status. At this moment, the converter needs to absorb active power and lagging reactive power from the grid, and the electric energy will be transmitted to the DC load through the converter from the grid. It is noteworthy that when the converter reaches Point B, it will realize unity power factor rectifier control; when it runs to Point A, the converter does not absorb active power from the grid, but it absorbs lagging reactive power from the grid.
2. When the end point of voltage vector $\dot{V}$ runs at BC trajectory, the converter is operating at the rectification status. At this moment, the converter needs to absorb active power and capacitive reactive power from the grid, and the electric energy will be transmitted to the DC load through the converter

from the grid. When the converter reaches Point B, the converter does not absorb active power from the grid, but it absorbs capacitive reactive power from the grid.

3. When the end point of voltage vector $\dot{V}$ runs at CD trajectory, the converter is operating at the active inverting status. At this moment, the converter transmits active power and capacitive reactive power to the grid, and the electric energy will be transmitted to from the converter DC side to the grid. When the converter reaches Point D, it can realize active inverting control of the unity power factor.
4. When the end point of voltage vector $\dot{V}$ runs at DA trajectory, the converter is operating at the active inverting status. At this moment, the converter transmits active power and lagging reactive power to the grid, and the electric energy will be transmitted from the converter DC side to the grid.

Obviously, to realize the four-quadrant operation of the converter, the key is the grid side current control. On the one hand, we can indirectly control the grid side current through controlling the converter AC voltage; on the other hand, we can directly control the grid side current of the converter through the closed-loop control of the grid-side current.

4.1.3.2 Classification and topological structure of a converter

1. Converter classification

With the development of converter technology, many types of converters have been designed. See Fig. 4.8 for the classification of converters.

Despite the variety of classification methods, the most basic is to classify the converters into the voltage source and the current source because these two types have their own characteristics in terms of the main circuit structure, pulse signal generation, and control strategy, etc., and show circuit duality. The other classification methods can be classified into the current source or voltage source converter from the perspective of the main circuit topological structure.

- Bidirectional AC/DC converter
  - Classification according to the DC energy storage form
    - Voltage source
    - Current source
  - Classification according to number of phases
    - Single - phase circuit
    - Three - phase circuit
    - Multi - phase circuit
  - Classification according to switching modulation
    - Hard switching modulation
    - Soft switching modulation
  - Classification according to the bridge circuit structure
    - Half - bridge circuit
    - Full - bridge circuit
  - Classification according to the modulation level
    - Two - level circuit
    - Three - level circuit
    - Multi - level circuit

FIGURE 4.8 Classification chart of converters.

2. Topological structure of voltage source converter (VSC)

The most distinctive topological characteristic of VSC is that the DC side uses a capacitor for DC energy storage, so the converter DC side shows voltage source characteristics of low impedance. The VSC has the following characteristics:

1. The AC side can be connected with the AC voltage source or load, and the AC output side is generally connected with inductance in series, which is mainly used to filter the grid-side current harmonic waves.
2. DC side is connected with a large capacitor in parallel, which is equivalent to the voltage source; the DC-side voltage basically has no pulse, and the DC circuit shows low impedance characteristics.
3. Due to the clamping effect of the DC voltage source, the AC-side output voltage waveform is a rectangular wave and has no relation with the load impedance angle. The AC-side output current waveform and phase are different at different situations of the load impedance.
4. When the AC side is under the resistance-inductance load, it needs to provide reactive power, and the DC-side capacitor functions as a buffer for reactive energy. To provide a channel for the reactive power fed back from the AC side to the DC side, each bridge arm power device of a VSC must be connected in a feedback diode in antiparallel.

The common topological structures of a VSC include the following types.

1. Single-phase half-bridge and full-bridge VSC topological structure. The single-phase half-bridge VSC main circuit topological structure is shown in Fig. 4.9A: it has two bridge arms; one arm consists of controllable power devices and an antiparallel diode; the other arm consists of two large enough capacitors connected in series. The junction of the two capacitors is the midpoint of the DC power source. The load or AC-side power source is

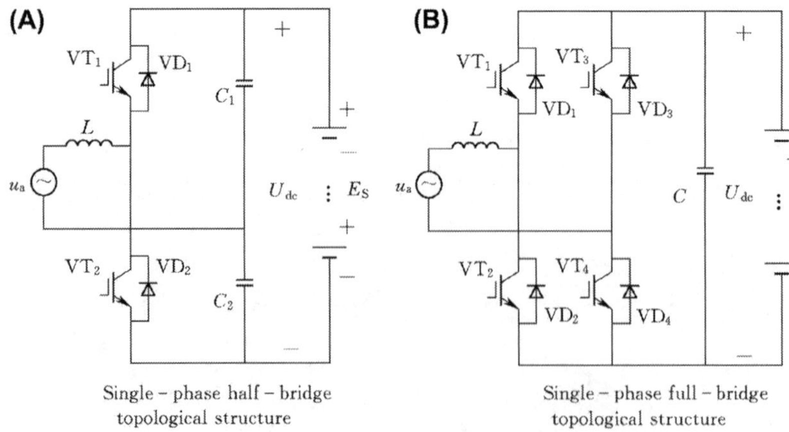

FIGURE 4.9 Topological structure of a single-phase VSC.

connected between the DC power source midpoint and the junction of the two power devices. The advantages of half-bridge circuit are that it is simple, uses fewer devices, and has low cost. The disadvantage is that the output AC voltage amplitude U_m is only $U_d/2$. Under the same AC-side circuit parameters, the half-bridge circuit DC voltage should be two times that of the full-bridge circuit to make the half-bridge circuit get the same AC-side current characteristics. Therefore, it has higher requirements on the voltage withstanding performance on the power devices. Moreover, to maintain midpoint potential of a capacitor of the half-bridge circuit as unchanged, the balance between the two capacitor voltages should be controlled in practice. Therefore, the half-bridge circuit is normally used in the low-cost and small power scenarios of only several kW.

See Fig. 4.9B for the topological structure of the single-phase full-bridge VSC main circuit. This topological structure uses the "Elbridge" structure with four power switches, and the load or AC-side power source is connected in H bridge, and the DC side only needs to be connected with a capacitor in parallel. The single-phase full-bridge circuit uses power devices as many again as the half-bridge circuit, but its voltage-use rate and control strategy are better than the half-bridge circuit. Therefore it is the most widely used in single-phase circuits.

2. Three-phase half-bridge and full-bridge VSC topological structure. See Fig. 4.10A for the topological structure of the main circuit of the three-phase half-bridge VSC. The AC side adopts the method of three-phase symmetrical connection without a neutral line, and it uses six power devices. This is the most common three-phase converter. The so-called three-phase bridge circuit is the three-phase half-bridge circuit. The three-phase half-bridge is more suitable for the three-phase grid balance system. When the three-phase grid is unbalanced, its control performance will deteriorate and even cause malfunction.

To overcome disadvantages of the three-phase half-bridge circuit, the three-phase full-bridge design can be used. Its topological structure is shown in Fig. 4.10B. Its characteristics are as follows: the public DC bus is connected with three independently controlled single-phase full bridges, and it is connected to the grid with a three-phase four-wire system through the transformer. Therefore, the three-phase full-bridge converter is actually composed of three independent single-phase full-bridge converters. When the grid is unbalanced, it will not seriously affect the converter control performance. Because the power devices that the three-phase full-bridge circuit needs are twice as the three-phase half-bridge circuit, the three-phase full-bridge is seldom used.

3. Three-level VSC topological structure. The converter topological structures mentioned before are conventional two-level topological structures. The disadvantage of such a topological structure lies in that when it is used

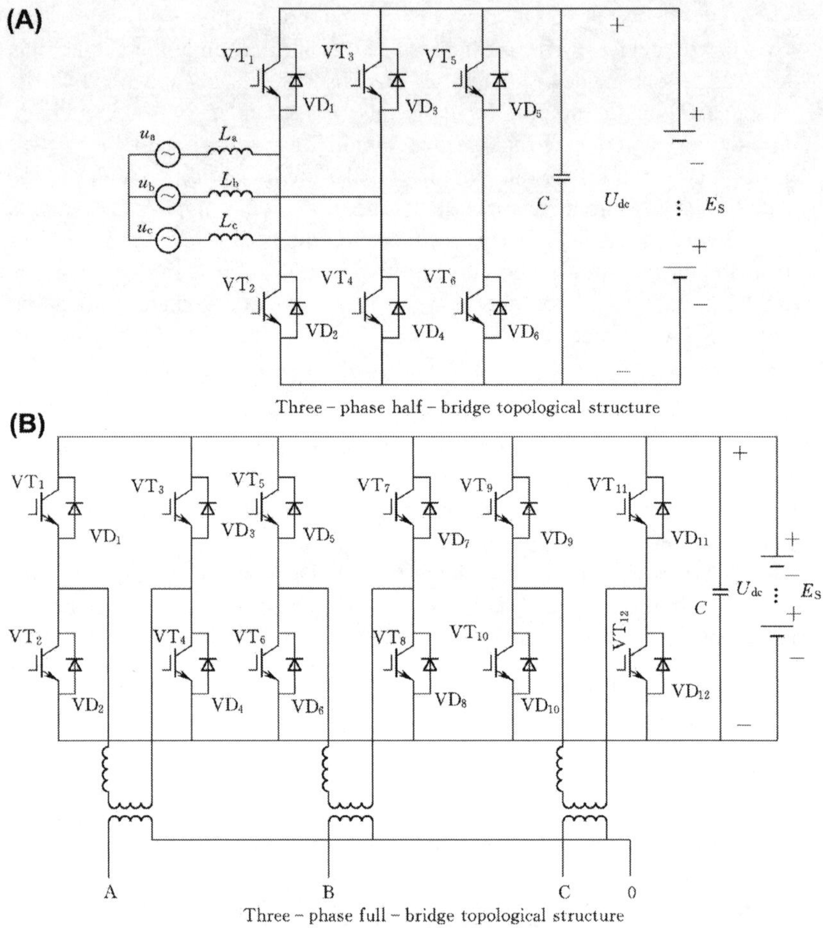

FIGURE 4.10 Three-phase VSC topological structure.

in the high voltage scenario, it needs to use high reverse-voltage power devices or several power devices connected in series. In addition, because the AC-side output voltage switches on the two levels, it will cause relatively high content of harmonic wave when the switching frequency is not high. To solve such problems, a multilevel circuit is designed. Fig. 4.11 shows the topological structure of a three-phase three-level VSC circuit. This circuit is also known as a midpoint clamping circuit. This circuit has three bridge arms, and the DC-side has two capacitors in series. Each half-bridge arm consists of two full-controlled power devices in series with an antiparallel diode. The midpoints of the two devices in series are connected by the midpoint of clamping diode and DC-side capacitor. The output phase voltage of this inverter circuit has a variety of levels, and its waveform is

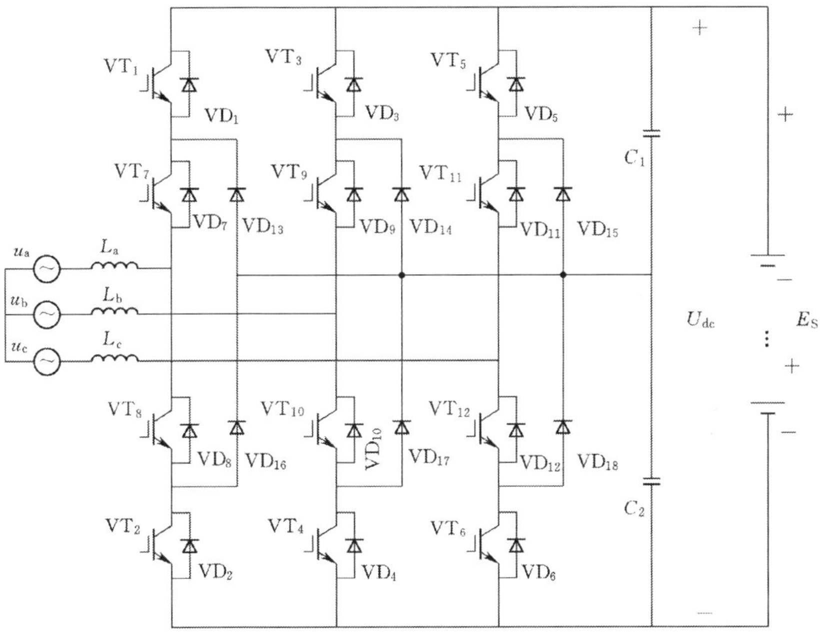

FIGURE 4.11 Topological structure of three-phase three-level VSC.

close to a sine wave. Moreover, with proper control, its output voltage harmonic can be significantly less than the two-level circuits. Obviously, the three-level circuit has effectively reduced the AC harmonic voltage and current while improving the voltage withstanding performance, and consequently, it improves the grid-side waveform quality. However, the disadvantage of the three-level circuit is that the required power switches are increased compared with the two-level circuit, and its control is relatively complicated.

To adapt to the high-voltage large power applications, and lower AC output voltage, multilevel VSC topological structures with more diode clamps can be designed by using the similar method of the three-level circuit.

3. Current source converter (CSC) topological structure

The most distinctive characteristics of CSC are that the DC side uses inductance for DC energy storage, so the DC side shows high impedance current source characteristics. CSC mainly has the following characteristics:

1. The AC side can connect the AC voltage source or load, and the AC output side is connected with a capacitor in parallel for absorbing energy stored in the load inductor during commutation.
2. The DC side links with a large inductor in series, which works as a current source; the DC side current basically has no impulse, and the DC loop shows high impedance characteristics.

3. The function of the switching device in the circuit is only to change the DC channel. Therefore the AC-side output current is the rectangular wave, and it has nothing to do with the load impedance angle. The AC-side output voltage waveform and phase are different due to the differences of the load impedance situations.
4. When the AC side is under the resistance-inductance load, it needs to provide reactive power, and the DC-side inductance serves as a buffer for reactive energy. Because the DC current does not feed back reactive energy, it does not need to connect an antiparallel diode with the switching devices like the VSC.

The CSC normally uses single-phase and three-phase topological structures. Fig. 4.12A and B show the single-phase full-bridge and three-phase

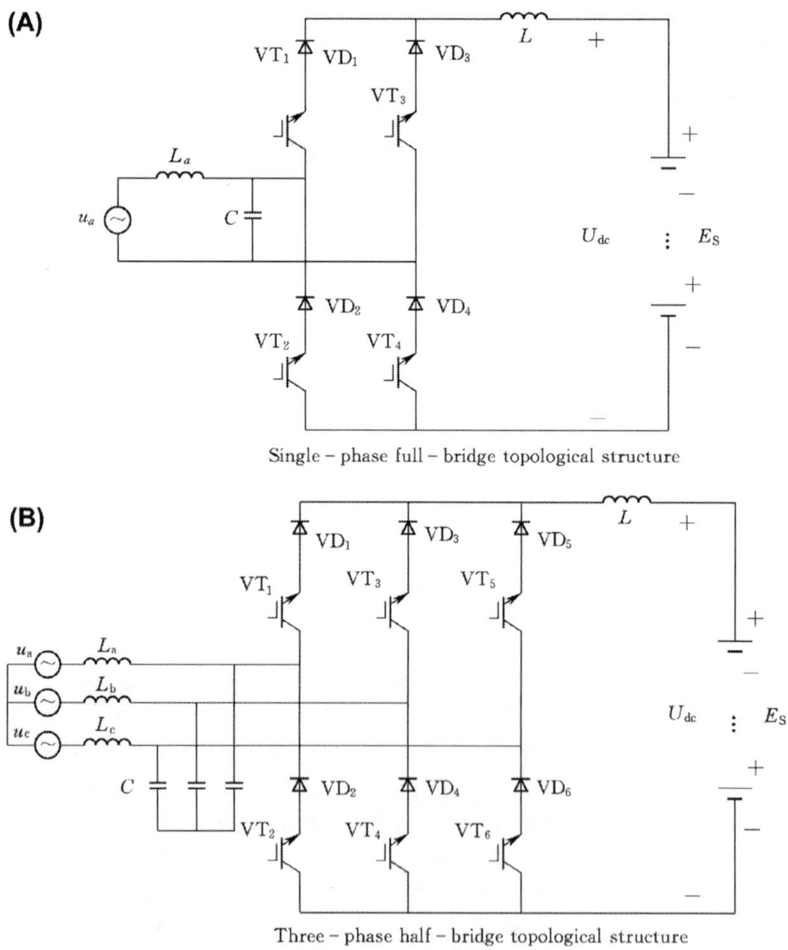

FIGURE 4.12 CSC topological structure.

half-bridge CSC topological structures, respectively. The AC-side inductor and capacitor constitute the LC second-order filter circuit, and the DC side connects the big inductor for energy storage. Generally, it needs to connect diodes in series with the CSC branch power switch in forward direction for the purpose of blocking the reverse current and improving the power switching tube's capability of withstanding reverse voltage.

At present the application of high-power converters is mainly VSC because the common electrical energy sources such as generators, grids, and batteries are VSCs, and compared with the energy storage inductors of CSC, the energy storage capacitors of VSC have obvious advantages in terms of energy storage efficiency and volume and price of energy storage devices. With the development and application of superconducting technology, especially the breakthrough development and functionalization of high-temperature superconducting technology, superconducting technology will address energy storage efficiency of energy storage inductors of CSC. Meanwhile the energy storage coil in the superconducting electric ESS has the current source characteristics. As a result, CSC will enjoy a wide range of applications in future superconducting energy storage. This section will mainly introduce the mathematical model of the most widely used three-phase half-bridge VSC.

4.1.3.3 Mathematical model of three-phase VSC

In the ESS, a bidirectional AC/DC converter normally adopts VSC. See Fig. 4.13 for the three-phase VSC main circuit topological structure. Suppose that the grid is an ideal voltage source; the three-phase circuit equivalent

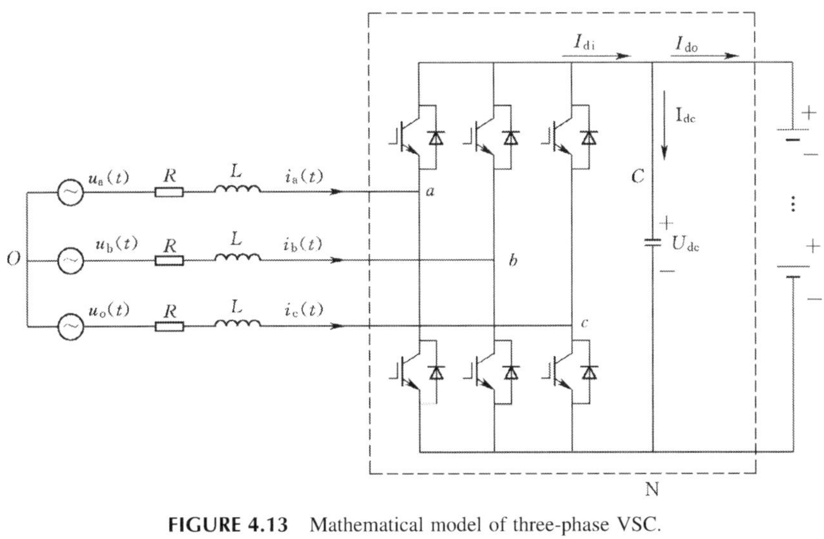

FIGURE 4.13 Mathematical model of three-phase VSC.

resistance equals the inductance. Ignore the switching device turn-on voltage drop and switching losses. Ignore effect of distribution parameters.

1. Mathematical model in the three-phase stationary coordinate system

The three-phase rectifying bridge switching function is defined as shown:

$$S_k = \begin{cases} 1, \text{Upper bridge arm switches on, lower bridge arm switches off} \\ 0, \text{Lower bridge arm switches on, upper bridge arm switches off} \end{cases} \tag{4.24}$$

Where, k = a, b, c, in normal operation, and either upper or lower bridge arm is switched on.

Choosing point O in Fig. 4.13 as zero potential, based on Kirchhoff current law and Kirchhoff voltage law, we can get this:

$$\begin{cases} u_a = Ri_a + Lpi_a + S_a U_{dc} + u_{NO} \\ u_b = Ri_b + Lpi_b + S_b U_{dc} + u_{NO} \\ u_c = Ri_c + Lpi_c + S_c U_{dc} + u_{NO} \end{cases} \tag{4.25}$$

$$C_p U_{dc} = S_a i_a + S_b i_b + S_c i_c - I_{do} \tag{4.26}$$

Where p is the differential operator.

In the three-phase balance system,

$$\begin{cases} u_a + u_b + u_c = 0 \\ i_a + i_b + i_c = 0 \end{cases} \tag{4.27}$$

Add up three phase in Eq. (4.25), substitute in Eq. (4.27), we can get:

$$u_{N0} = -\frac{S_a + S_b + S_c}{3} U_{dc} \tag{4.28}$$

Substituting Eq. (4.28) into Eq. (4.25), combined with Eq. (4.26), we can get the following:

$$\begin{cases} Lpi_a = u_a - Ri_a - \left(S_a - \dfrac{S_a + S_b + S_c}{3}\right) U_{dc} \\ Lpi_b = u_b - Ri_b - \left(S_b - \dfrac{S_a + S_b + S_c}{3}\right) U_{dc} \\ Lpi_c = u_c - Ri_c - \left(S_c - \dfrac{S_a + S_b + S_c}{3}\right) U_{dc} \\ CpU_{dc} = -I_{do} + S_a i_a + S_b i_b + S_c i_c \end{cases} \tag{4.29}$$

When the switching frequency is much higher than the grid frequency, the switching function S_k can be substituted by the percentage d_k (duty cycle) of the switch-on time of the upper bridge arm in a switching cycle, and we can get the mathematical model of state space of a converter in the three-phase stationary coordinate system:

$$Zpx = Ax + u \tag{4.30}$$

Where $Z = \text{diag}\,[L \quad L \quad L \quad C]$, $x = [i_a \quad i_b \quad i_c \quad U_{dc}]^T$, $u = [u_a \quad u_b \quad u_c \quad -I_{do}]^T$

$$A = \begin{bmatrix} -R & 0 & 0 & -\left(S_a - \dfrac{S_a + S_b + S_c}{3}\right) \\ 0 & -R & 0 & -\left(S_b - \dfrac{S_a + S_b + S_c}{3}\right) \\ 0 & 0 & -R & -\left(S_c - \dfrac{S_a + S_b + S_c}{3}\right) \\ S_a & S_b & S_c & 0 \end{bmatrix}$$

2. Mathematical model in the synchronous rotating coordinate system

Using the equations in Section 4.1.1 to make the rotating coordinate transformation in Eq. (4.30), we can get the converter mathematical model in *dq* coordinate system:

$$Zpx = Ax + u \tag{4.31}$$

Where $Z = \text{diag}\,[L \quad L \quad C]$, $x = [i_d \quad i_q \quad U_{dc}]^T$, $u = [u_d \quad u_q \quad -I_{do}]^T$

$$A = \begin{bmatrix} -R & wL & -d_d \\ -wL & -R & -d_q \\ d_d & d_q & 0 \end{bmatrix}$$

Extending and transforming Eq. (4.31), we will get the voltage equation of a converter in the two-phase synchronous rotating *dq* coordinate system:

$$u_d = Ri_d + L\frac{di_d}{dt} - \omega_e Li_q + u_{dl} \tag{4.32}$$

$$u_q = Ri_q + L\frac{di_q}{dt} + \omega_e Li_d + u_{ql} \tag{4.33}$$

$$u_{dl} = d_d udc$$
$$u_{ql} = d_q udc$$

Where u_{dl}, u_{ql} is the front-end voltage of the converter.

Here position d axis of the synchronous coordinate system at the space vector, i.e., in the rotating coordinate system, and d axis coincides with u_d, $u_q = 0$, so the equation of the front-end voltage of converter is this:

$$u_{dl} = -\left(Ri_d + L\frac{di_d}{dt}\right) + wLi_q + u_d \tag{4.34}$$

$$u_{ql} = -\left(Ri_q + L\frac{di_q}{dt}\right) - wLi_d \tag{4.35}$$

The cross-coupling term is defined as shown:

$$u_{dlc} = wLi_q + u_d \tag{4.36}$$

$$u_{qlc} = -wLi_d \tag{4.37}$$

The PI regulator output voltage of the control system is defined as this:

$$u'_{dl} = -\left(Ri_d + L\frac{di_d}{dt}\right) \tag{4.38}$$

$$u'_{ql} = -\left(Ri_q + L\frac{di_q}{dt}\right) \tag{4.39}$$

From Eqs. (4.32)–(4.39), we can see that we can control input power transformation energy by controlling input current with different control methods, and accordingly control the DC-side output voltage. Therefore, dual closed-loop PI modulation is normally adopted to realize the converter control mentioned. The outer ring uses the constant power or constant voltage to control according to the control objective, while the inner ring uses AC input current to control. The role of the outer ring is to protect stability of the control objective, while the current inner ring is to improve the dynamic performance of system and realize the current limiting protection. The output adjusted by the outer ring is the reference value of inner ring input current. After getting the current error by comparison, we make PI regulation the current error to reduce mutation of current in the dynamic process. After getting the regulated two-phase voltage in dq coordinate system, we transform to the ABC coordinate system or $\alpha\beta$ coordinate system with the inverse transformation formula in Section 4.1.1, and we adopt proper PWM technology to generate corresponding six-channel driving impulse to control on and off of three-phase rectifying bridge Insulated Gate Bipolar Translator (IGBT).

4.1.4 Bidirectional DC/DC converter principle and mathematical model

4.1.4.1 Principle overview

Transforming the uncontrolled input DC voltage into the controlled output DC voltage is called DC/DC transformation. The so-called DC/DC converter is

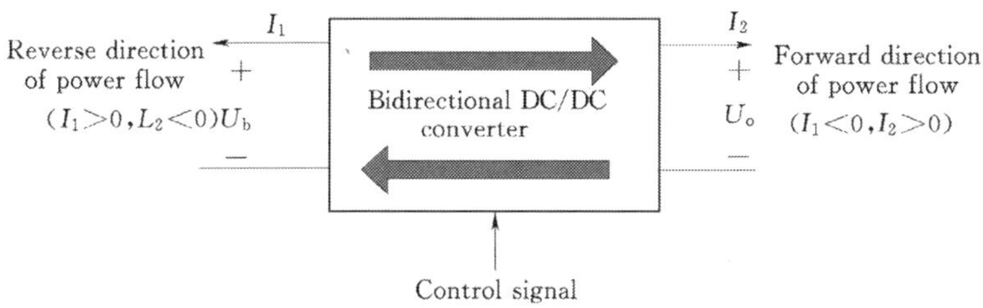

FIGURE 4.14 Schematic diagram of the two ends of a bidirectional DC/DC converter.

DC/DC converter dual-quadrant operation. Its input and output voltage polarity will remain unchanged, but the direction of input and output current can be changed. Functionally, it is equivalent to two unidirectional DC/DC converters. The status of converter output can change in the first and second quadrant of the V-I plane. Exchanging the converter input and output can still realize the voltage transformation function. Power can not only flow from the input end to the output end, but it also can flow from the output end to the input end. Fig. 4.14 is the schematic diagram of the two ends of the bidirectional DC/DC converter. From the perspective of basic converter topology, the bidirectional switch can be used to replace the unidirectional switch to realize bidirectional flow of energy.

4.1.4.2 Converter classification and topological structures

Based on whether there is electric isolation between input and output, the DC/DC converter can be divided into isolated and nonisolated converters. The following paragraphs will briefly introduce the two basic topological structures.

1. Basic topological structures of a nonisolated converter

The typical nonisolated bidirectional DC/DC converter has six basic topological structures: buck converter, boost converter, buck–boost converter, Ćuk converter, Zeta converter, and SEPIC converter. They are basic elements of bidirectional DC/DC converters. Because the body diode of existing controllable power devices (e.g., Power MOSFET, IGBT) have problems such as storing too much electrical charge and poor reverse recovery, more research is being carried out on soft switching technology in the nonisolated converter. Fig. 4.15 shows the basic topological structure of the nonisolated buck–boost converters.

2. Basic topological structure of isolated converter

In practice, the switching power supply normally needs electrical isolation between input and output because of reasons such as voltage level–based transformation, security, isolation, and system connection in parallel or series.

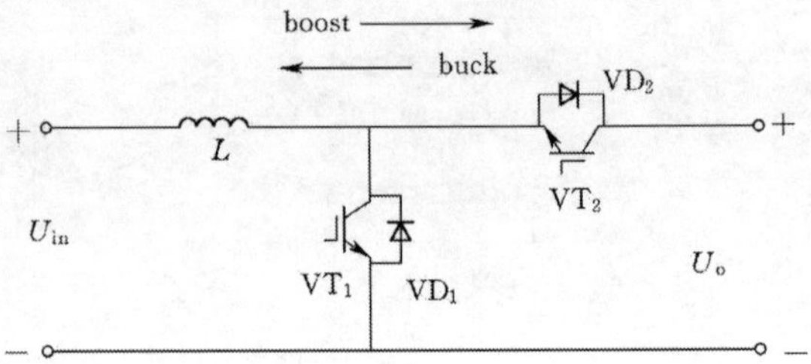

FIGURE 4.15 Basic topological structure of nonisolated bidirectional DC/DC converter.

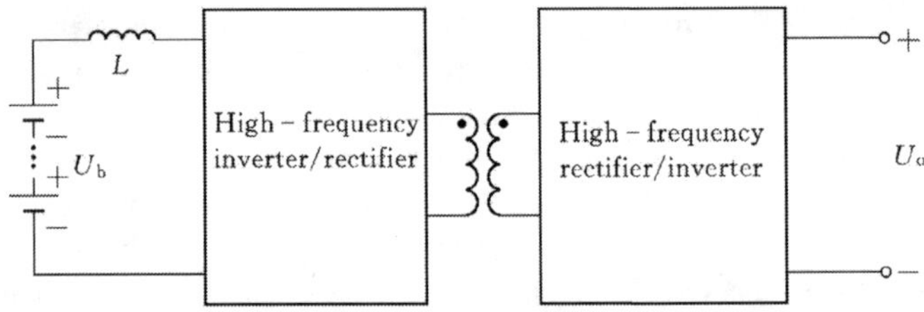

FIGURE 4.16 Basic topological structure of isolated bidirectional DC/DC converters.

Therefore a high-frequency transformer can be added to the nonisolated DC/DC converter to get the isolated DC/DC converter. Fig. 4.16 shows the basic form of an isolated buck–boost converter. The high-frequency rectifier/inverter unit and high-frequency inverter/rectification unit can be composed of full-bridge, half-bridge, or push-pull circuit topological structures.

4.1.4.3 Working principle of bidirectional buck–boost converter

The ESS requires bidirectional flow of energy. The DC/DC converters should have the buck–boost bidirectional exchange functions, i.e., buck–boost chopper circuit. Therefore this section will introduce the principle of the buck–boost chopper circuit in detail. See Fig. 4.16 for the equivalent circuit of the buck–boost bidirectional DC/DC converter of the ESS. In the figure, L stands for chopper inductance, C_1 is the DC bus capacitance, C_2 is filter capacitance, VT_1, VT_2 stand for IGBT of the buck-booster chopped wave of the ESS, and VD_1, VD_2 stand for the fly–wheel diode. If L is high, C_1 is high too.

1. Buck circuit control principle

In Fig. 4.17, VT_1 and VD_2 constitute the buck chopper circuit. During the voltage step-down, the ESS is charging, as shown in Fig. 4.18. The AC-side power supply after rectification serves as the input power supply U_d, U_E is the battery voltage, and U_0 is the output voltage.

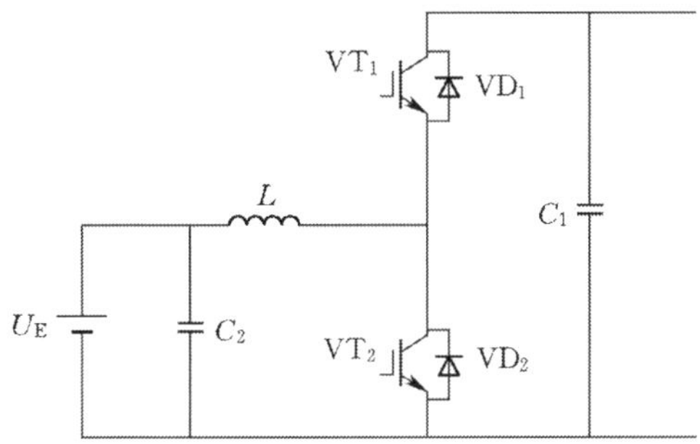

FIGURE 4.17 Equivalent circuit of bidirectional DC/DC converter.

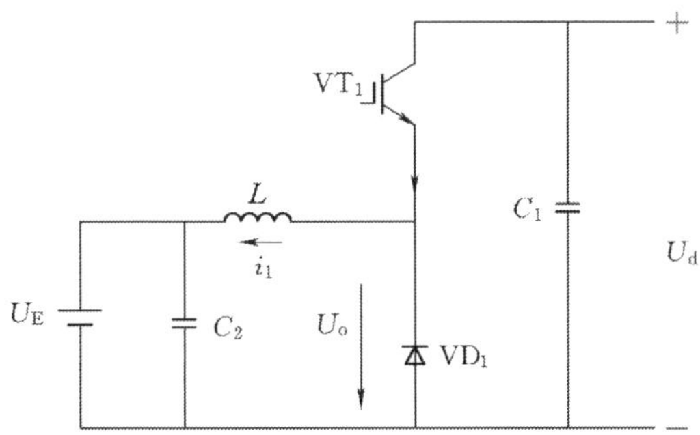

FIGURE 4.18 Equivalent circuit of buck circuit.

When VT_1 is switched on, the DC bus charges the energy storage device, output voltage $U_0 = U_d$, and inductive current i_1 rises according to the exponential curve. When VT_1 is switched off, inductance L continues flowing through the diode VD_2, the output voltage is close to zero, and inductive current declines in the exponential curve. To make the current continual and impulse small, it normally connects the inductor with higher L value in series.

When a cycle ends, switch on VT_1 and repeat the cycle again. When the circuit is stable, the average output voltage is as follows:

$$U_o = \frac{t_{on}}{t_{on} + t_{off}} U_d = \frac{t_{on}}{T} U_d = DU_d \tag{4.40}$$

Where t_{on} is the time that VT_1 is in the status of switching on; t_{off} is the time that VT_1 is in the status of switching off; T is the switching cycle, and D is the duty ratio, also known as conduction ratio.

From Eq. (4.40), we can know $D = t_{on}/T \leq 1$, and the maximum output voltage average is U_d; if the duty ratio D is reduced, U_o reduces accordingly. Therefore the circuit is called a buck chopper circuit.

2. Boost circuit control principle

In Fig. 4.16, VT_2 and VD_1 form the booster chopper circuit. During the boosting of voltage, the ESS discharges, as shown in Fig. 4.19, the DC-side battery serves as the input supply U_E, and it outputs DC bus voltage U_d after the voltage boosting.

When VT_2 is switched on, the energy storage device charges inductor L, and the charging current is basically maintained at i_1, while the voltage of capacitor C_1 supplies the DC bus. Because C_1 is high, the bus voltage U_d is kept as a constant. Suppose the time that VT_2 is in the status of switching on is t_{on}, then the energy accumulated at the inductor L is $U_E I_1 t_{on}$. When VT_2 is switched off, U_E and L charge capacitor C_1 together to maintain the bus voltage constant. Suppose the time that VT_2 is in the status of switching off is t_{off}, then the energy discharged by inductor L is $(U_d - U_E)\ I_1\ t_{off}$. When the circuit is under stable operation, the energy accumulated and discharged in the cycle T is the same:

$$U_E i_1 t_{on} = (U_d - U_E) i_1 t_{off} \tag{4.41}$$

Here it is simplified:

$$U_d = \frac{t_{on} + t_{off}}{t_{off}} U_E = \frac{T}{T - t_{on}} U_E = \frac{1}{1 - D} U_E \tag{4.42}$$

Similarly, in Eq. (4.42), t_{on} stands for the time when VT_2 is in the status of switching on, and t_{off} is the time when VT_2 is in the status of switching off, T is the switching cycle, and D is the duty ratio. Because $0 < D \leq 1$, $1/(1-D) \geq 1$, the output voltage is higher than the input voltage, and therefore the circuit is called a booster chopper circuit.

In Fig. 4.17, if VT_1 and VT_2 are switched on at the same time, the supercapacitor will be short circuited. Therefore we must prevent such a situation.

A boost—buck control system is a dual closed-loop control system, and the outer loop is the voltage control loop. The sampled output voltage is subtracted

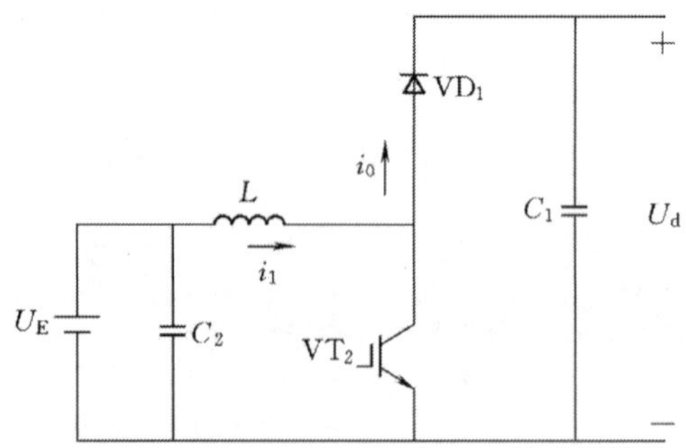

FIGURE 4.19 Equivalent circuit of booster circuit.

from the voltage set value, and the voltage loop PI operation is performed according to the voltage error signal. Then, we get the given signal of the inductor current. The inner loop is a current control loop, and the sampled inductor current is subtracted from the inductor current set value. The current loop PI operation is performed according to the current error signal, and then we get the duty cycle signal of the switching element and output to the IGBT switching element of the main loop of the converter.

According to the differences of the output voltage modulation methods, the chopper circuit has three control methods:

1. The method keeping the switching cycle T unchanged and modulating the switching-on time t_{on} is PWM, also called pulse widening modulation.
2. The method keeping the switching-on time t_{on} unchanged and modulating the switching cycle T is called frequency modulation.
3. The method keeping t_{on} and T adjustable and changing the duty ratio is called mixed modulation.

The first modulation is the most widely used.

4.1.5 Typical topological structure of an ESS

The BESS consists of the battery management system and energy conversion devices. The BESS has the following two typical topological structures according to the different combination of the energy conversion devices.

4.1.5.1 Monotier topological structure

The monotier topological structure of the single-tier converter ESS that only uses the AC/DC converter is shown in Fig. 4.20. The bidirectional AC/DC

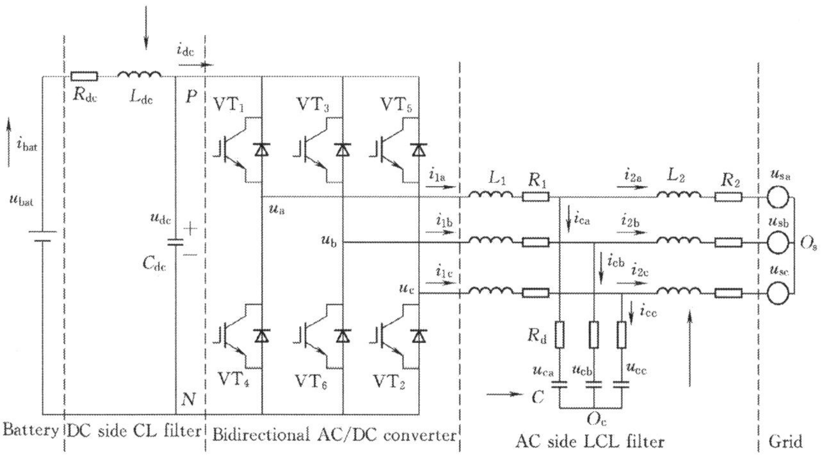

FIGURE 4.20 Topological chart of the monotier on-grid system.

converter DC side is directly connected with the energy storage device, while the AC side is connected with the power source (the power source can be the grid or distributed power source) or load system. When the ESS is charging, AC/DC is under rectification state, and the AC-side power source charges the energy storage devices through the three-phase full-controlled rectification bridge (if the ESS powers the load system independently under the off-grid condition, the AC side is connected with the load system, and additional charging unit is needed to charge the system.). When the ESS is discharging, AC/DC is under the inversion state (active inverter or reactive inverter), and the ESS supplies the grid or the load system through the thee-phase full-controlled rectification bridge.

The monotier conversion topological technology has the following characteristics:

1. The circuit structure is simple, but the energy conversion efficiency is high, and total system loss is small.
2. The cost of equipment is lower, and it is easier to realize standard manufacturing.
3. The control system is simple, and a single AC/DC controller can realize active and reactive unified control.
4. The control strategy is simple, and it is easier to realize control in practice. It is easy to switch the two modes between the on-grid and off-grid and connect with the upper level monitoring systems and realize various advanced control strategies.
5. The DC side has frequency-doubled low-frequency ripple and high-frequency switching ripple, the LC filter features high difficulties in design, the battery-side ripple is large, the precision of control is lower, and the time for conversion between charge and discharge is long.
6. The DC-side voltage features a narrow scope. When we design the single machine with large-capacity, the battery pack requires several in series and parallel connection, which increases the difficulties of battery grouping; when a single battery is replaced because of fault, the performance of the whole battery group will be reduced.
7. When the AC or DC side has failure, the battery side will be subject to surge current for a short time; consequently the battery service life will be reduced.

4.1.5.2 Two-tier topological structure

The topological structure of the two-tier conversion ESS that uses AC/DC converter and DC/DC converter concurrently is shown in Fig. 4.21. The bidirectional AC/DC converter AC side is connected with the power source (the power source can be the grid or other distributed power source) or load system, and the AC side is connected with the ESS through the

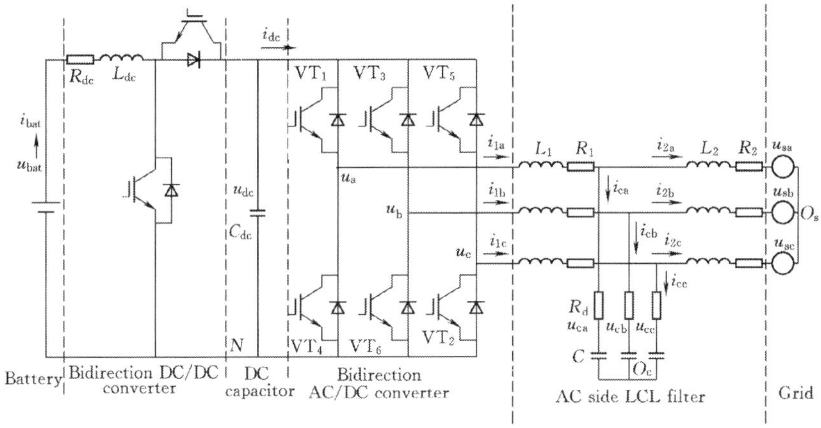

FIGURE 4.21 Topological chart of the two-tier on-grid system.

DC/DC converter. The DC/DC converter has functions of conversion of current and modulation of voltage, and it can directly control the discharge current and bus voltage on the AC side, and then control the input and output active power; when it is on-grid, the AC/DC converter realizes the conversion of the system and the grid power; when it is off-grid, the AC/DC converter provides the system voltage and frequency support.

The two-tier conversion topological technology has the following characteristics:

1. The circuit structure is relatively complicated, energy conversion efficiency is low, and the whole system loss is slightly higher.
2. The equipment costs are high, and it is not easy to realize unified modular manufacturing.
3. The control system is complicated, and the active power control of the system is realized through the DC/DC converter controller, and the system reactive control is realized through the AC/DC converter controller; it is difficult to match the upper level monitoring systems and realize advanced control strategies.
4. The control strategy is relatively complicated and realized through the AC/DC and DC/DC control strategies. Coordination is needed between the AC/DC strategy and the DC/DC strategy. When the DC side has several DC/DC circuits, it is difficult to realize the two-mode switching between the on-grid and the off-grid.
5. There is no need for complex LC filters on the DC side. The ripple on the battery side is small, the control accuracy is high, and the time of charge-discharge conversion is short.

6. When we design the single machine with large-capacity, it can be realized with several DC/DC on the DC side. Each unit of DC/DC can be connected with an independent battery pack. There is no need to connect several battery packs in series and parallel, which reduces the difficulty of configuration of battery packs. A single-cell battery will not reduce the overall system performance index after being replaced due to failure.
7. When there is a fault on the AC side or on the DC side, the presence of a DC/DC circuit section can effectively protect the battery, prevent the battery from incurring an inrush current, and prolong the service life of the battery.

4.2 Grid-connected operation control technology

When the ESS is connected to the grid, it directly adopts the grid frequency and voltage for support, and the ESS is operating under the current source mode. It controls the input and output power, DC-side voltage or charging and discharging current. Different control methods can be adopted according to the topological structures of the grid-connection system of the ESS. The monotier topological conversion system controls the active and reactive power system through the AC/DC converter. We know from Section 4.1.3 that the ESS grid-connected control is realized with the dual closed loops. The inner controls the input and output current of the energy storage, while the outer loop can choose constant power (*P*/*Q*) control or constant voltage control. The two-tier topological grid-connected system can have the following two combined control modes: one makes the AC-side output power as the main control target to realize constant power output of the AC side, and AC/DC converter adopts the *P*/*Q* control method to control the input and output power of the ESS, while the DC/DC converter uses the constant voltage control method to control the DC bus voltage and maintain the DC bus voltage stability; the other makes the DC-side voltage the main control target to keep the DC-side voltage stable, and the AC/DC converter uses the constant voltage control method to maintain the DC bus voltage constant, while the DC/DC converter adopts the constant current control method to control the charging and discharging power of the DC-side ESS. The following paragraphs will introduce the specific control methods of AC/DC converter and DC/DC converter in different control combinations.

4.2.1 AC/DC converter control

An AC/DC converter uses the dual closed-loop control. The inner loop controls the input and output current of the ESS, while the outer loop can adopt the *P*/*Q* control or constant voltage control according to the control targets.

4.2.1.1 P/Q control

The goal of the AC/DC converter *P/Q* control is to maintain the active and reactive power of the ESS close to their reference values. When the ESS is connected to the power grid, it directly uses the grid frequency and voltage as support. Based on the active and reactive reference values given by the upper-level controller, the energy storage converter can realize active and reactive power control according to the *P/Q* controls strategy. Its active power controller and reactive power controller can adjust the active and reactive power output, respectively, and maintain the output power of the ESS stable according to the active and reactive output power with given reference values. See Fig. 4.22 for the *P/Q* control block diagram.

In the diagram, P_{ref} and Q_{ref} stand for given power reference values; P and Q are the measured values of power; i_{dref} and i_{qref} are reference values of the dq axis components of the AC-side current; i_d and i_q are actual values of the AC-side current dq axis components; u_d, u_q are actual values of the inverter output voltage dq axis components; u_{dl}, u_{ql} are reference values of the inverter output voltage dq axis components; L is the AC-side coupling inductance; θ is the initial phase angle of voltage.

To realize the control mentioned, the first is to conduct active and reactive decoupling and use the coordinate transformation formulas introduced in Section 4. To transform the voltage and current components of the three-phase ABC coordinate system output by the AC/DC converter to components of the synchronous rotating dq coordinate system, and make q axis voltage component $u_q = 0$, the inverter output power can be written as shown:

$$\begin{cases} P = u_d i_d + u_q i_q = u_d i_d \\ Q = u_d i_q - u_q i_d = u_d i_q \end{cases} \tag{4.43}$$

The difference between the given reference values of the power P_{ref}, Q_{ref} and the actually measured values P and Q provides reference values i_{dref} and i_{qref} for the inverter output current under function of the PI regulator.

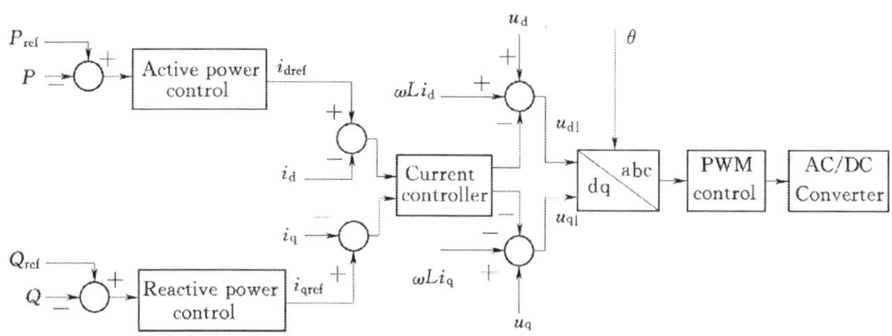

FIGURE 4.22 *P/Q* control block diagram of AC/DC converter.

The difference between the output current reference value and the actual value i_d, i_q provides the reference components for the output voltage of inverter under the function of the PI regulator. Meanwhile, we can calculate *dq* axis voltage coupling components $\omega L i_d$ and $\omega L i_q$ according to the outlet filter inductance parameter L of the inverter, get the inverter output voltage reference values u_{dl}, u_{ql} by superposition, and then transform them to the three-phase *ABC* coordinate components through the coordinate transformation to control the inverter.

4.2.1.2 Constant voltage control

The purpose of the control of the AC/DC converter constant voltage is to maintain the ESS DC bus voltage value around the reference values. The outer loop uses the DC voltage PI modulation to maintain the battery voltage constant. The given voltage values are the allowable voltage values of the energy storage batteries; the actual values are DC bus voltage values that are measured with measurement components. After amplitude limiting of the DC voltage outer loop modulation, the output value is as the given value of the inner loop active current. See Fig. 4.23 for the constant voltage control block diagram.

In Fig. 4.23, U_{ref} is the DC voltage reference value; U_{dc} is the DC voltage measured value; i_{dref}, i_{qref} stand for AC-side current *dq* axis component reference values; i_d, i_q are actual values of AC-side current *dq* axis components; u_d, u_q are actual values of inverter output voltage *dq* axis components; u_{dl}, u_{ql} are reference values of the inverter output voltage *dq* axis components; *L* is the AC-side coupling inductance; *3* is the voltage initial phase angle.

The reference value of the reactive current of the inner loop is calculated by the active current reference value and power factor:

$$i_{qref} = \frac{\sqrt{(1 - \lambda^2)}}{\lambda} i_{dref} \tag{4.44}$$

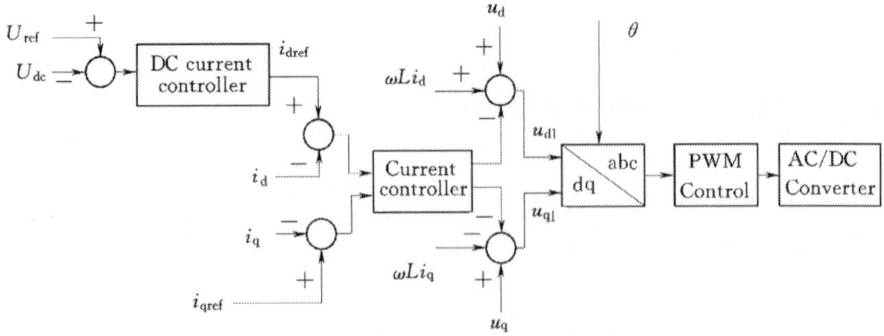

FIGURE 4.23 Constant voltage control block diagram of AC/DC converter.

The difference between the given reference value of DC voltage U_{ref} and the actually measured values U_{dc} provides reference values i_{dref} for the inverter output d axis current under function of PI regulator. The reference value of q axis current is calculated with Eq. (4.44). The difference between the output voltage reference value and current actual value i_d and i_q provides the reference component for the inverter output voltage under the function of the PI regulator. Meanwhile, we can calculate dq axis voltage coupling components ωLi_d and ωLi_q according to the inverter outlet filter inductance parameter L, get the inverter output voltage reference values u_{dl}, u_{ql} by superposition, and then transform them to the three-phase ABC coordinate components through the coordinate transformation to control the inverter.

4.2.2 DC/DC converter control

DC/DC converter control is to realize control of DC-side input and output active power and can use constant voltage control or constant current control according to the difference of control targets.

4.2.2.1 Constant voltage control

See Fig. 4.24 for the DC/DC converter constant voltage control block diagram. In the diagram, U_{ref}, U_{dc} stand for the reference value and measured value of the DC bus voltage; I_{ref}, I_{dc} are the reference value and measured value of the charging and discharging current at the DC energy storage device side.

The control principle of Fig. 4.24 is as follows: The difference between the DC bus voltage reference value U_{ref} and actual value U_{dc} can provide reference value I_{ref} for the charge and discharge current at the energy storage side under the function of the PI regulator. When $U_{ref} > U_{dc}$, the energy storage device discharges and makes the bus voltage rise; when $U_{ref} < U_{dc}$, the energy storage device charges and makes the bus voltage decline. Under the function of the PI regulator, the difference between the DC charge and discharge

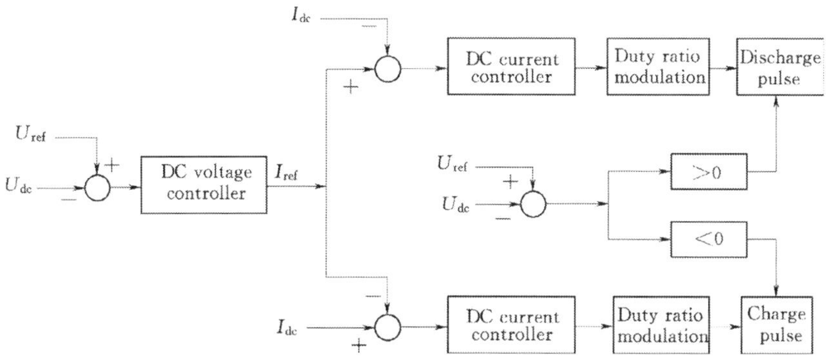

FIGURE 4.24 DC/DC converter constant voltage control block diagram.

current reference value I_{ref} and the actual value I_{dc} outputs the modulation depth and compares with the triangle carrier phase to make duty ratio modulation and output control impulse. To make smooth switching between charge and discharge, normally two controllers are used in parallel to switch the charge and discharge impulse according to the voltage value differences and to control the DC/DC converter.

4.2.2.2 Constant current control

See Fig. 4.25 for the constant current control block diagram of the DC/DC converter. In the diagram, I_{ref}, I_{dc} stand for the reference value and measured value of the charge and discharge current at the DC energy storage device side.

The control principle of Fig. 4.25 is as follows: Under the function of the PI regulator, the difference between the DC energy storage side charge and discharge current reference value I_{ref} and actual value I_{dc} outputs the modulation depth and compares with the triangle carrier phase to make duty ratio modulation and output control impulse. When the current reference value $I_{ref} > 0$, the ESS discharges; when the current reference value $I_{ref} < 0$, the ESS charges. To make smooth switching between charge and discharge, normally two controllers are used in parallel to control switching of the charge and discharge impulses to control the DC/DC converter according to the voltage reference value signs.

4.2.3 Island detection

4.2.3.1 Analysis of the island effect

The island effect is a phenomenon when the power supply system of the power company stops working because of failure or outage maintenance, the grid-connected new energy generation system or ESS fails to detect the outage and immediately cut off its connection with the grid, but it forms a self-powered island powering the ambient consumers that cannot be controlled by the power company.

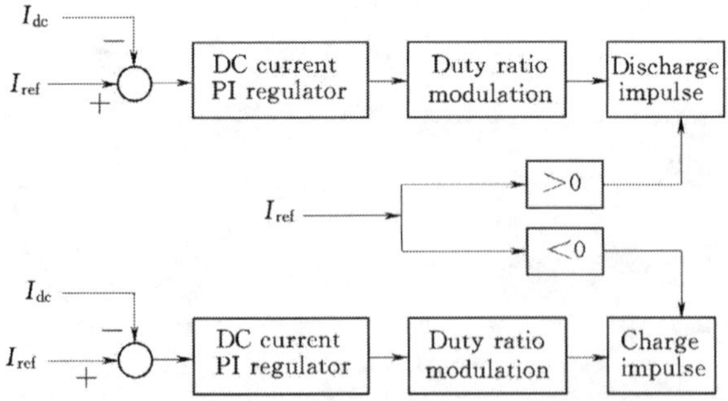

FIGURE 4.25 Constant current control block diagram of the DC/DC converter.

Generally, the island effect may have adverse impacts on equipment of the whole power distribution system and the customer premise equipment:

1. Endanger the life safety of power maintenance personnel.
2. Affect the protection switch action procedure of the power distribution system.
3. The instability of the power supply voltage and frequency in the island area will damage the electric equipment.
4. The out of sync of voltage phase caused by power restoration will lead to surge current, which may cause tripping again or cause damage to the PV system, load, and power supply system.
5. The single-phase power supply of the power system may cause the out-of-phase power supply of the three-phase load system.

Obviously, as a safe and reliable grid-connection inversion device, the energy storage converter must have the capability of rapidly detecting an island and immediately cutting off connection with the grid; and the protection against an island shall coordinate with the grid-side line protection system.

When the inverter is connected to the grid, the output voltage is controlled by the grid; what the inverter can control is the input current, including the current amplitude, phase, and frequency. The frequency and phase should be the same as the grid voltage, and in practice, it is realized through the public coupling point voltage zero-crossing point (ZCP) synchronization, and the amplitude is adjustable according to the actual systems. The study of island detection technology does not need to care about how the internal control of the inverter is realized, but it should care about the output characteristics of the inverter power supply. Therefore when studying the island detection technology, the inverter power can serve as an equivalent controlled current source with adjustable amplitude, with frequency and phase tracing the grid, as shown in Fig. 4.26.

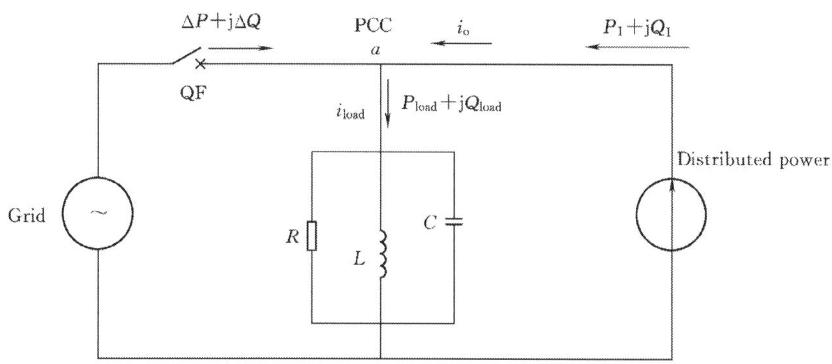

FIGURE 4.26 Typical island system model.

In Fig. 4.26, when the grid-connection switch QF is closed, the inverter system provides power $P_I + jQ_I$ to the point of common coupling (PCC), the power consumed by the load system is $P_{load} + jQ_{load}$, and the power that the grid provides to the load system is $\Delta P + j\Delta Q$. At this moment, the amplitude and frequency of the PCC voltage is determined by the grid. Where we have the following:

$$\Delta P = P_{load} - P \tag{4.45}$$

$$\Delta Q = Q_{load} - Q \tag{4.46}$$

When $\Delta P \neq 0$, $\Delta Q \neq 0$, if the grid is switched off, the voltage, frequency, and amplitude of the PCC will change. Therefore ΔP and ΔQ are called power nonmatching extent.

4.2.3.2 Passive island detection

The passive island detection is to judge whether island phenomenon occurs by directly detecting the changes of output voltage, frequency, phase, or harmonic wave of the inverter. This method does not need additional hardware circuit or separate protection relay. In detection, it has no impact on the output energy quality or interference on the grid. This method is characterized by a simple principle, easy realization, and fast detection speed. But when the grid-connected system output power is balanced with the local load power, the passive detection method will lose the capability of detecting an island and have a high nondetection zone (NDZ). The passive island detection mainly has the following methods:

1. Over/undervoltage protection and over/underfrequency protection detection methods

IEEE Recommended Practice for Utility Interface of Photovoltaic (PV) Systems (IEEE Std. 929.2000) stipulates that the grid-connected system must have the overvoltage protection (OVP), undervoltage protection (UVP), overfrequency protection (OFP), and underfrequency protection (UFP) and the corresponding protection threshold. The OVP/UVP/OFP/UFP detection is a method that stops the inverter grid connection when the voltage amplitude and frequency of the PCC exceed the normal range. When the inverter works, the working range of voltage and frequency shall be set up reasonably and allow normal fluctuation of grid voltage and frequency. For 220 V/50Hz, the working range of voltage and frequency are normally $194\text{V} \leq U \leq 242\text{V}$, $49.5\text{ Hz} \leq f \leq 50.5\text{ Hz}$. If the voltage or frequency drift reaches the islanding detection threshold, the island can be detected. However, when the local load of the inverter is close to its output power, the voltage and frequency drift will be very small or even zero. Therefore this method has the undetectable zone. This method features a good economy performance, but it is enough for OVP/UVP/OFP/UFP island detection because the undetectable zone is large.

2. Harmonic detection method

Harmonic detection method judges whether there is island by monitoring the PCC total harmonic distortion (THD) and whether THD exceeds the preset threshold. In the grid-connected inverter system, the PCC harmonic comes from two aspects, as shown in Fig. 4.26: one is the distortion of voltage of point a caused by the grid voltage distortion; the other is the distortion of voltage of point a caused by the current harmonic of the high-frequency switch. Normally the grid voltage harmonic distortion is smaller and stays unchanged. When the inverter system is connected to the grid, its output current harmonic will follow into the grid through a point of PCC. As the grid impedance is small, the THD of voltage of point a is normally lower than the threshold value (the THD of grid-connected inverter is smaller than 5% of the rated current). When the grid is switched off, the current harmonic from the inverter system flows to the load system. Because the load impedance is much higher than the grid impedance, the inverter output current will generate a relatively high-voltage harmonic at "a" point (the product of harmonic and load impedance). By detecting harmonics or changes of the voltage harmonic, we can effectively detect the occurrence of an island.

Theoretically, the voltage harmonic detection improves reliability of the island detection to some extent. However, in practice, due to the nonlinear load and other elements, the grid voltage has relative high harmonics, the THD of output current of the grid-connected inverter shall meet the standard of below 5% of the rated current, the operation threshold of harmonic detection cannot be easily decided, and it may cause incorrect operation because the nonlinear load is connected or the switch acts, so this method has some limitations.

3. Phase jump detection (PJD)

PJD is to detect an island by monitoring the grid-connected inverter output voltage and current phase difference changes. When the inverter system is connected to the grid, voltage of "a" point of PCC is clamped by the grid voltage, and the output current and voltage of the inverter system should be the same as the grid to enable the system to work under the unity power factor mode. In such a case the load current is decided by the voltage of "a" point, i.e., by the grid voltage. When the grid is disconnected, there will be an island that a PV grid-connected generation system independently powers the load system. In such a case, voltage of "a" point is decided by output current i_0 and load impedance Z. Because of the function of locked-phase loop, i_0 and "a" point voltage synchronize at the ZCP. Between ZCPs, i_0 follows the internal reference current of the system and will not have mutation. Therefore, for the nonresistance load, phase of "a" point voltage will have mutation. Accordingly, we can use the phase mutation detection method to detect island.

The PJD method is simple and easy to realize. But when the load impedance angle is close to zero, i.e., the load is close to be resistive, the method will fail because of limitations of the preset threshold. Normally the phase mutation detection shall be used in combination with OVP and UVP detection methods to reduce the undetectable zone and improve reliability of the island detection.

4.2.3.3 Active island detection

The active island detection refers to the method making certain disturbance of the output power, frequency, or phase by controlling the inverter. When the gird works normally, such disturbance will not be detected because of the balance function of the grid. When the grid fails, the disturbance delivered by the inverter will accumulate rapidly and exceed the allowable range, consequently triggering the island detection circuit. This method has a high precision and small undetectable zone, but it has complicated control and reduces the quality of the output energy of inverter. At present the antiislanding strategy of a grid-connected inverter normally adopts a passive detection method and an active detection method in combination.

1. Alternate current disturbances (ACD)

ACD is an active islanding detection method. It applies periodic current disturbance to the grid-connected inverter to break the power balance between the inverter and load system and reach the goal of detecting an island. For an inverter controlled by current source, it will change the output active power by reducing the grid-connected inverter output current after a certain duration. When the inverter is connected to the grid, its output voltage is kept constant at the grid voltage. When the grid is powered off, the output voltage of the inverter is decided by the load system. At the current disturbance time, when the output current amplitude changes, the voltage of load will change accordingly. When the voltage reaches the UVP range, it may detect occurrence of an island. The current disturbance method is normally used together with the OVP and UVP detection methods.

The periodic current disturbance detection method (ACD) features a simple principle, easy realization, and small current harmonic. However, the periodic changes of the grid-connected current will have a relatively high effect on the inverter system output power and have high requirements on the system dynamic performance. When multiple inverter systems are connected to the grid, the effect of a single inverter current disturbance on the PCC voltage will be weakened, reducing reliability of island detection.

2. Active frequency drift (AFD) detection

AFD is a common active disturbance detection method. It detects an island by making the inverter output current frequency have a certain drift. In grid-connected operation, each periodic inverter output current initial phase traces the grid voltage phase. Therefore, small current deviation will not have obvious effect on the normal operation of the system. When an island occurs, the PCC voltage frequency will change due to the current deviation and generate certain feedback to consequently increase the voltage frequency drift, form a continuous frequency changing trend, and finally cause the output voltage and current to break the frequency protection threshold and realize the purpose of antiislanding.

In the event of resistive load, the PCC voltage waveform may follow the output current waveform of the inverter when the grid trips; when it exceeds the preset threshold value, the AFD can detect occurrence of islanding. In case of inductive load, because the inductive load can make the PCC voltage frequency on the rise trend after the grid is disconnected, it will exceed the frequency protection threshold faster and consequently detect an island. But in case of capacitive load, because the capacitive load can make the PCC voltage frequency on the decline trend after the grid is disconnected, if the frequency decline trend counteracts the rise trend of the inverter output current, the detection method will fail.

3. Slip-mode frequency shift (SMS) detection

SMS is an active island detection method. It controls the output current of the inverter and causes it to have a certain phase difference with the PCC voltage to detect islanding when the PCC frequency deviates from the normal rage after grid voltage loss. In normal circumstances, the inverter phase angle response curve is designed within the vicinity of the system frequency range, and the inverter phase angle increases faster than RLC load under unity power factor. When the inverter is connected to the distribution grid, the distribution grid provides a fixed reference phase angle and frequency to make the inverter working point stable at the working frequency. When islanding occurs, if the inverter output voltage frequency has small fluctuation, the inverter phase response curve will increase the phase errors and reach a new steady point. The new status point frequency will exceed the OFP/UFP threshold, and the inverter will shut down because of frequency error. This method actually shifts frequency by shifting the phase. It is as simple as the AFD and does not need additional hardware. It has high reliability of island detection and similar weakness as AFD method, i.e., the possibility of failing to detect an island will increase with growing load quality factor.

4. Frequency jump (FJ) detection

FJ detection is an improved method of AFD and similar to the impedance measurement method. FJ detection adds dead band in the output current waveform (not all periods), and the frequency vibrates under the preset mode. For example, when adding the dead band in the fourth period, the inverter current causes frequency mutation in normal circumstances, but the grid prevents its fluctuation. When the island forms, the FJ method detects whether the island phenomenon occurs by examining whether the vibration mode of the inverter output voltage frequency meets the present mode. This method has the following advantages: if the vibration mode is mature, FJ method is effective to prevent an island when a single inverter is working; but with multiple inverters working, if the frequency drift directions are different, the efficiency and effectiveness of detecting an island will be lowered.

4.2.4 Low-voltage ride through

LVRT requires that when the grid fails or disturbance causes a grid connection point voltage decline, the ESS can be ensured to operate and stay connected with the grid and provide certain reactive power to the grid at certain voltage decline scope and duration until the grid voltage restores, i.e., "riding through" the low voltage period (zone).

The ESS LVRT is normally required at the high-voltage, high-capacity energy storage station. The energy storage converter PCS device should have the LVRT capability. The technical requirements on the energy storage converter LVRT capability are shown in Fig. 4.27.

1. The ESS has the capability of ensuring on-grid operation for 625 s after the PCC voltage declines by 20% of the rated voltage.
2. When the ESS can recover the voltage to 85% of the rated value within 2 s after the PCC voltage drops, the energy storage converter can ensure continuous on-grid operation.

For different faults of the grid, the following requirements are imposed on the ESS LVRT:

1. When the grid experiences three-phase short circuit that causes the PCC voltage drop, and the voltage of different lines of the ESS falls in the area of the voltage contour line and above in the diagram, the ESS must guarantee on-grid continuous operation; when the voltage of any line of the

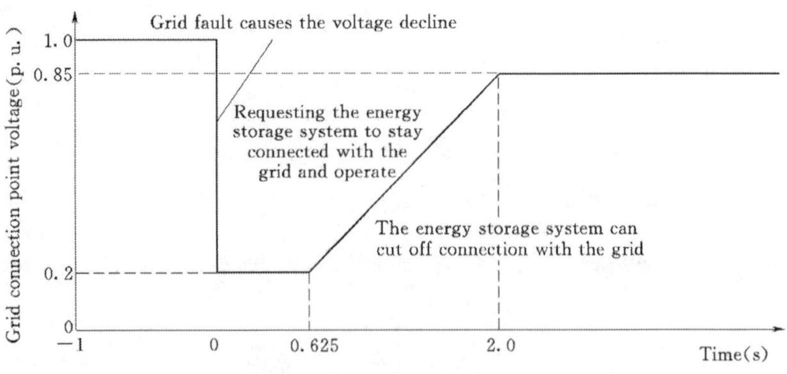

FIGURE 4.27 LVRT requirements on an energy storage inverter.

ESS PCC is lower or partially lower than the voltage contour line in the diagram, the ESS is allowed to be off the grid.

2. It is the same when the grid faces two-phase short circuit that causes PCC voltage drop.
3. When the grid faces single-phase earthing short circuit that causes the PCC voltage decline, and the voltage of different phases of the ESS PCC fails in the contour line and above area in the diagram, the ESS must guarantee on-grid continuous operation; when the voltage of any phase of the ESS PCC is below or partially below the contour line in the diagram, the ESS is allowed to be off the grid.

4.3 Off-grid operation control technology

In off-grid operation, the ESS independently powers the load system or serves as the main power supply in the microgrid, and it shall provide voltage and frequency support for the load system and maintain stability of the microgrid voltage and frequency. Therefore, in the monotier transformation topology or two-tier transformation topology, the AC/DC converter shall control the AC-side voltage and frequency, and it adopts the V/f control method. In the two-tier transformation topology, the DC/DC converter mainly controls the voltage of the DC bus and the charge and discharge current of the DC side, and it adopts the constant voltage or constant current control as in the on-grid system. See Section 4.2.2 for details; it will not be discussed here. The basic thinking of V/f control is that the outlet voltage amplitude and frequency will remain unchanged no matter how the ESS output power changes. It mainly provides strong voltage and frequency support for the off-grid operation of the ESS and has the follow-up characteristics of certain load power. There are several V/f control methods as follows according to the differences of the adopted control variables and modes.

4.3.1 V/f control

4.3.1.1 Voltage single closed-loop control

The voltage single closed-loop control only makes closed-loop control to a type of voltage that can adopt the voltage effective value, voltage instantaneous value, or voltage vector after coordinate transformation decoupling. See Fig. 4.28 for the voltage effective value single closed-loop control block diagram. In the diagram, U_{CA}, U_{BC}, U_{AB} stand for the measured voltage signals of the three-phase lines; U_{ref} is the reference value of the voltage effective value; f is the given frequency instruction. According to the measured voltage instantaneous value, we can get the effective values of the voltage of three phases through the Fourier transform and compare them with the reference values, and the difference outputs the three-phase voltage modulation depth

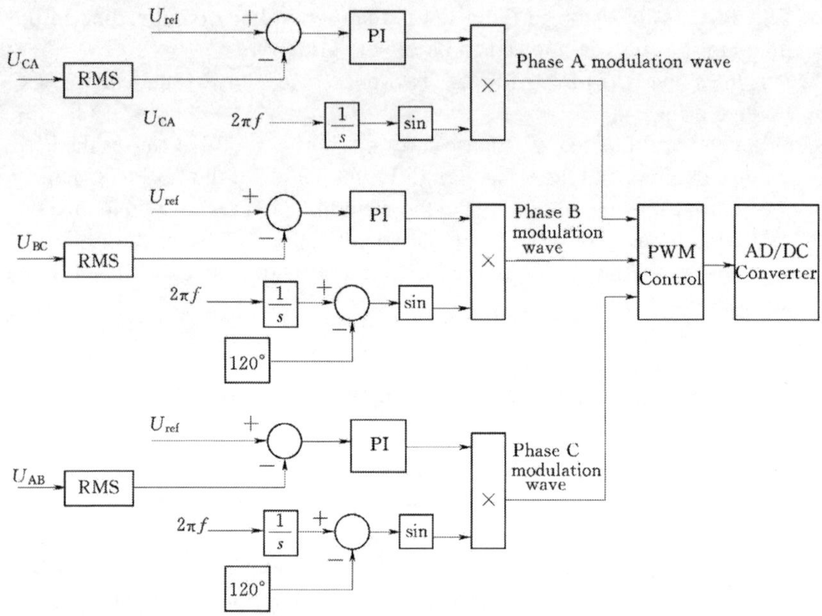

FIGURE 4.28 Block diagram for the voltage effective value single closed-loop control.

values after PI modulation. Multiply them and the phase of the three-phase voltage, and we get the three-phase modulating wave. We choose proper carrier signals to generate control impulse with the PWM algorithm to control the AC/DC converter. With the existence of the effective value closed loop, when the PI parameter modulation is proper, the output voltage effective value will change with the given value.

The control of frequency adopts the open-loop control. We get the phase of the modulation wave after the frequency instruction integral and make the modulating wave frequency as the given frequency to ensure the output voltage at the given frequency and the phase difference of the ABC three phases at 120 degrees.

See Fig. 4.29 for a block diagram of the voltage instantaneous value single closed-loop control. Where U_{CA}, U_{BC}, U_{AB} stand for the measured voltage signals of the three-phase lines; U_{ref} is the voltage reference value; f is the given frequency instruction. Multiplying the given voltage value and the given voltage phase, we get the voltage instantaneous given value. Compare the measured voltage instantaneous value with the voltage instantaneous given value, and we get the modulating wave output with PI modulation. We choose the proper PWM control algorithm to generate the control impulse to control the AC/DC converter. As the voltage instantaneous given value is the sine signal, it can make the output voltage maintain at sine through the closed-loop control to guarantee the quality of output energy.

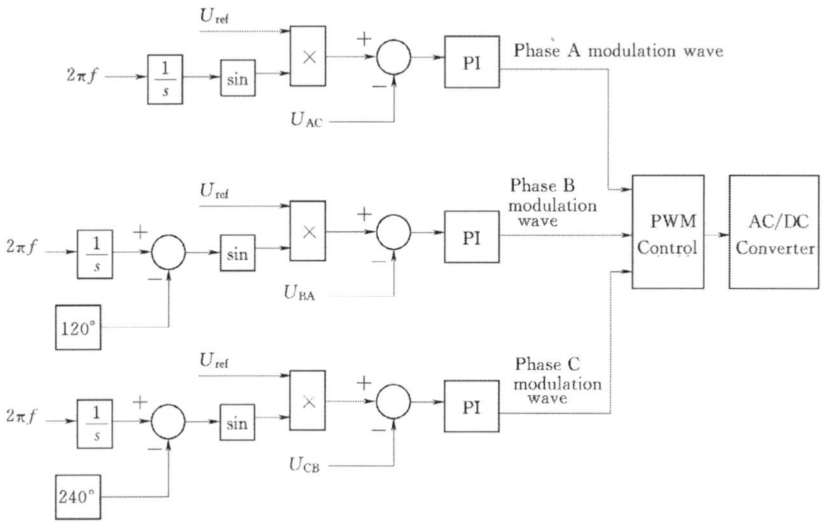

FIGURE 4.29 Voltage instantaneous value single closed-loop control block diagram.

The control of frequency adopts the open-loop control, and we get the modulation wave phase after the frequency instruction integral and make the modulating wave frequency as the given frequency to ensure the output voltage at the given frequency. The phase difference of the ABC three phases is 120 degrees.

From the preceding, we can know that the ESS can only collect the information of voltage at the inverter port to make voltage modulation through adjusting the inverter modulation factor when using the voltage single closed-loop control.

4.3.1.2 Voltage dual closed-loop control

The voltage dual closed-loop control makes the inner loop and outer loop control the two different voltage values. See Fig. 4.30 for the voltage effective value instantaneous value dual closed-loop control block diagram. In the diagram, U_{CA}, U_{BC}, U_{AB} stand for measured voltage signals of the three phases; U_{ref} is the voltage reference value; f is the given frequency instruction. According to the measured voltage instantaneous value, we can get the effective values of the voltage of three phases through the Fourier transform and compare them with the given effective values to output the reference value of the voltage instantaneous given value with PI modulation. Multiplying it and the given phase, we get the voltage given value, and we compare it with the actual voltage instantaneous value to get the modulating wave value output after PI modulation. We choose the proper PWM control algorithm to generate control impulse to control the AC/DC converter. The frequency still uses the open-loop control method.

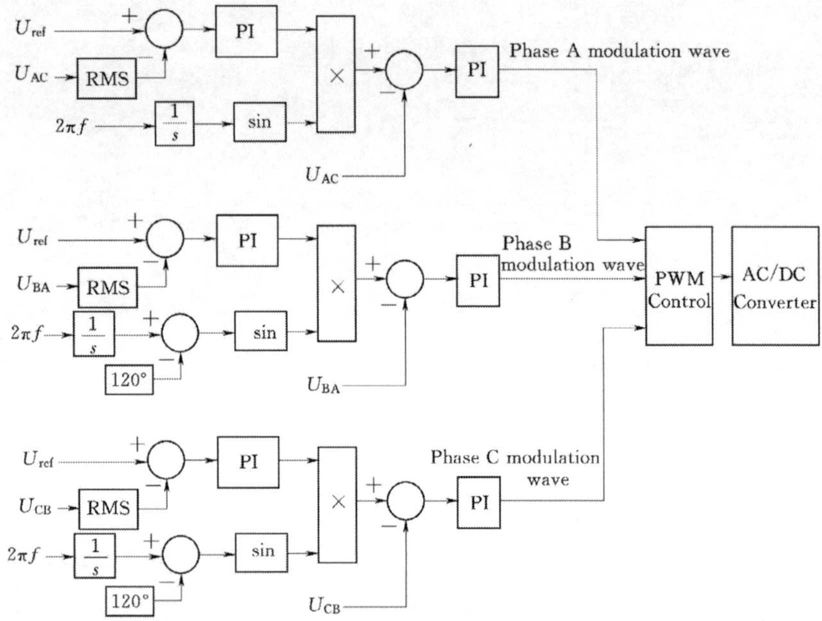

FIGURE 4.30 Dual closed-loop control block diagram of voltage effective value instantaneous value.

This control mode can control the voltage effective value and instantaneous value as well as enabling the output value to maintain the sine characteristics. On the basis of guaranteeing the output voltage effective value at the given value, it can prevent the energy quality decline caused by the nonlinear load.

4.3.1.3 Voltage and current dual closed-loop control

The voltage and current dual closed-loop control makes the output voltage the outer loop control and makes the filter inductance current the inner loop control, as shown in Fig. 4.31. In the diagram, U_{ref} is the given voltage reference value; u_{dref}, u_{qref} are dq components of the voltage reference value; i_{dref}, i_{qref} are reference values of the dq axis components of the AC-side current; i_d, i_q are the actual values of the AC-side current dq axis components; υ_d, υ_q are actual values of the inverter output voltage dq axis components; υ_{sd}, υ_{sq} are reference values of the inverter output voltage dq axis components; L_s is the AC-side coupling inductance; f is the given frequency instruction; w is the voltage initial electrical angle, and *3* is the voltage phase angle.

This control strategy adds the current inner loop based on the voltage closed loop to realize effective control of the output voltage effective value and the waveform of the output current. The voltage outer loop control provides voltage support for the AC side, while the inductance current inner loop control can quickly trace the load changes to improve the dynamic response speed.

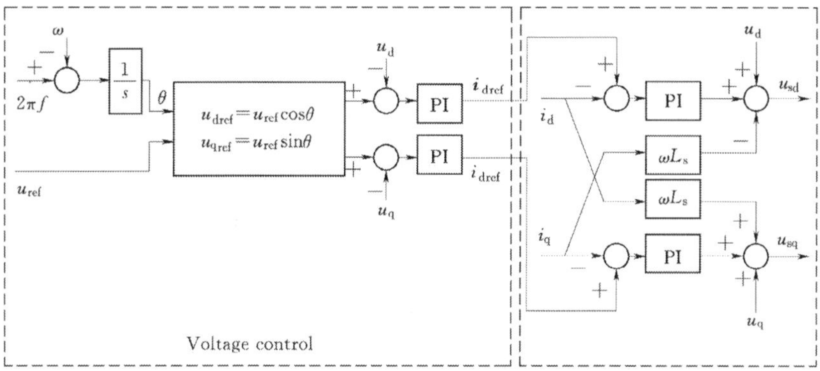

FIGURE 4.31 Voltage and current dual closed-loop control block diagram.

4.3.2 Black start control

Black start refers to the process of finally restoring a whole electric system to operation without relying on the external transmission network by starting the unit with self-start function to drive the units without self-start function and gradually expand the system recovery scope after the power supply system is "black" because of power failure.

Many studies have been conducted in China and abroad on the black start, but mostly on the traditional grid. The black start of new energy generation and ESS is mainly used in the circumstance that the new energy generation and the ESS constitute a microgrid. This section will mainly introduce the black start of a new energy generation microgrid that includes the ESS.

When the microgrid needs to restart because of system outage in the off-grid operation, users request the microgrid to have the black start function in the off-grid operation to improve the power supply reliability. A microgrid is mostly applied in the medium- and low-voltage distribution networks, and its internal power supply includes a large number of power supplies of inverter type. These electric devices have more flexible control and rapid response, but much lower overload capacity, fault ride-through capability, and unit generation capacity than the traditional rotating mechanical power supplies. Therefore they have higher requirements on the black start control of microgrid. As one of the key execution units of the microgrid black start process, the microsource of the black start should have both the voltage source output characteristics and black start function. Therefore the ESS is of course the first choice of the black start power supply.

Under the microgrid off-grid black start, the ESS adopts V/f control and establishes the voltage and frequency of the microgrid system. In the process of V/f start, if direct full-voltage start and the system has inductive load or isolation transformer start, the system will have high surge when the reactor or isolation transformer suddenly adds high voltage, resulting in the inverter-type

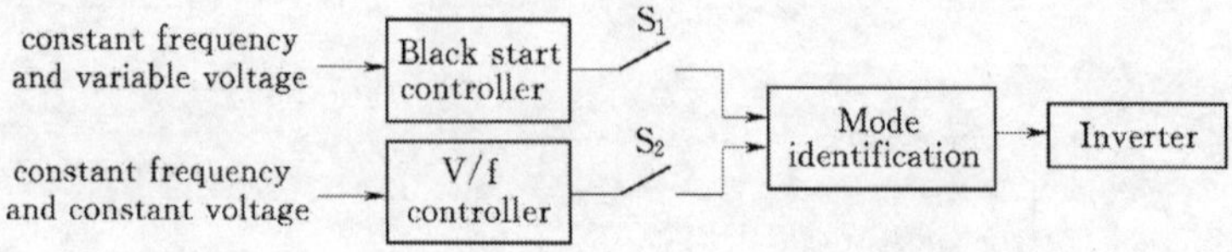

FIGURE 4.32 The schematic diagram of a black start control switch.

voltage source overcurrent protection. Therefore it adopts the soft start control with constant frequency and variable voltage. In the several cycles at the very beginning of start, the modulation depth is gradually increased from zero to the value corresponding to the rated voltage, and the voltage on the isolation transform/inductive load/capacitive load gradually increases to reduce current changes of the reactor or transformer. When the system voltage and frequency are established and reach the rated value, it is concluded that the black start succeeds, and the control is switched to the V/f control with constant voltage and frequency. See Fig. 4.32 for the switching of the black start.

The ESS black start also needs to address the power supply of the control system in case of outage. The conventional solution is to use UPS to power the control system. However, the UPS power supply has the following problems: (1) Provision of UPS will lead to cost increase; (2) UPS has a limited duration of guaranteeing power supply, and the capacity declines with the use of time; (3) If the UPS battery exhausts, we need to charge it with a diesel generator or other charge equipment before it can be used again. Therefore there will be many problems to use UPS as the power supply in case the microgrid faces the risk of a long-term outage. As a result, the black start of the ESS adopts the self-powered mode from the battery unit of the ESS to provide the temporary emergency power supply for the microgrid system through the wide transformation voltage, guarantee the black start power supply, and effectively solve weakness of UPS power supply.

4.3.3 Multimachine parallel coordinated control

When multiple ESSs are connected in parallel, there are mainly two strategies for coordination and control of the different systems: master—slave control and peer-to-peer control. This section will introduce these two control strategies and analyze the typical parallel systems. The control strategies here in refer to the control relationship among parallel systems; after the coordination and control strategy, corresponding control technology of the stand-alone introduced in Sections 4.2 and 4.3 can be adopted according to the relationship and status.

4.3.3.1 Master—slave control

The master—slave control method is to adopt different controls to various ESSs and give them different functions. One (or several) ESS(s) serve(s) as the main power supply to detect the electric quantity in the grid, adopt corresponding modulation means according to the grid operation, and control the other "slave" power supply output by means of the communications line to reach the purposes of maintaining the power balance of the whole system and the voltage frequency stable at the rated value.

The master–slave control strategy is mainly used when the ESS is in the status of an island. When the ESS is on-grid, the voltage and frequency are supported and modulated by the grid because the total capacity of the ESS is small compared with the grid. The ESSs can only control the output power according to the actual need. When the ESS is operating in the situation of an island, the system disconnects from the grid. In such a case, one or several ESSs need to serve as the grid to maintain the voltage and frequency stable if the microgrid, where the ESS is connected, needs to maintain the internal voltage and frequency stable. In such a case the ESS is called the main power supply or reference supply. When the ESS with multimachine in parallel is cut off from the grid, one plan is to use an ESS as the main supply that uses V/f control and provides reference voltage and frequency; the rest are slave systems that adopt P/Q control; the other plan is to use multiple ESSs in synchronicity and show a single power supply characteristic, and the slave modular uses the P/Q control.

A common master–slave control structure is shown in Fig. 4.33. In the system, the main power supply adopts the V/f control inverter to generate stable sine voltage continuously, and the slave power supplies adopt the P/Q control inverter to follow the distributed power of the control center. They maintain a good power equalization effect through the communications distribution power.

The master–slave control has many advantages. For example, the inverter does not need phase-locked loop for synchronous control, and it has a good load equalization effect. The impedance of connection line does not affect the load equalization, and it is convenient for system escalation; the energy

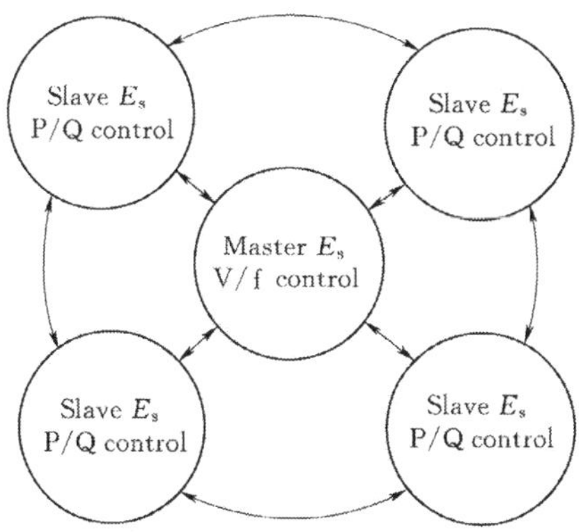

FIGURE 4.33 Master–slave control schematic diagram.

equality of the system is controlled well; if adopting the $N+1$ operation mode (i.e., adding an additional power supply to ensure that the system can maintain the functional integrity after losing any power supply), the system liability and stability can be enhanced. However, the master–slave control has some disadvantages. As the main power supply is specified, the whole system controls the other power supplies through the main power supply. Therefore the main power supply is required to have a certain capacity. What is more, when the main power supply fails, it will affect the whole system. Therefore most systems do not realize real redundancy. Moreover, all the master–slave control technologies need communication interconnection, and the system reliability depends on the communication reliability to some extent. The communication equipment increases the system costs and complexity.

4.3.3.2 Peer-to-peer control

Peer-to-peer control means to take the same control method for all ESSs, and literally, these ESSs are "equal" to each other and have no relation of subordination. All ESSs participate in the active and reactive modulation with the preset control mode to maintain the system voltage frequency stable. During the off-grid operation, ESSs under the peer-to-peer control strategy should participate in the voltage and frequency modulation and adopt the V/f or droop control technology. In the mode of parallel connection without junctor, the ESSs in parallel are connected with each other through the output-end AC bus, and the commonly used is frequency voltage droop control technology. The so-called droop control mainly refers to the electric inverters simulating the active power-frequency (P–f) curve and reactive power-voltage (Q–V) curve modulation characteristics of the traditional grid, and making the system voltage and frequency modulation by means of decoupling the droop characteristics between P–f and Q–V. By examining the ESS output-end voltage and frequency, and comparing them with the given reference values, it modulates the output active and reactive power according to the droop characteristic curve and controls the output voltage and frequency of the ESS. There are mainly two droop control methods for inverters: one is similar to the traditional synchronous generator modulation and adopts P–f and Q–V adjustment ratio control mode; the other adopts the P–V and Q–f antiadjustment control mode. Though these two methods are different in form, they share similar basic principles. They only make choice of the droop control strategy according to different line parameter characteristics.

Suppose the active power and reactive power on a line flow from point A to point B; see Fig. 4.34.

$$\begin{cases} P = \dfrac{U_1}{R^2 + X^2}[R(U_1 - U_2 \cos \delta) + XU_2 \sin \delta] \\ Q = \dfrac{U_1}{R^2 + X^2}[-RU_2 \cos \delta + X(U_1 - U_2 \cos \delta)] \end{cases} \tag{4.47}$$

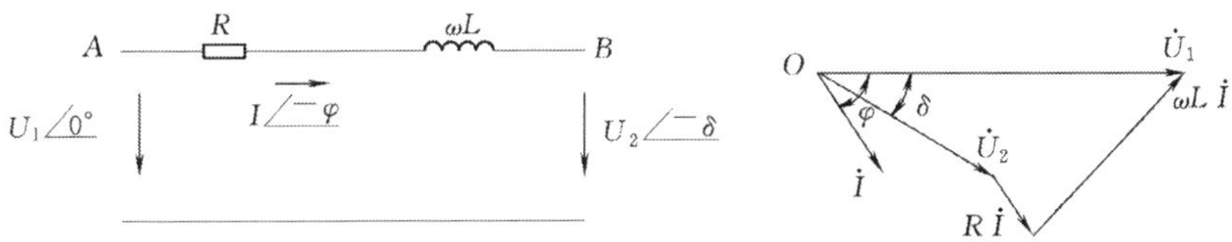

FIGURE 4.34 Power flow and vector schematic diagram.

For a high-voltage line, the inductive reactance value is much higher than the resistance ($X \gg r$); when δ is smaller, it can be approximately deemed that $\sin\delta \approx \delta$, $\cos\delta \approx 1$, so the equation can be written as shown:

$$\begin{cases} P = \dfrac{U_1 U_2}{X}\delta \\ Q = \dfrac{U_1^2 - U_1 U_2}{X} \end{cases} \tag{4.48}$$

This shows that the active power is related to the power angle, while the voltage difference is related to the reactive power. Therefore the active power control can be realized though the power angle control, i.e., the frequency control can be achieved through the power angle while the voltage difference can be directly controlled by the reactive power.

For a medium- or low-voltage distribution line, the inductive reactance is close to or much lower than the impedance in the line parameters, so the equation can be written as follows:

$$\begin{cases} P = \dfrac{U_1^2 - U_1 U_2}{R} \\ Q = \dfrac{U_1 U_2}{R}\delta \end{cases} \tag{4.49}$$

From the equation, we can see that the difference of voltage is related to the active power, while the power angle is related to the reactive power, i.e., the frequency is related to the reactive power. Therefore, we can choose reasonable droop control mode according to the actual situation of the lines.

Fig. 4.35 shows thc typical frequency droop characteristic curve and the voltage droop characteristic curve.

The droop control process is described as follows: as shown in Fig. 4.35A, suppose the currently adopted droop curve is *y*, when the microgrid operation frequency is f_3, the ESS outputs active power P_1, and the ESS operates at point A of the frequency droop characteristic curve y; if the active power provided by the ESS increases, for example changing to P_2, the droop control can make the operating point of the ESS to move from A to E along curve *y*. In such a case the outlet frequency of the ESS declines to f_4; if the frequency is too low,

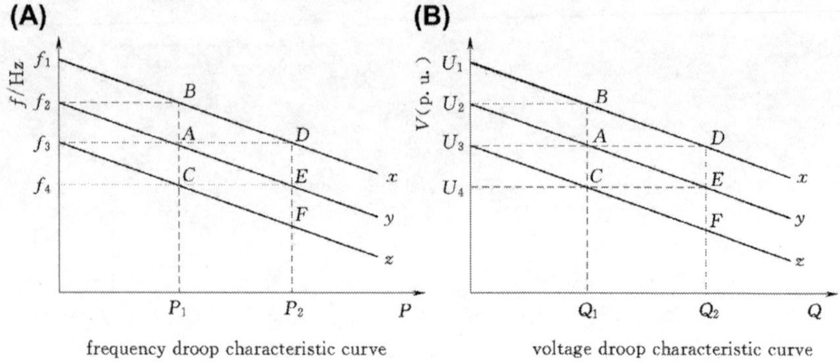

FIGURE 4.35 Schematic diagram of the droop control strategy.

it needs to give out orders through the communications line to shift the droop curve upward in parallel, adopt curve *x*, and adjust the outlet frequency to f3; then the ESS operates at point D of the frequency droop characteristic curve *x*. It is the same control process for the microgrid reactive and voltage control, as shown in Fig. 4.35B. So we can see the ESS under the function of the droop control has the same operation characteristics as the conventional power supply in the traditional grid.

1. Traditional droop control

When the output impedance of the inverter shows inductive characteristics, the basic droop control equation is this:

$$\begin{aligned} \omega &= \omega_0 - mP \\ U &= U_0 - nQ \end{aligned} \tag{4.50}$$

Where ω, ω_0 is the inverter output angle frequency and initial angle frequency; U and U_0 is the inverter output voltage amplitude and initial amplitude; M and n are the droop factors of the active and reactive power.

In practice, frequency control can be used to replace the phase angle control because the frequency signal is easier to measure.

2. Frequency voltage antidroop control

The antidroop control equation is as follows:

$$\begin{cases} P_s^* = K_v(v_{ref} - v_q) \\ q_s^* = K_\omega(\omega_{ref} - \omega) \end{cases} \tag{4.51}$$

Where $\omega_{ref} = \omega^*$, $v_{ref} = v^*$; when the output impedance mainly shows resistive characteristics, the load equalization effect of this control method is acceptable. But when the line inductance is at the same order of magnitude as the converter output filter inductance, the energy quality will drop significantly, mainly reflected in the voltage disturbance. The impact on the energy quality mainly comes from LC filter circuit. The LC filter is composed of the line inductor and the capacitor of AC side of the converter. Moreover, this method does not work when connected to higher voltage levels, and in such a case the traditional droop equation is usually used.

The droop control uses the local measured grid status variables as the control parameters, and it realizes redundancy. The reliable operation of the system does not rely on communications. This control strategy has many ideal characteristics, for example extensibility, modular, redundancy, and flexibility. With the droop control strategy, when an ESS is out of operation due to failure, the rest of the ESSs can continue to run unaffectedly, showing high system reliability; it realizes "plug and play." When the system needs to expand, it only needs to set the same control strategy in the newly added ESS to access to the system. There is no need to adjust the other modules, and it is not subject to the geographic location. The installation and maintenance are easy, and the operation in parallel is more reliable. However, this method has the disadvantages of frequency and amplitude deviations and slow transient response; because of the mismatch between the impedance of inverter output side and the impedance of load bus line or errors in the voltage/current inductor measured values, it is easy to cause circulation in inverters; the resistance value of the low-voltage distribution network is much higher than the reactance value ($R \gg X$) and different from the power transmission network. Therefore the effect of the line impedance must be taken into account. When the operation status of three-phase microgrid changes (for example initiative island operation), the control mode should be adjusted accordingly, especially when the linear and nonlinear load exist simultaneously, and the droop control strategy could not solve such problems. Therefore the droop control, i.e., parallel connection without interconnection line, is seldom used in practice. The widely used are still the products connected in parallel with interconnection line. Given the existing hardware development level and control technology, the control with an interconnection line has obvious advantages than that without an interconnection line in terms of the current sharing response speed and stability.

If the peer-to-peer control strategy is adopted, the ESS only needs to measure the electric quantities at the output port to independently participate in the voltage and frequency modulation and does not need to know the status of operation of the other ESSs and has no communications in the whole process. Moreover, when an ESS is out of operation because of failure, the rest of the ESSs can continue to run unaffectedly, showing high system reliability. When the system needs to add new energy storage units, it only needs to set the same control strategy in the newly added units to access to the system without alteration of the rest of the system, realizing "plug and play" to facilitate system expansion. The peer-to-peer control strategy also has some disadvantages: the overall energy quality of the system (voltage, frequency, harmonic) is not as stable as that of the master–slave control strategy.

4.3.3.3 Analysis of typical parallel systems

1. Parallel system I

Fig. 4.36 shows a parallel system consisting of multiple AC/DC converter monotier transformation topology, where multiple AC/DC converters are connected in parallel at the AC side. As shown in the diagram, when the system runs off-grid, it can use master—slave control or peer-to-peer control. When adopting the master—slave control, the system uses one or several AC/DC energy storage converters to serve as the main power supply and adopts the V/f control to maintain the system voltage and frequency; the rest of the AC/DC converters adopt P/Q control, and the main power supply collects the total power signals to divide equally. In the case of peer-to-peer control, all AC/DC converters adopt V/f or droop control, and they participate in the system voltage and frequency modulation.

2. Parallel system II

Fig. 4.37 shows the system composed of two-tier transformation topology of multiple AC/DC converters connected in parallel with DC/DC converters. As shown in the diagram, when the system runs off-grid, it can adopt

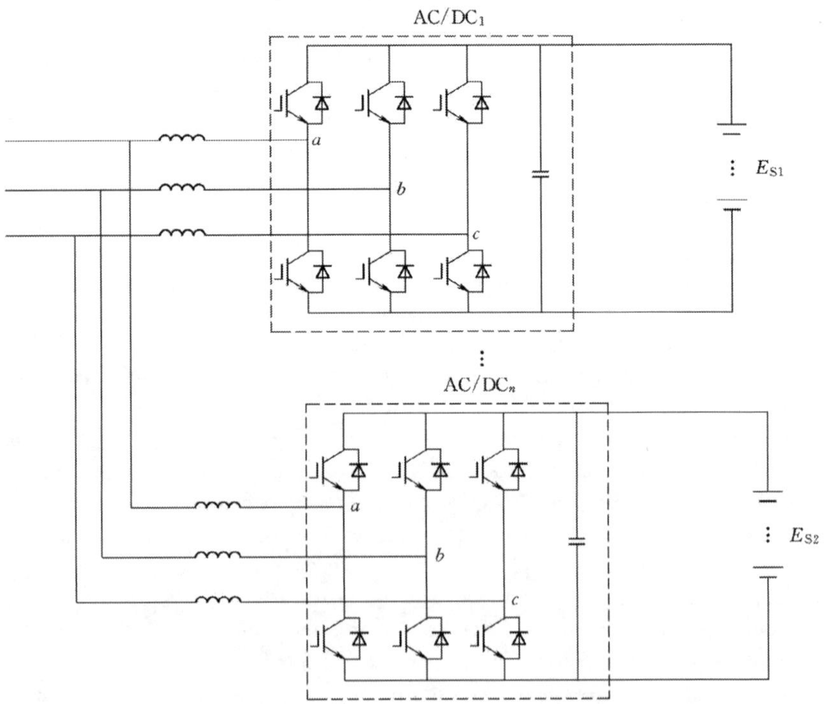

FIGURE 4.36 Topological diagram of parallel system I.

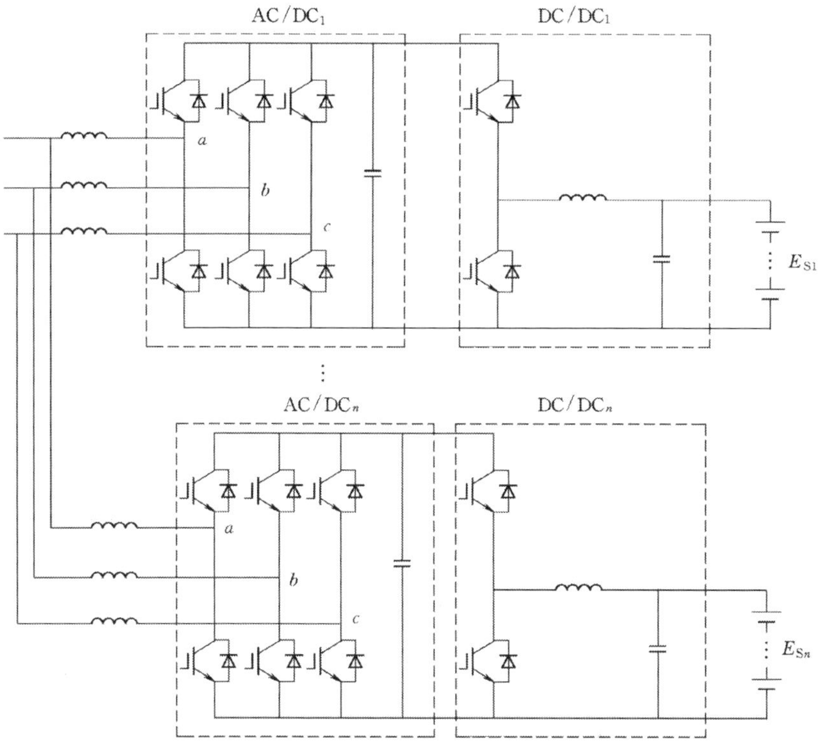

FIGURE 4.37 Topological diagram of parallel system II.

master—slave control or peer-to-peer control. In the case of master—slave control, one or several ESSs serve as the main power supply to maintain the system voltage and frequency. The AC/DC converter of the ESS that serves as the main power supply adopts the V/f control, while DC/DC converter adopts the constant voltage or constant current control; the AC/DC converters of the rest of the slave ESSs adopt the P/Q control, and the main power supply collects the total power signals to divide equally, and DC/DC converters adopt the constant voltage or constant current control. In the case of peer-to-peer control, the AC/DC converters of all ESSs adopt V/f or droop control, while DC/DC converters adopt the constant voltage control or constant current control.

3. Parallel system III

Fig. 4.38 shows the system composed of one AC/DC converter connected with multiple DC/DC converters in parallel; multiple DC/DC converters are connected in parallel at the AC side. When the system shown in the diagram runs off-grid, the common AC/DC converter must adopt V/f control to provide

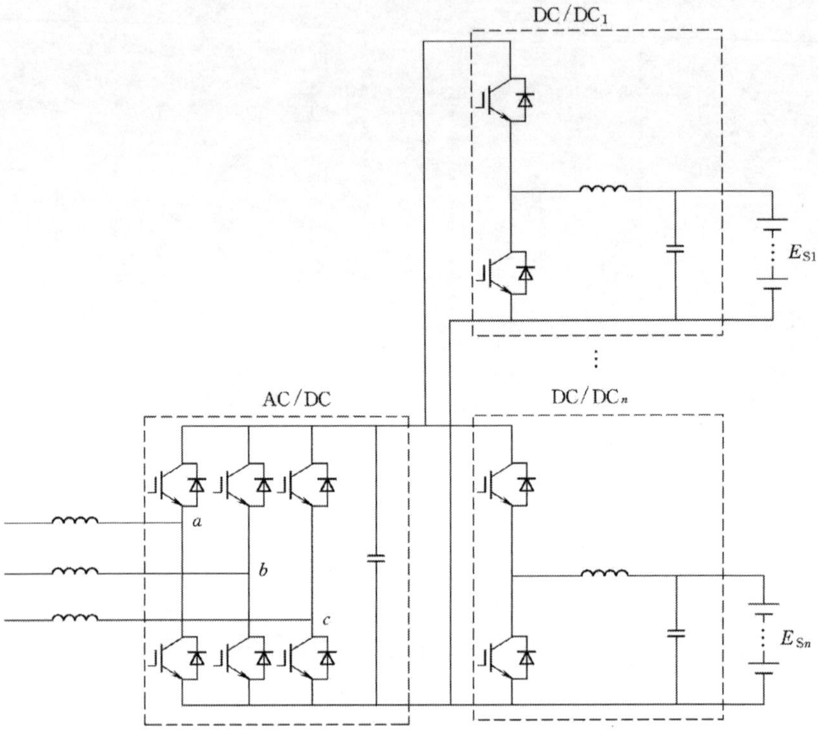

FIGURE 4.38 Topological diagram of parallel system III.

system voltage and frequency support, while the choice of coordination and control strategy is made in the DC/DC converters, and the DC/DC converters in parallel can adopt master—slave control or peer-to-peer control. In the case of peer-to-peer control, the DC/DC converters adopt constant voltage or constant current control; in the case of master—slave control, one or several DC/DC converters adopt the constant voltage control, while the rest of the DC/DC converters adopt the constant current control, and the main controller collects the total current signals to divide equally.

From the foregoing exposition, we can see it is difficult to make a unified control strategy that is applicable to all systems, as each control method has its advantages and disadvantages. One of the disadvantages of the master—slave control is that the system stability is subject to the quantity of slave modules, and the relationship between the quantity of parallel modules and system stability and dynamic response is very complicated. Moreover, there are differences in relationship for different parallel control methods. It is one of the main directions for the future studies to clarify the relationship. The master—slave control does not realize redundancy, but it is easier to be applied in practice. In contrast, the parallel systems with droop control technology realize redundancy, but the system voltage and frequency will change with changes of the load and cannot guarantee the energy quality. And the existing dynamic compensation methods also cannot not generate ideal results. Therefore, the energy quality control of the parallel ESSs that adopt the droop control needs further studies. In conclusion, a single or combined control strategies shall be adopted to guarantee reliability and stability of multiple ESSs in parallel according to the actual application conditions and load type and other elements.

4.4 Dual-mode switching control technology

The dual-mode switching of the ESS mainly refers to the switching between on-grid operation and off-grid operation. From Sections 4.2 and 4.3, we can know that in the on-grid mode, the AC/DC converter adopts the P/Q control or constant voltage control; in the off-grid mode, the AC/DC converter adopts V/f control. However, in the on-grid or off-grid mode, the DC/DC converters are working at constant voltage or constant current status and do not change with the system operation mode changes. Therefore the on-grid and off-grid switching of the ESS mainly refers to the switching of the AC/DC converter control mode.

4.4.1 Control of switching from on-grid to off-grid

The process of the ESS switching from on-grid to off-grid is to switch the control mode of the AC/DC converter from P/Q mode or constant voltage control to the voltage/frequency (V/f) control. This switching mainly happens when the grid suffers the planned outage or abrupt failure, and the ESS is required to continuously power the load. It requires seamless switching, keeping the important and sensitive load powered when the grid is in outage and the other distributed power supplies in the microgrid do not trip; after switching the PCS controlled voltage and frequency are stable.

When the large power grid needs to cut off the ESS and the external grid because of an abrupt failure or artificial needs, the AC/DC converter should immediately change the control strategy to realize smooth switching from on-grid to off-grid. In such a case, the converter detects the grid voltage phase before the switching and makes it the initial value of the voltage phase controlled by the VSC in the off-grid mode, and it switches the converter to the V/f voltage source control mode at the moment when the on-grid switch is switched off.

Thus we can get the control logic and procedure of the converter transforming from on-grid to off-grid:

1. monitoring the off-grid instruction of "island" status information;
2. once the off-grid requirement is confirmed, the converter changes to V/f control, and the output voltage phase traces the external grid voltage phase;

3. giving out the instruction of switching off the on-grid switch;
4. delaying and waiting for the on-grid switch to be switched off reliably;
5. the converter adopts the V/f control with the standard voltage and frequency as the reference.

See Fig. 4.39 for the control flow chart.

Initiative switching from on-grid to off-grid: When the grid needs to power off for planned maintenance, the controller can forwardly transfer to the off-grid mode after receiving the blackout instruction to ensure the power supply continuity of the microgrid load and normal operation of the microgrid. In the process of initiative switching from on-grid to off-grid, the on-grid switch is artificially switched off under the normal circumstance of the grid. When the ESS receives the initiative off-grid instruction, it traces the grid voltage amplitude and phase before switching off. At the moment of switching off, to maintain the load voltage unchanged, the converter control mode is switched to the voltage frequency control; the voltage effective value and frequency adopt the standard values (380 V/50 Hz) of the distribution network, and the output voltage phase should continue the load voltage phase before the grid switches off.

Passive switching from on-grid to off-grid: When the grid fails, the ESS can rapidly identify the failure and switch to the off-grid operation mode.

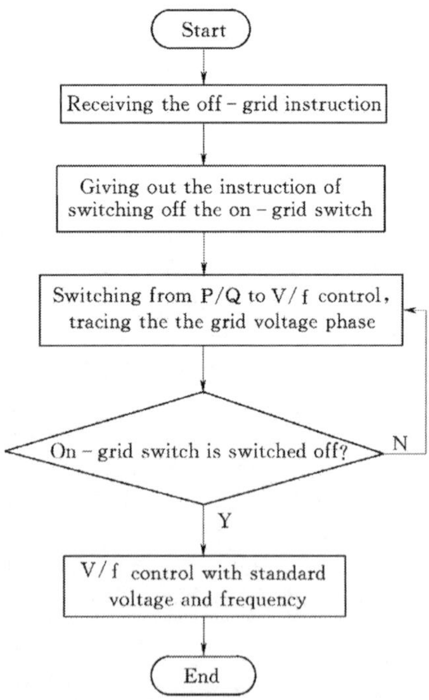

FIGURE 4.39 Procedure switching from on-grid to off-grid.

The switching time should be short enough to minimize effect of the grid failure to the microgrid load and microsource. To realize smooth nonimpact switching, one needs to rapidly and precisely detect the grid failure, and the converter should rapidly switch from the on-grid working mode to the off-grid working mode. To realize the grid detection precision and rapidness, the system should adopt the combined frequency detection and amplitude detection method to identify the grid failure. In the passive switching, the inverter rapidly starts after the grid is powered off because of inherent delay of the grid failure detection module. Thus the load voltage will not be as smooth as in the process of initiative switching and will decline for a short time.

4.4.2 Synchronization control of the switching from off-grid to on-grid

The switching of the energy storage converter from off-grid to on-grid is mainly the process of the AC/DC converter changing from the V/f control mode to the P/Q control mode or constant voltage control mode. The process of the ESS switching from the off-grid to the on-grid is called "synchronization" and is controlled by the dedicated synchronization device.

As in the off-grid power supply working mode, the energy storage converter output voltage is synchronous with the built-in reference signal of the system; but the energy storage converter cannot get the synchronous standards when the grid is power off. And when the grid is recovered, the output voltage amplitude, frequency, and phase of converter can be inconsistent with the grid. Therefore when the on-grid switch is closed, the system must make the converter output voltage amplitude, frequency, and phase match the grid voltage through the phase-locked loop trace and control. Otherwise, there will be high voltage differences that will cause the on-grid surge current and endanger the converter when the on-grid switch is closed. Considering effect of the line connection inductor and others, one needs to control the rising speed of the grid current to avoid the overvoltage spikes at the load end or possible surge current on the load.

Thus we can get the control logic and procedure of the converter transforming from off-grid to on-grid:

1. Inspect whether the grid where the converter is located meets the on-grid conditions.
2. Make the phase-locked trace to the grid voltage keep the converter output voltage consistent with the grid voltage in terms of amplitude, frequency, and phase.
3. Close the on-grid switch.
4. Gradually increase the converter power control to the given rated power value.

See Fig. 4.40 for the control flow chart.

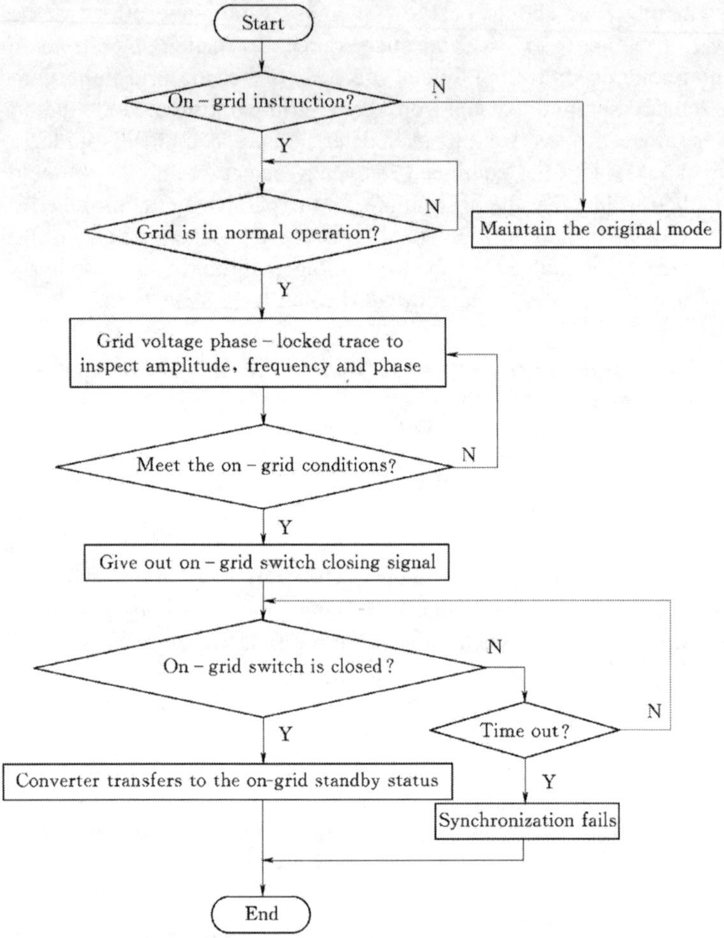

FIGURE 4.40 Procedure of switching from off-grid to on-grid.

The upper level monitoring system gives out the on-grid instruction. When the ESS receives the instruction, the converter operates with the V/f control, the voltage instruction is the distribution network voltage effective value, and the frequency instruction is 0.1Hz lower than the gird frequency for the on-grid modulation. In such a case the converter output voltage is consistent with the distribution voltage, and the frequency is 0.1Hz lower than the distribution network frequency for synchronization detection. The energy storage converter receives the grid voltage and frequency signal from the synchronization device and adjusts the output voltage and frequency according to the grid voltage and frequency to keep consistent with the grid voltage frequency. The synchronization device inspects the output voltage, frequency, and phase of the distribution network and energy storage converter, respectively. When the energy storage converter output voltage meets the on-grid requirements when comparing with the distribution network voltage, it gives out the signal of switching on the on-grid switch. The on-grid requirements are that the voltage difference is less than 5%, the frequency difference is less than 0.15Hz, and the phase difference is less than 5 degrees. After the synchronization device detects that the synchronous on-grid succeeds, the ESS transfers to the on-grid standby status, and it waits for the power or voltage control instruction from the monitoring system. If this procedure is not completed within the designated time, it will be deemed as "synchronization fails."

4.5 Case study

4.5.1 Test system

Focusing on the operation control technology of an ESS described in this chapter, the following will explain the research results of the distributed ESS laboratory of the Renewable Energy Research Center of China Electric Power Research Institute as the example. Fig. 4.41 is the schematic diagram for the distributed generation/energy storage and microgrid system. The whole

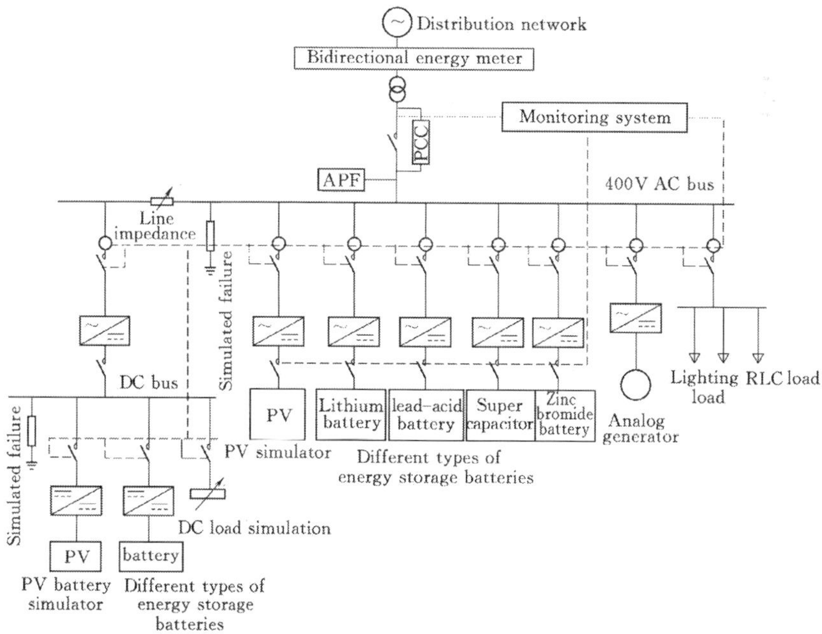

FIGURE 4.41 Distributed generation/energy storage and microgrid system diagram.

laboratory mainly includes a variety of energy storage systems, DC microgrid system, AC/DC hybrid system, multilevel nested distributed generation systems, circuit simulation, adjustable dummy load, and lighting load, etc. On the basis of the existing test platform and the theoretic study of this chapter in combination, this section carries out studies of the DC microgrid operation control technology, AC/DC microgrid coordination and control technology, hybrid ESS control, improved droop technology−based multimachine parallel connection technology, and combined PV/diesel/storage integrated operation control technology, among others. The energy storage batteries of the testing platform include 40k-W/75-kWh iron−lithium battery, 30-kW/60-kWh iron−lithium battery, 40-kW/100-kWh lead−acid battery, 25-kW/50-kWh zinc−bromide battery, two 50-kW × 10-s supercapacitors, and the energy conversion devices include two 100-kW and four 50-kW AC/DC energy storage converters (PCS), two 100-kW DC/DC converters, a 100-kW photovoltaic simulator, and a 30-kW diesel generator and some testing and lighting load.

4.5.2 Function verification

The microgrid operation control technology described in this chapter has been experimentally verified on the distributed generation/energy storage and microgrid test platform of the Renewable Energy Research Center, including the charge-discharge test of two-tier ESS, off-grid black start test, multimachine parallel connection test, on-grid and off-grid seamless switching control test, and hybrid energy storage operation control test.

1. Two-tier ESS charge and discharge test

The charge and discharge full-power test of two-tier ESS is conducted at the platform, as shown in Fig. 4.42. The test uses a 100-kW PCS and a 100-kW DC/DC converter and the lithium battery as the energy storage device. The 100-kW charge and discharge switching test of the two-tier ESS with rated power is carried out. Fig. 4.42A shows the AC voltage of the three phases and AC current waveform of Phase A when the ESS is switching from charge to discharge at the 100-kW rated power. Fig. 4.42B shows the AC voltage of the three phases and AC current waveform of Phase A when the ESS is switching from discharge to charge at the 100-kW rated power. Fig. 4.43 shows the DC-side voltage waveform of the DC side of DC/DC converter during the switching of the rated power.

2. Off-grid black start test

In some extreme circumstances, for example complete outage because of failure in the initiative islanding transit or microgrid instability, the self-start and independent power supply characteristics of the distributed generation should be made of good use to black start the microgrid to ensure that the load

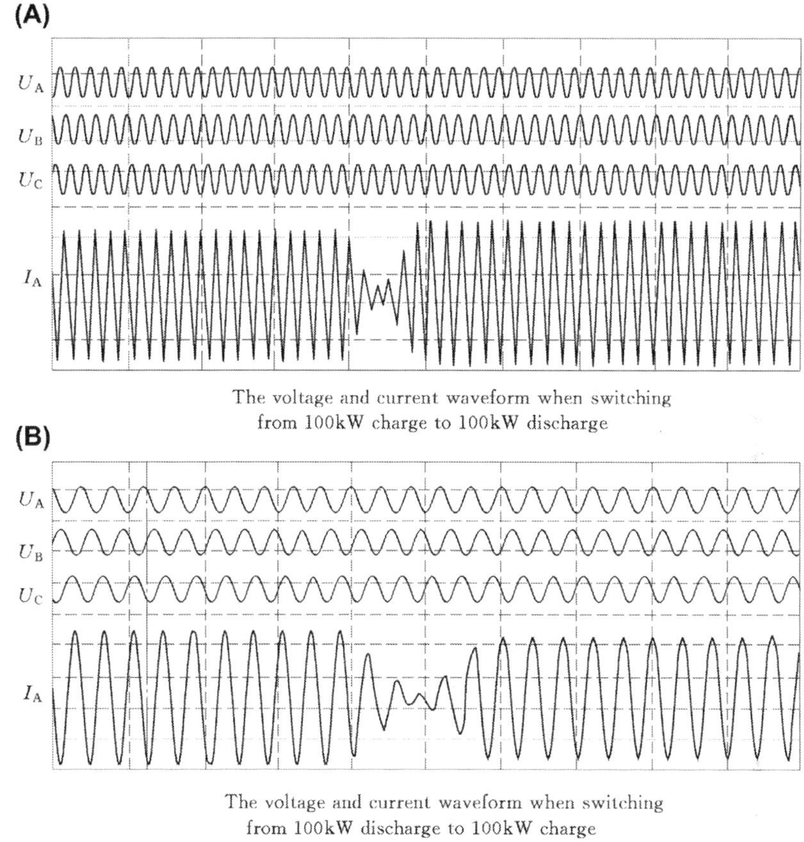

FIGURE 4.42 AC-side experimental waveform of the ESS in switching between 100-kW full-power charge and discharge.

is repowered. Start tests are carried out for the two-tier energy storage power supply system with impact load and 20-kW resistive load, respectively.

Fig. 4.44A shows the start waveform with impact load (7.5-kW electrical machinery), and (B) shows the experimental waveform with impact load and 20-kW resistive load in combination.

3. Multimachine parallel connection test

As a single PCS has limited capacity, the expansion of system is difficult when the system load increases. It is necessary to adopt multiple converters to run in parallel. Tests and studies are carried out in light of the problems such as uneven power distribution and large circulation in the parallel operation. The test uses two 50-kW PCS and a 100-kW PCS in parallel. For batteries, a 40-kW/100-kWh lead–acid battery and 40-kW/75-kWh iron–lithium battery are chosen.

(A)

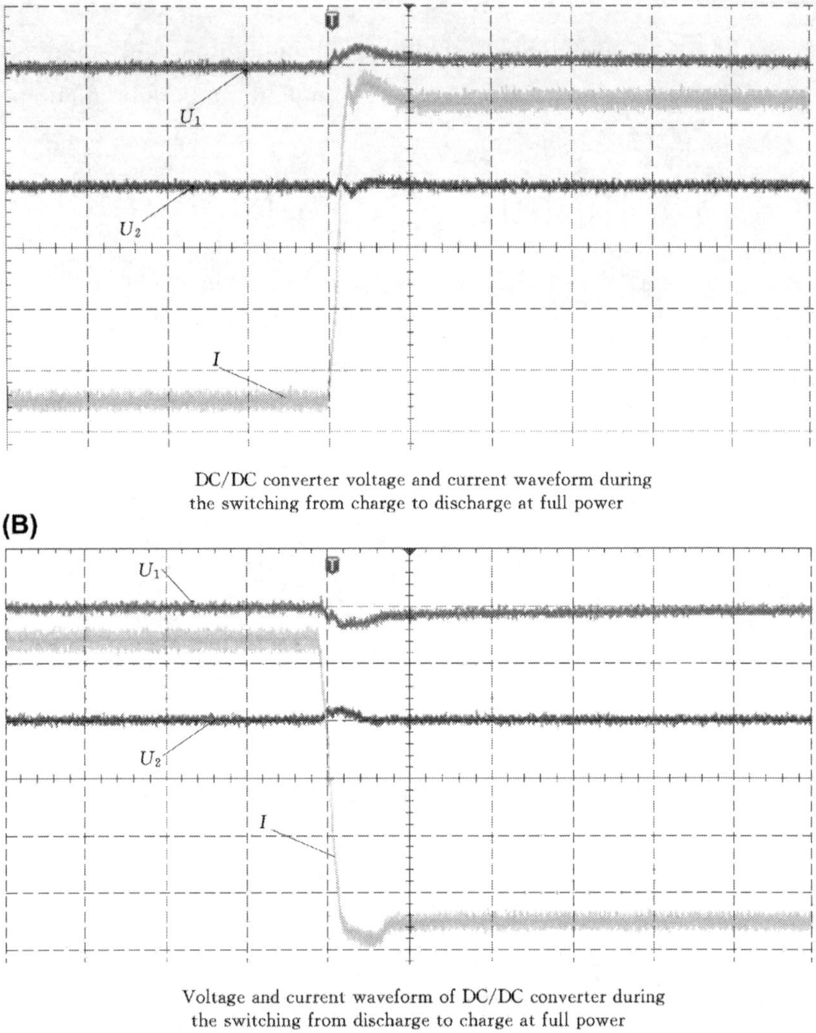

FIGURE 4.43 DC-side waveform of the ESS during switching between charge and discharge at 100-kW full power.

Fig. 4.45 shows two 50-kW PCS and 100-kW PCS operating in parallel, and input of 36-kW adjustable RLC load. Then the power of these three PCSs is (A) 10.6 kW, (B) 10.4 kW, and (C) 15 kW. The operating conditions are AC bus voltage 220V and frequency 49.85 Hz. First cut off C, and the power of A and B immediately increases and divides equally; then C is connected, so the power of A and B decreases, and the three PCSs share the power equally and operate steadily.

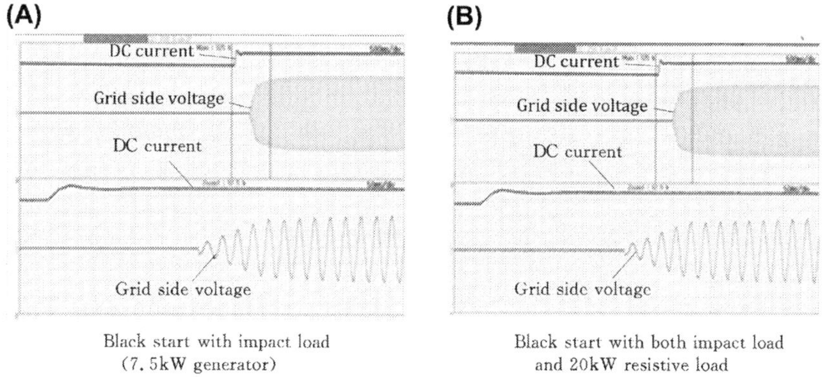

FIGURE 4.44 Off-grid black start test of energy storage system with different load.

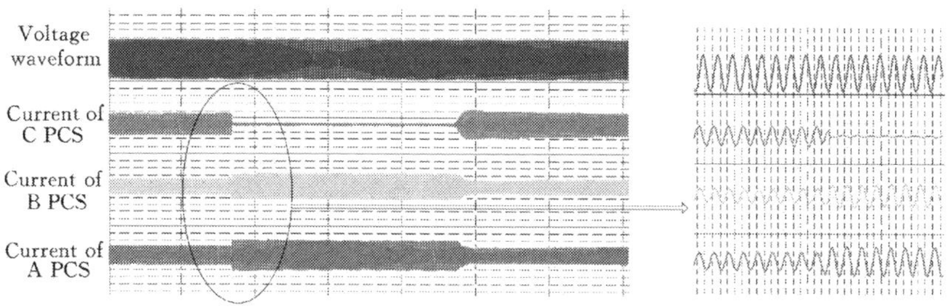

FIGURE 4.45 Three PCSs operating in parallel, one of them cut off and connected.

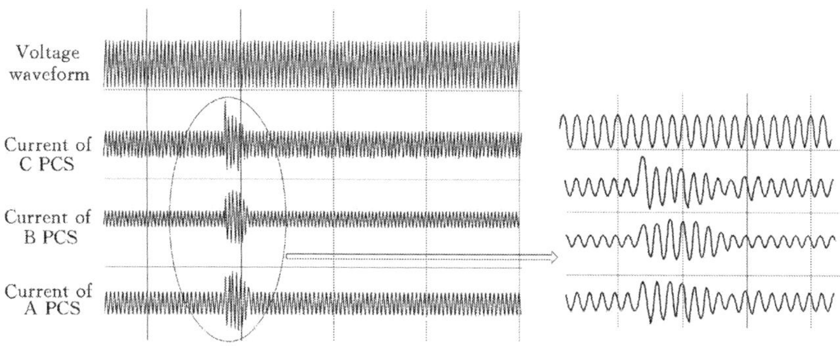

FIGURE 4.46 A, B, and C PCSs with resistance operating in parallel, starting the motor.

Fig. 4.46 shows (A, B, and C) three PCSs with 36-kW resistive load operating in parallel. With input of motor load, the surge current is small, the voltage fluctuation is low, and the three PCSs realize power equalization and stable operation when impact load is connected.

U_1 is the high-voltage side voltage, amplitude at 700 V; U_2 is the low-voltage side voltage, amplitude at 280 V; I is the current value, switching between 140 and 140 A.

4. Control test of seamless switching between on-grid and off-grid

For seamless switching between on-grid and off-grid, when the grid fails, the ESS can quickly identify and promptly switch to the off-grid operation mode. The switching time should be short enough to minimize effect of the grid failure on the internal load and power supply of the power supply system. See Fig. 4.47 for the switching process.

Off-grid to on-grid synchronization control: in the process when the energy storage converter transfers from off-grid to on-grid, the control mode is changed from V/f control to constant power control. Before grid connection, first, the energy storage converter must make the output voltage match the grid voltage in terms of amplitude, frequency, and phase through phase-locking loop control. Otherwise the high voltage differences generated during switching of the on-grid switch will cause high on-grid surge current and endanger the safety of the converter. See Fig. 4.48 for the switching process.

5. Hybrid energy storage test

This book carries the hybrid energy storage test with a zinc–bromine battery and supercapacitor as an example. The zinc–bromine battery has 40 ms voltage runaway during the switching between charge and discharge, and the supercapacitor is connected in parallel to effectively provide instant power support and safe and stable operation of the system.

A 25-kW/50-kW-h zinc–bromine battery and 50-kW × 10-s supercapacitor are connected in parallel to carry out the test of switching from

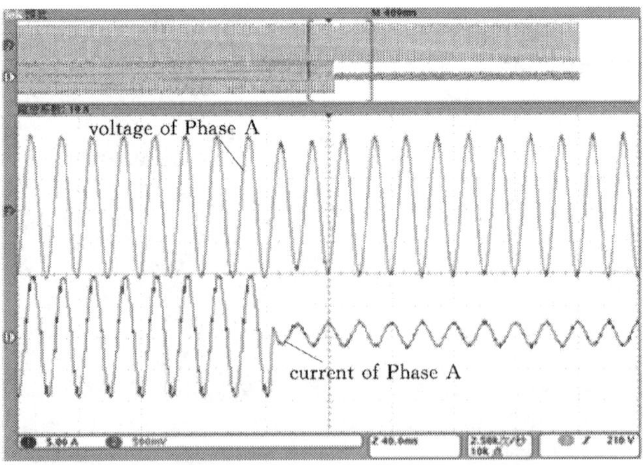

FIGURE 4.47 Waveform during the switching from on-grid to off-grid.

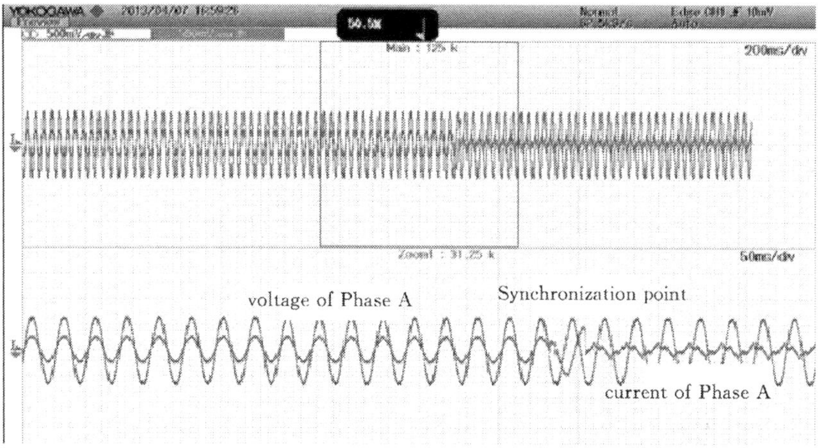

FIGURE 4.48 Waveform during the switching from off-grid to on-grid.

15-kW charge to 15-kW discharge. Fig. 4.49 shows the waveform during the switching process when a zinc–bromine battery is adopted alone. From the diagram, we can see the voltage declines temporarily at the moment of switching and remains low for some time.

Fig. 4.50 shows the experimental waveform of the hybrid ESS during the power switching. From the waveform, we can see the DC voltage is relative smooth at the moment of switching and recovers to the original level rapidly after the switching.

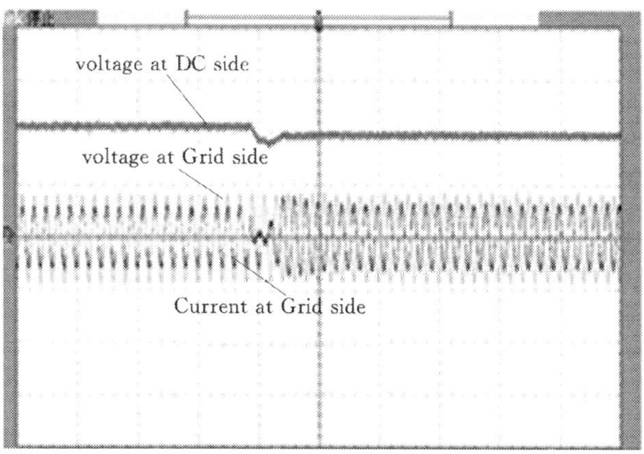

FIGURE 4.49 Experimental waveform of a zinc–bromine battery during switching from 15-kW charge to 15-kW discharge.

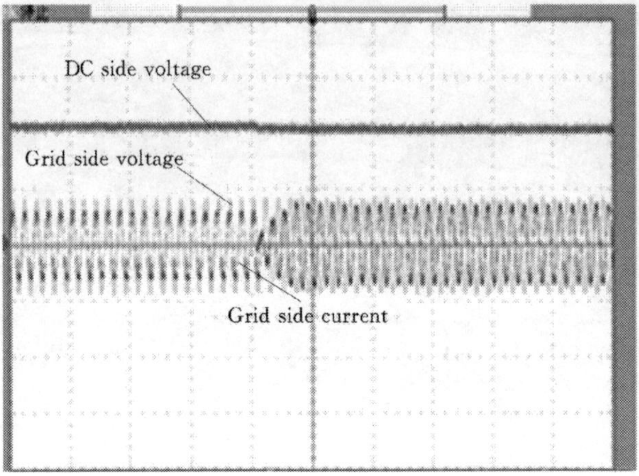

FIGURE 4.50 Experimental waveform of the zinc–bromine battery and supercapacitor connected in parallel during the switching from 15-kW charge to 15-kW discharge.

References

[1] Wang C, Xia J, Yang J, et al. Modern control technology of motor. 1st ed. Beijing: China Machine Press; 2006. p. 1–26.

[2] Wang Z, Huang J. Power electronics technology. 4th ed. Beijing: China Machine Press; 2000. p. 100–65.

[3] Zhang X. A study of PWM rectifiers and their control technology. Hefei University of Technology; 2003. p. 8–44.

[4] Xu J. A study of three-phase voltage source PWM rectifiers control technology. Huazhong University of Science and Technology; 2004. p. 11–27.

[5] Xiong Y. A stody of current source multi-level converter topology and control strategy. Zhejiang University; 2004. p. 11–6.

[6] Xu H. A study of high power Bi-directional DC-DC converter topological structure and its analysis theory. Graduate University of Chinese Academy of Sciences; 2005. p. 43–7.

[7] Zhang F. A study of Bi-directional DC-DC converters. Nanjing University of Aeronautics and Astronautics; 2004. p. 5–10.

[8] Tong Y, Wu T, Jin X, et al. A study of Bi-directional DC-DC converters. Proceedings of the CESS 2007;27(13):81–6.

[9] Sang B, Yang B, Li G, Tao Y. A study of distributed generation and microgrid applications of the lithium battery energy storage system. Power Electronics Technology 2012; 46(10):57–9. 00.

[10] Ye Z, Kolwalkar A, Zhang Y, et al. Evaluation of anti-islanding schemes based on non-detection zone concept. IEEE Transactions on Power Electronics September 2004;19(5).

[11] Wang L, Sun X, et al. PV power generation technology of distributed generation system. 1st ed. Beijing: China Machine Press; 2010. p. 213–21.

[12] Zheng J, Wang Y, Li X, et al. Microgrid smooth switching control method and strategy. Automation of Electric Power Systems 2011;35(18):17–24.

[13] Zhang J, Huang W. Microgrid operation control and protection technology. 1st ed. Beijing: China Electric Power Press; 2010. p. 41–122.

[14] Xing H, Jin X, Lin M. Micro-grid off-grid black-start optimization ControlScheme. Transactions of China Electrotechnical Society 2013;28(4):182–9.

[15] Tang XS, Deng W, Li N. Energy storage based renewable energy micro-grid operation control technology. Electric Power Automation Equipment 2012;32(3):99–103.

[16] Zhang Q, Peng C, Chen Y. A control strategy for parallel operation of multi-inverters in microgrid. Proceedings of the CESS 2012;32(25):126–31.

Chapter 5

Integrated ESS application and economic analysis

Chapter outline

Grid-scale Energy Storage Systems and Applications. https://doi.org/10.1016/B978-0-12-815292-8.00005-8

In recent years, a series of technical breakthroughs in electrochemical energy storage have been witnessed, for example, those in sodium–sulfur cells, flow cells, lithium-ion batteries, and lead–acid batteries. More than 200 energy storage demonstration projects, at megawatt level and above, have been set up across the world, indicating huge potential for further expanded application of energy storage technologies. Battery energy storage systems (BESSs) boast high energy conversion efficiency, flexible design, quick charge–discharge shift, and freedom from restrictions on site. Therefore, such systems can play an important role at different sections of an electric system.

Many developed countries, particularly the United States, have invested heavily in large energy storage power stations, such as those located in California and Pennsylvania. These stations, for power generation and distribution, integrate user-end and distributed energy sources with microgrids and large-scale renewable energy sources as well as auxiliary services. Japan is the global leader in developing energy storage technologies. It has prioritized developing integrated storage technologies for wind and photovoltaic power and has launched a number of demonstration projects. Meanwhile, it has made remarkable achievements in the deployment of small user-side PV storage hybrid systems. China is relatively a late starter in research of large-scale energy storage systems (ESSs) and energy storage power stations. However, China has witnessed rapid development in early-stage application, such as system integration and function verification, and it has built up several large-scale battery-based energy storage power stations.

Following the increased construction of energy storage power stations, the architectures of ESSs and the system integration approaches have become known. BESSs have a complicated structure that requires quite specific integration technologies. Therefore, this chapter will introduce integrated ESS designs, typical ESS application in power systems, and methods for analyzing benefits from ESSs under single function mode based on its application in typical scenarios, as well as analysis of comprehensive efficiency of ESSs in the Chinese electricity market.

A BESS usually consists of a battery for energy storage, battery management system (BMS), power conversion system (PCS), energy storage monitoring system, and low-voltage access switch or step-up transformer (See Fig. 5.1). The energy storage monitoring system responds to instructions from superior system to conduct real-time processing, analysis, graphic display, storage and query of data from energy storage subsystems, and dispatches power distribution. PCS links the battery with the grid for bidirectional conversion between DC and AC. It responds to instruction from the energy storage monitoring system, charging and discharging in designated operation mode as well as interacting with BMS to ensure a secure and stable operation of the BESS. BMS monitors the battery operation, collects voltage information, balances the current and temperature of the battery, as well as protects it. The battery is the energy storage unit of the BESS.

5.1 Integration design for ESSs

5.1.1 Definition

An energy storage unit consists of PCS, battery pile (BP), and BMS. The power of an energy storage unit is according to the capacity of the whole unit and can be 50, 100, or 250 kW.

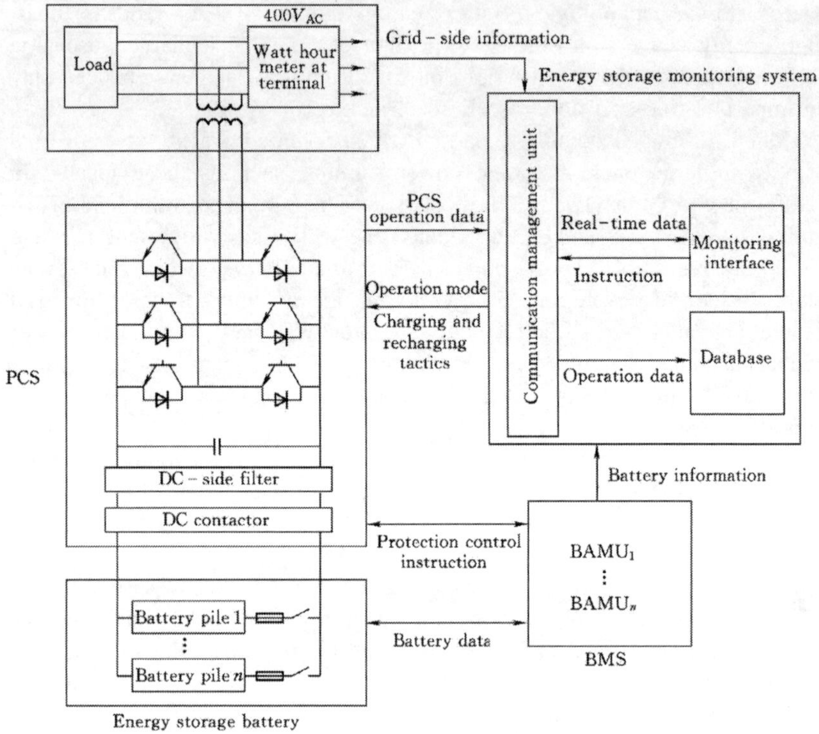

FIGURE 5.1 Composition of BESS.

An energy storage branch is composed of a storage unit and a low-voltage access switch. Under specific application scenarios, a step-up transformer can be used to access a higher voltage level, such as 10 kV. An energy storage branch of a high-power BESS usually employs a power of 250 or 500 kW, while that of a low-power BESS usually has a power of 50 or 100 kW.

An energy storage circuit has energy storage branches placed parallel to each other, together with a step-up unit and a monitoring unit. Each circuit employs a power of 1 or 2 MW. An energy storage branch is the smallest unit in a circuit. The step-up unit is composed of low-voltage-side circuit interrupter, grid-side circuit interrupter, step-up transformer, and its monitoring and protection devices. The monitoring unit collects information about the PCSs and BMSs in the energy storage branches and the operation of step-up unit, and it sends the information to the central monitoring system.

5.1.2 BP application technology

5.1.2.1 Design methods

BP is the core component of an energy storage power station. It consists of a number of individual cells configured in a series, parallel, or a mixture of both.

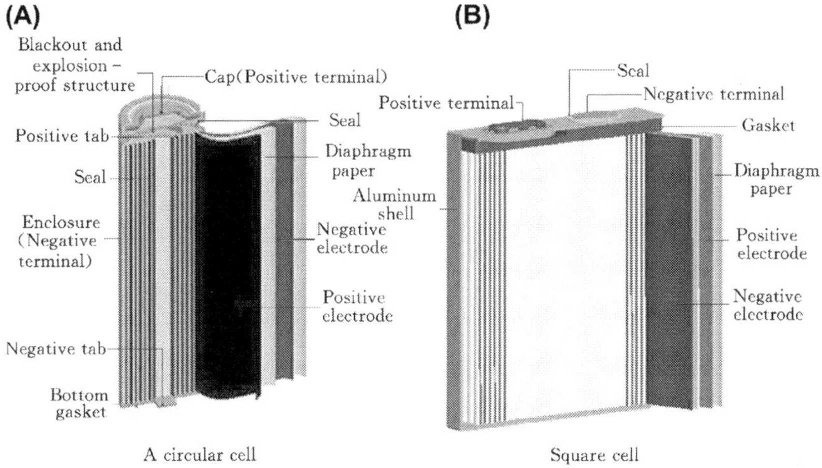

FIGURE 5.2 Cell.

Currently, the capacity of an individual battery ranges from several ampere hours to several hundreds of ampere hours, while the total capacity of an energy storage power station may range from several to tens of megawatts. Therefore, a power station may need tens of thousands of individual cells. Scientific BP design will not only improve the reliability and security of the BP but also facilitate the monitoring, management, and maintenance of the power station.

BP formulation process can be like this: Cell → Unit → Module → Series → Pile.

A cell is composed of electrode, electrolyte, and diaphragm. It is an independent closed system and the smallest component of a BP. Please refer to Fig. 5.2 for the typical structure of a cell.

A number of cells are configured in parallel (See Fig. 5.3) into a unit to meet a certain capacity requirement for a BP. The number of cells in parallel depends on the capacity of individual cells as well as the criteria for safety guarantee. To avoid average current control, the fewer the cells in parallel, the better.

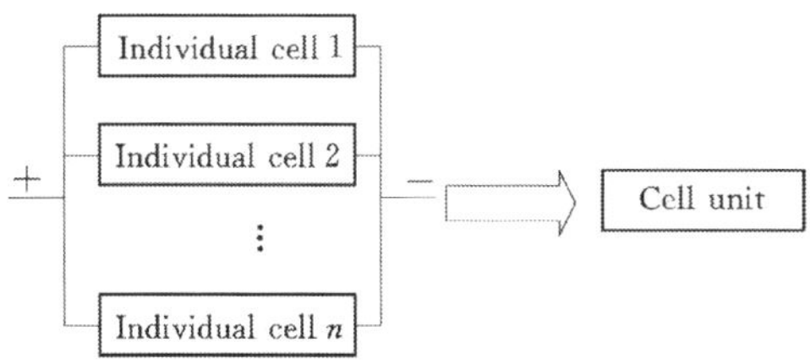

FIGURE 5.3 Cells in parallel.

A module consists of a number of units configured in series (See Fig. 5.4). The number of units should be carefully determined to guarantee the management of the module and match with the number of interfaces on the battery module management unit (BMMU). Usually, each BMMU controls one module, both configured in a box, which is called a battery box. Currently, most of the battery modules consist of 12 units, a few with eight or 10. However, the number of units can be adjusted according to the number of interfaces on the BMMU to meet the demand for working voltage.

A number of modules are configured into a series to meet certain system power (working voltage/current) demand (See Fig. 5.5). In the series, the number of modules should be enough to reach the required working voltage of the energy storage converter. On the other hand, it should be minimized to avoid inconsistency in battery capacity during charging and discharging. Among the existing battery-based ESSs and those under development, the nominal voltage of the battery series ranges from 500 to 800 V. A series consist of 17, 20, or 25 modules.

A number of battery series are configured into a pile to meet capacity demand of an ESS (See Fig. 5.6). The number depends on the system's capacity, redundancy rate, and working mode.

Fig. 5.7A is the BP design for a Chinese energy storage power station, where 300 kW × 5 h lithium iron phosphate batteries were configured. The BP consists of three parallel series; each consists of 20 modules in series; each module consists of 10 individual 200 Ah lithium iron cells in series.

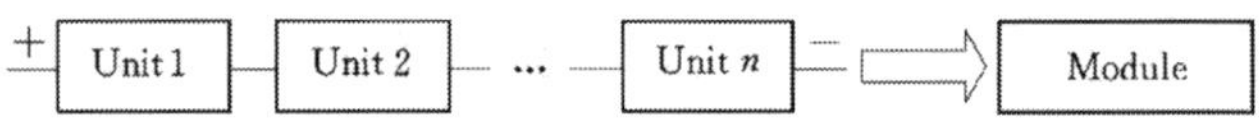

FIGURE 5.4 A module.

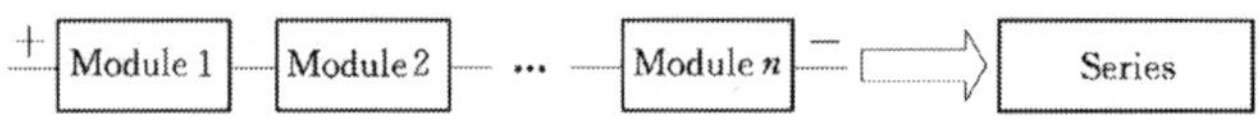

FIGURE 5.5 Battery series.

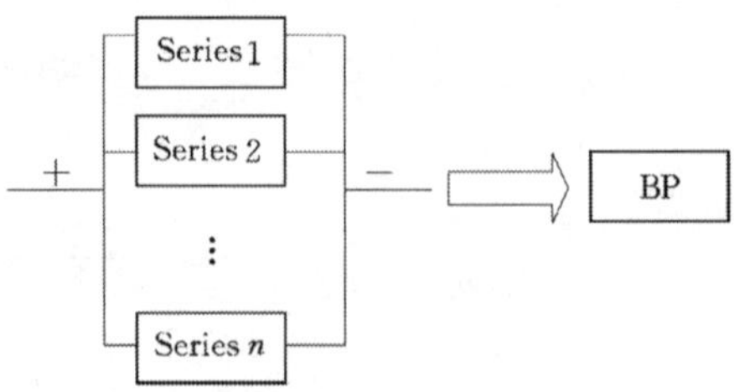

FIGURE 5.6 BP.

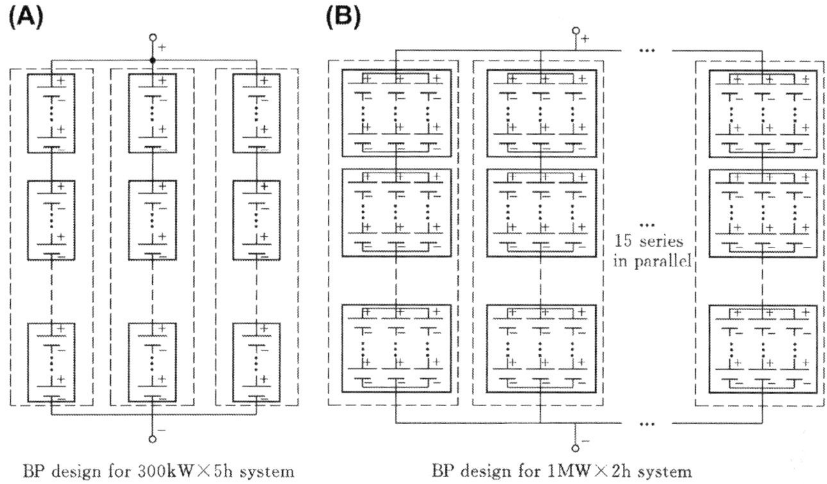

FIGURE 5.7 BP design for demonstration Chinese energy storage power stations.

In total, 600 cells were used. Fig. 5.7B is another BP design model, where 1 MW × 2h lithium iron phosphate batteries were configured. The BP consists of 15 parallel series; each series consists of 20 modules in series; each module consists of 12 units in series; each unit consists of 3 cells in parallel.

5.1.2.2 Evaluation indicators

1. Consistency. It refers to the sustained performance of the cells that compose the battery units, series, and piles under specific load, charging, and discharging conditions. It usually includes two coefficients: $R = (\text{Xmax} - \text{Xmin})$ and δ (standard deviation coefficient). For the cells in a BP, R should not be higher than 4%, and δ should be at E level.

$$R = \frac{U_{\text{max}} - U_{\text{min}}}{U_{\text{p}}} \times 100\% \tag{5.1}$$

Where U_{max} is the maximum voltage of an individual cell; U_{min} is the minimum voltage of an individual cell; U_p is the average voltage of an individual cell.

Coefficient δ ranges from A to F levels. Please refer to Lithium Ion Batteries for Electric Vehicles (QC/T 743—2006) for calculation methods for δ. It is used to analyze cell and module consistency.

For analysis on cell consistency, taking the 24-cell unit as an example, δ in discharge capacity is calculated using the following formula:

$$\delta = \sqrt{\frac{\sum_{n=1}^{24} (C_n - \overline{C})^2}{23}} \tag{5.2}$$

$$C_\delta = \frac{\delta}{\overline{C}} \tag{5.3}$$

Where C_n is the capacity of the nth individual cell; $\overline{C}$ is the average capacity of the 24th individual cell.

To analyze module consistency, taking the 10-cell module as an example, δ in discharge voltage under simulated working conditions is calculated using the following formula:

$$\delta = \sqrt{\frac{\sum_{n}^{10} (U_n - \overline{U})^2}{9}} \tag{5.4}$$

$$U_\delta = \frac{\delta}{\overline{U}} \tag{5.5}$$

Where U_n is the discharge cutoff voltage at the mth discharge stage of the nth individual cells; $\overline{U}$ is the average discharge cutoff voltage at the mth discharge stage of the 10th individual cells.

Consistency of different modules at different stages can be analyzed based on data on discharge at different stages.

Coefficient δ is rated at A, B, C, D, E, and F levels. See Table 5.1.

2. Discharge capacity. Considering the inconsistency between individual cells, the discharge capacity of a unit may reduce to some extent. Also, the manufacturing techniques and operation conditions, such as temperature and current flow, may impact the discharge capacity. Therefore, the discharge capacity should be evaluated based on all these factors. So far, there is no unified standard. So, it should be evaluated according to specific conditions and evaluation methods provided by the battery manufacturers.

TABLE 5.1 δ **levels.**

Level	A	B	C	D	E	F
δ (%)	≤ 1.5	2 ± 0.5	3 ± 0.5	4 ± 0.5	5 ± 0.5	≥ 5.5

3. Cycle life. The service life of a battery includes storage life and cycle life. The storage life refers to the period during which the battery's performance deteriorates from the best level to a certain lower level under some idle and storage conditions. For example, under the New Generation of Vehicles (PNGV) and Freedom CAR programs of the United States Department of Energy, the batteries' storage life is expected to be 15 years. Cycle life refers to the period during which the battery's performance deteriorates from the best level to a certain lower level under operating conditions.

Currently, due to brevity in energy storage research, there exist only a few methods for testing the service life of batteries for energy storage, with limited results. Most current service life tests target power batteries. Different power batteries are tested differently. For example, those for electric vehicles are different from those for hybrid electric vehicles. Generally, battery manufacturers for vehicles adopted one of the two standards to judge the service life of batteries: the discharge quantity is less than 80% of the rated capacity of the battery; the actual internal resistance reaches 150% of the rated internal resistance.

The battery cycle life is also affected by the working conditions, such as the depth and speed of discharge, the temperature of the battery, the charging and discharging tactics, and the alternating current quality.

5.1.2.3 Factors that impact the performance of a battery pack

Limited by connection and integration techniques, the battery pack cannot compare with individual cells in terms of specific energy and power. What is more, a battery pack is susceptible to overcharge, overdischarge, overcurrent, and overheat, which may be followed by capacity attenuation, shortened service life by several or more than 10 times compared to individual cells, cost upsurge for use and maintenance of battery system, as well as hazards such as a fire. Currently, the cycle life of a single lithium-ion cell is 3000 times or above, while that of a pack configured using such cells is 1500 times at most. This is due to inconsistency between individual cells and the difference caused by configuration modes, connection techniques, and operation conditions.

1. Inconsistency between individual cells. Variance exists in the initial performance of individual cells that compose a battery pack. Such differences will multiply in the charge, discharge, and natural discharge processes. For example, if the cells have different capacities, for those with large capacity, the charge and discharge may be insufficient, while for those with small capacity, it may overcharge and overdischarge. Some cells may suffer long-term insufficient charging, which will lead to passivation, while others may suffer long-term overdischarge that will lead to cell damage. These conditions will speed up the capacity attenuation and loss of cycle

life of the battery pack and therefore make conditions for battery management too complicated.

2. Configuration mode. Different configuration modes result in varied consumption rates of a battery's internal resistance and capacity.

"Configuration in parallel and then series" is good for self-balance of capacity among the cells and connection stability. However, charge and discharge control should be made to prevent current unbalance in multiple circuits. A small battery system can adopt the mode to reduce capacity loss.

3. Connection techniques. The connectors play a critical role in a battery system. How to prevent the connectors from eroding remains a problem in the development of connection techniques. If the connectors are oxidized after long-term use, the resistance on the contact surface will increase and cause a voltage drop, leading to unbalance of battery voltage, lower efficiency, and safety pitfalls. Currently, enclosing the outer surface of the battery cells with film is the most common solution to insulate the connectors from corrosion.
4. Operation conditions. Reasonable charge mode, discharge power, and depth are critical factors that guarantee sustained stable operation of the battery pack. Both overcharge and overdischarge will have great impact on the service life of the battery pack. Long-term high-level charge and discharge will result in great capacity attenuation of the battery pack.

Temperature is also a major influence on the electrochemical performance of a battery. Extremely high or low temperature undermines the performance of the battery. Therefore, in improper temperature conditions, charge and discharge should be avoided to protect the ESSs from permanent capacity attenuation and safety pitfalls.

5.1.3 Typical ESS design

5.1.3.1 Branch ESS

Power in such type of ESS ranges from 30, 50, 100, 250–500 kW, and discharge duration from 2 to 4 h. It connects with a 400-V low-voltage grid through a grid-connected switch cabinet.

The system consists of an energy storage unit. See Fig. 5.8 for the topological structure of a 100-kW branch ESS.

5.1.3.2 Circuit ESS

This ESS has power ranging from 200 kW to 1 MW and discharge duration ranging from 2 to 4 h, and its branches have power ranging from 100, 250–500 kW. It connects to a 400-V low-voltage grid through a grid-connected switch cabinet, or 10 or 35-kV grid through a step-up transformer unit.

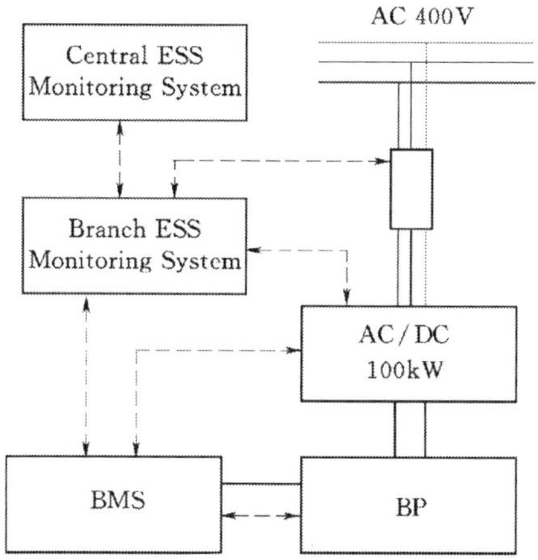

FIGURE 5.8 Topological structure of branch ESS.

The system consists of several parallel branches. It can be connected to the ESS of a medium-voltage grid through a step-up transformer unit. See Fig. 5.9 for the topological structure of a 500-kW circuit ESS.

5.1.3.3 Power station ESS

This ESS has power ranging from several to tens of megawatts, typically 10 and 20 MW, and discharge duration ranging from 15 min to 4 h. The system

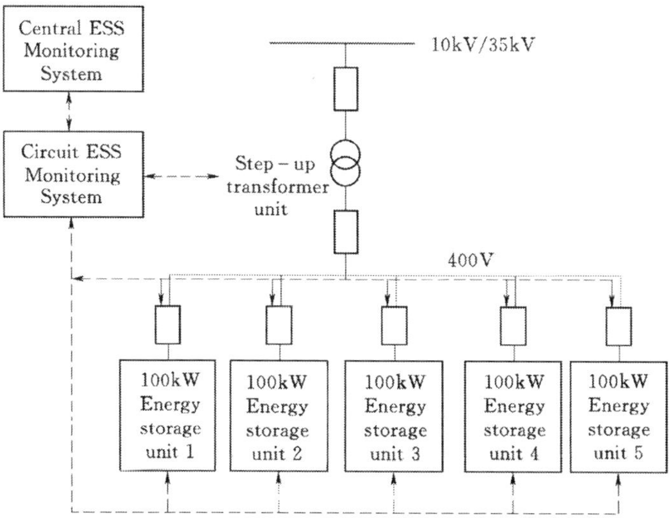

FIGURE 5.9 Topological structure of a 500-kW circuit ESS.

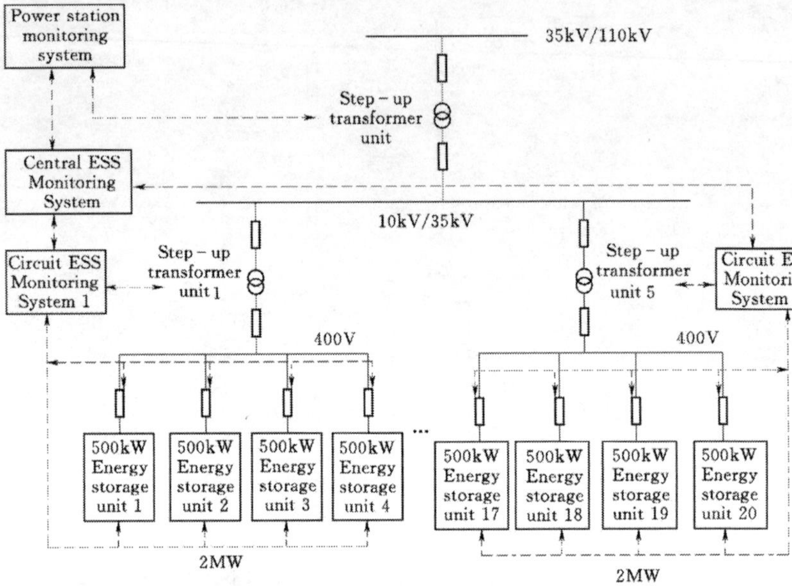

FIGURE 5.10 Topological structure of power station ESS.

circuits have 1 or 2 MW power. The system connects to a 10-kV or above (35 and 110-kV) grid.

The system is composed of several energy storage circuits in parallel. It converges the currents through a high-voltage bus and can be connected to a higher voltage grid after an upgrade. See Fig. 5.10 for the topological structure of this kind of system.

5.1.3.4 Examples for system integration design

5.1.3.4.1 Connection modes

One example is a 10-MW/40-MWh lithium–iron battery storage system. The energy storage unit has a power capacity of 500 kW/2 MWh. It consists of ten 1-MW/4-MWh energy storage circuits. Each circuit is connected to the 400-V bus through two 500-kW/2-MWh branches and then to a 35-kV bus through a 0.4/35-kV step-up transformer. See Fig. 5.11.

5.1.3.4.2 Design for energy storage unit

The energy storage unit consists of a 500-kW/2-MWh BP, a 500-kW PCS, and BMS. As Fig. 5.12 indicates, the PCS includes a set of 500-kW AC/DC converters and 10 sets of 50-kW DC/DC converters. The BP consists of ten 50-kW/200-kWh series in parallel. Each series is configured with a set of battery alley management units and is controlled by a 50-kW DC/DC converter for charge and discharge.

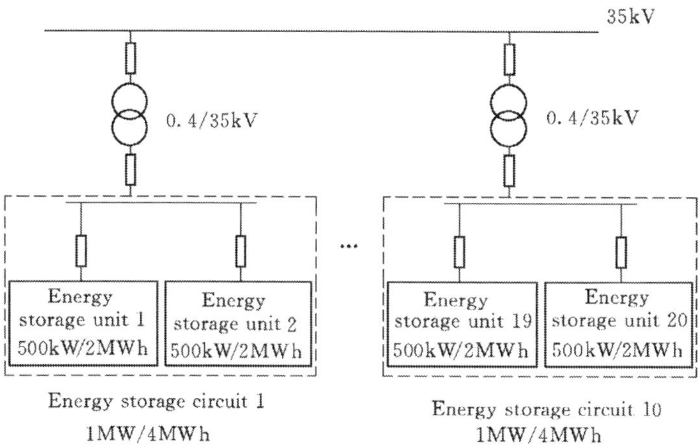

FIGURE 5.11 Connection modes of a 10-MW/40-MWh lithium–iron battery storage system.

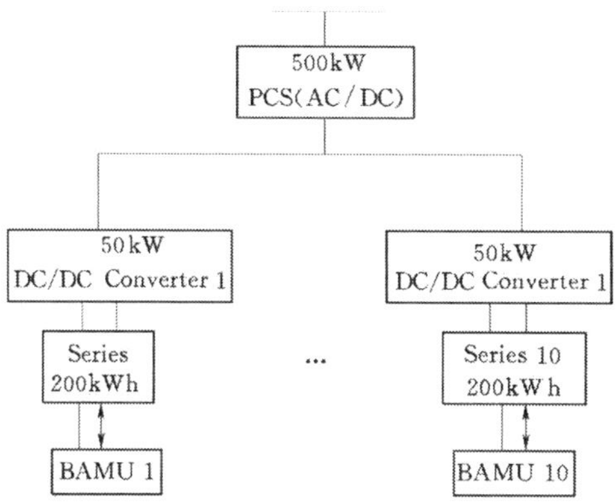

FIGURE 5.12 Topological structure of an energy storage unit.

Each 50-kW/200-kWh series can consist of 25 25.6-V/320-Ah modules, with working voltage ranging from 500 to 730 V, and nominal voltage of 640 V. Each module consists of 8 3.2-V/320-Ah units in series. Each unit consists of a number of individual cells according to their capacity. For example, a unit may include 16 3.2-V/20-Ah cells, or 8 40-Ah cells.

5.1.3.4.3 Energy storage monitoring system design

Based on the capacity and performance requirements, 10-MW/40-MWh battery-based ESS should be equipped with an appropriate monitoring system. See Fig. 5.13 for the structure of such monitoring system, consisting

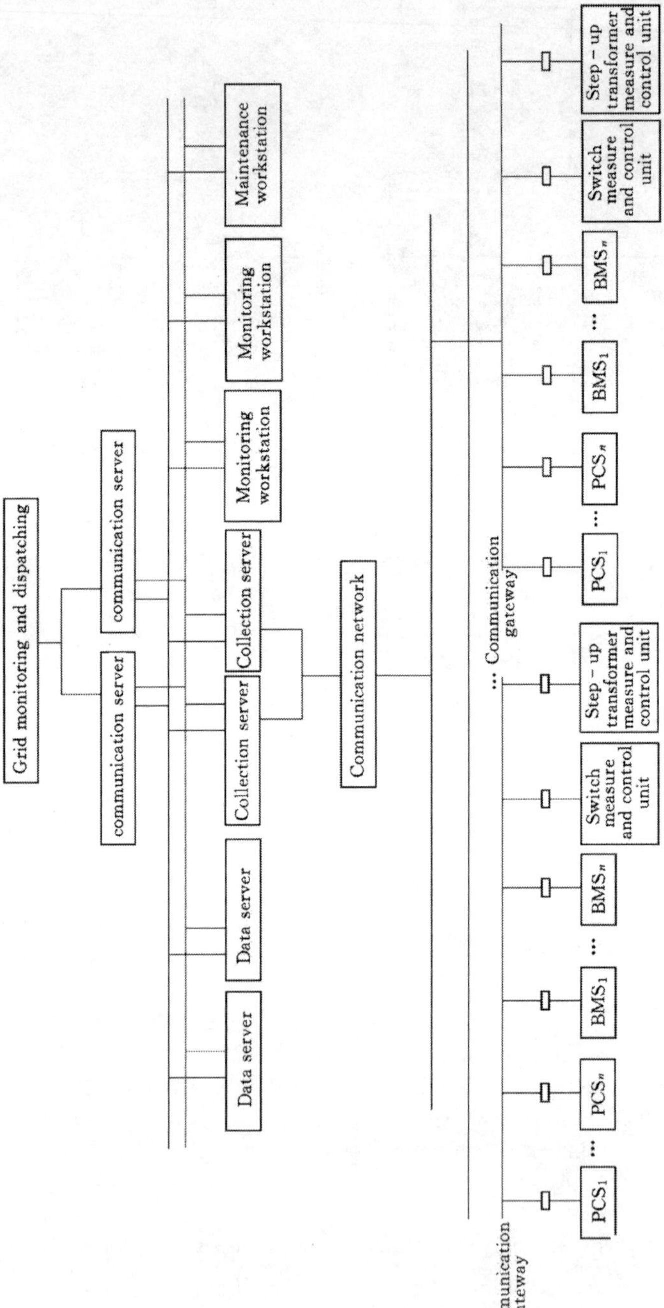

FIGURE 5.13 Structure of the monitoring system in an energy storage power station.

of a power station monitoring system and an onsite monitoring system. See Fig. 5.13.

An onsite monitoring system consists of onsite monitoring unit, BMS, converter, and other onsite controllers. It monitors the operation of several circuits (the PCS, BP, grid-connected switches, and step-up transformers), collecting relevant information and developing reports on a power station's monitoring system. The power station's monitoring system has a front management unit, real-time/historical information server, data server, operators' workstation, and engineers' workstation. It exchanges data with an onsite monitoring system and remote grid dispatching/monitoring system, as well as receives dispatching instructions to optimize power allocation.

5.1.4 ESS reliability evaluation indexes

As mentioned earlier, ESS consists of an energy storage battery, BMS, PCS, energy storage monitoring system, low-voltage access switch or step-up transformer, and integrated system. An integrated system has a connector cable, feeder cable, and bus. The reliability of an ESS can be evaluated on the following indexes: energy loss expectation, average fault frequency index, average fault duration index, average availability index, and average unavailability index. Next is a specific introduction to calculation methods for these indexes.

5.1.4.1 Energy loss expectation

$$ELE = \sum_{i=1}^{m} \lambda_i D_i P_i \tag{5.6}$$

Where λ_i is the fault frequency of the ith component, times/year; D_i is the fault duration of the battery pack in case of failure of the ith component, h; P_i is the power loss of the battery pack in case of failure of the ith component, MW; m is the total number of the components of the system; ELE is the total energy loss of the energy storage power station, MWh/year.

5.1.4.2 Average fault frequency index

$$AFFI = \frac{1}{n} \sum_{i=1}^{m} \lambda_i S_i \tag{5.7}$$

Where λ_i is the fault frequency of the ith a component, times/year; S_i is the number of battery pack faults caused by failures of the ith component; n is the total number of battery packs; m is the total number of the system components; AFFI is the average fault frequency of the battery packs in a power station, times/year.

5.1.4.3 Average fault duration index

$$AFDI = \frac{1}{n}\sum_{i=1}^{m}\left(\lambda_i \sum_{j=1}^{S_i} d_{ij}\right) \tag{5.8}$$

Where λ_i is the fault frequency of the ith component, times/year; d_{ij} is the fault duration of the jth battery pack in case of failure of the ith component, h, $j = 1, 2......$; S_i is the number of battery pack faults caused by failures of the ith component; n is the total number of battery packs; m is the total number of the components of the system; $AFDI$ is the average fault duration of the battery packs in a power station, h/year.

5.1.4.4 Average availability index

$$AAI = \frac{8760 - AFDI}{8760} \tag{5.9}$$

Where AAI is the average availability of the power station on the basis that all the battery packs are in normal operation.

5.1.4.5 Average unavailability index

$$AUI = 1 - AAI = \frac{AFDI}{8760} \tag{5.10}$$

Where AUI is the average unavailability of the power station on the basis that the battery packs are operating abnormally.

5.2 ESS integration under typical application mode

5.2.1 New energy power generation side

Wind and PV power generation is fluctuant and intermittent. Therefore, large-scale grid connection is vulnerable to problems, such as unstable grid operation and improper local allocation. When the wind and PV power exceeds expected levels, there may be adverse impact on the grid that may lead to serious mishaps. Related researches indicate that when the wind power generation unit's installed capacity takes less than 10% of the total installed capacity, traditional power grid technologies can be adopted and hydro power and gas-fired generating units can be added to boost the security of the power grid; when it is raised to 20% or higher, the grid's peak load regulation and security will be greatly challenged.

The application of large ESS on the new energy power generation side helps stabilize the wind or PV power output. By storing and stabilizing the output, it will minimize the damage caused by the fluctuation of wind and PV power generation on the grid's stability and the impact caused by the intermittency on the load allocation. It can also provide quick active support for enhancing the grid's frequency regulation capacity and guaranteeing reliable connection of the wind and PV power with a standard power grid. Therefore, the application of ESS can help optimize the power source structure of the whole system. This chapter will give an introduction to its application in wind power generation systems.

5.2.1.1 Technical requirements

In a wind power plant, the ESS can be used to cut down power transmission, time shift, make up the difference between predicted and actual output, provide frequency support for the power grid, and restrict power fluctuation. See Table 5.2 for specific technical requirements.

1. Cutdown power transmission. In case of any limitation on the transmission lines, the ESS can store the electric power generated by the wind power plant; when the transmission line capacity is satisfied, the ESS can release the electric power. This mostly happens in remote wind power plants, where transmission capacity closely reaches its limit. The ESS can be able to respond within a minute and give play to its other functions when the power transmission is not limited. Its working cycle is related to wind power generation curve. In this case, the ESS can increase the power generation and consumption while improving its delivery capacity. It can also prevent huge upgrading and renovation costs on the transmission lines. Such an ESS has a capacity of several to hundred MW and discharge duration of several hours.
2. Time shift. The ESS can store the wind power at load valley and release it at load peak. The function of the ESS should be given full play. Power should be purchased and delivered to the ESS only in case of low power consumption and insufficient power generation by the wind power plant. The ESS should be able to respond within a minute and execute its other functions concurrently. The stored energy can be valued based on the transaction duration in the power market. It can generate economic efficiency by avoiding costs of peak load regulation. Such an ESS can have a capacity of several to hundred MW and discharge duration of several hours.
3. Make up for the difference between predicted and actual output. The ESS's purpose is to store electric power when the actual power output of the wind power plant exceeds the amount specified in the bidding documents. It also releases electric power when the actual power output is below the specified amount. These should be real-time processes. Similarly, the functions of the ESS should be given full play to store surplus power at nonpeak hours. The ESS should be able to respond within

TABLE 5.2 Technical requirements for energy storage under different application scenarios.

Scenarios	Use of ESS in a wind power plant				
Technical requirements	Cut down power transmission	Time shift	Make up the difference between predicted and actual output	Provide frequency support for the power grid	Restrict power fluctuation
ESS power (MW)	2–200	10–500	2–200	2–200	2–50
Voltage at connection point (kV)	10–35	10–35	10–35	10–35	10–35
Full power discharge duration	5–12 h (excluding large-capacity compressed air energy storage), depending on energy storage type			10–30 min	Several seconds
Energy/discharge	50–120 MWh			0.2–25 MWh	10–200 MJ
Discharge cycle	Make optimization based on wind power output curve			24 times/year 1 time/day	Continuous triangular wave, 90 charge and discharge cycles/h
System response duration	<1 min	<1 min	<20 ms	<20 ms	<20 ms
Economic efficiency	Make comprehensive analysis on basis of cost/benefit calculation, market rules, and power price				

20 min and simultaneously execute its other functions. Such an ESS can have a capacity of several to hundred MW and discharge duration of several hours.

4. Provide frequency support to the power grid. The ESS works to alleviate the imbalance between power generation and load that may occur when the wind power output suddenly reduces by more than 50% or the wind power permeability reaches above 20%. The ESS can have the ability of immediately detecting disturbance and respond within 20 min to input effective power for up to 30 min. It is described as a 15-min full power discharge, one time/day, 24 times/year. Such an ESS can have a capacity of tens of MW and discharge duration of tens of minutes.
5. Restrict power fluctuation. The ESS can be able to give follow-up response to fluctuations experienced during wind power output, for example, 90 times/hour. Such an ESS can have a capacity of several to tens of MW and discharge duration of tens of seconds. In fact, it is difficult to count the cycle frequency of the ESS. Therefore, it can be designed with a service life of 15 years, and its cycle frequency depends on wind resources, congregation effect of wind power plants, and demand for grid connection in related areas. Moreover, wind power plants' erratic power change may result in erratic change of energy during each cycle. Therefore the cycle frequency of the ESS at full energy should be clarified. For example, it may be 1000 times. The design of the ESS should take into account anticipated operation conditions, such as aging following repeated circulation over time, to make sure the system will meet the demand for energy till the end of its service life. For example, for a 10-MWh ESS, the design should be able to ensure that the system will be able to transmit 10 MWh energy over a service life of 15 years.

5.2.1.2 Connection mode

To stabilize the power change of a specific wind power plant, the ESS should be installed close to the plant; to support a larger power grid, the ESS can be installed anywhere to connect with the power transmission and distribution networks.

Take a 10-W/40-MWh BESS as an example. See Fig. 5.11 for the design of the system.

Step-up transformer should be designed with a capacity of 1.2 MW, 1.2 times as much as that of the energy storage converter, a normal capacity of 1250 kVA, and a transformation ratio of 0.4/35 kV. The monitoring and protection systems (temperature control system) of the step-up transformer should be configured in the step-up transformer unit. A 0.4-kV breaker and 35-kV grid-connected breaker should be equipped with control and communication units, and the monitoring and protection systems (TV, TA, and protectors) should be installed in the breaker cabinet.

5.2.1.3 Operation mode

ESS installed at the new energy power generation side can be operated under the following several modes, each independent from each other and not necessarily multifunctional:

1. Power variation rate limitation. Power variation rate is limited to a certain range relying on real-time measurement of the output power of a wind power plant. The ESS starts charging when the output power of the wind power plant climbs sharply; it discharges when the output power drops sharply. This aims at meeting the needs of the grid for a certain power variation rate. Under such a mode, the ESS should be installed close to the wind power plant.
2. Grid power variation support. The ESS charges and discharges power according to the grid's dispatch instructions to make extra adjustment for the grid. Under such a mode, the ESS does not target a certain wind power plant. Therefore, it is not necessary for the ESS to be installed near the wind power plant.
3. Voltage regulation. The ESS works on reactive power absorption or release according to real-time need for regulation of transmission voltage.
4. Frequency regulation. The ESS can charge or discharge power according to the signal received each second. Under such a mode, the ESS can maintain a long-term operation at a set State of Charge value.
5. Low-voltage ride-through. The ESS works in power electronic equipment for low-voltage ride-through. Most wind power plants have power electronic equipment. So there is no need for extra support.

5.2.1.4 Electrical interface

1. Interface standard. The system should keep to the standards on safety, electrical design, electrical connection, harmonics, DC injection, and isolation.
2. Breaker. The breaker should be able to cut off full-power operation, through manual or remote operation, and it should be able to adapt to short-circuit current and normal pulse value.
3. Contactor. The system should include a contactor that responds to operation mode.

5.2.1.5 Communications and data management

1. Communications. The system should meet the requirements for multiple communication modes, such as Ethernet, Modbus, and CAN. The selection for communication modes is based on demand of the power grid or by the grid company. Typically, distributed network protocol, namely IEC 61850, is adopted for control interface.

2. Integration interface. The system should meet the grid power communications standards and stick to the ESS Communications/Distributed Power Communications Standards. The system may receive instructions from the wind power plant's control center or superior dispatching unit according to the ownership and operation rights. The results of in situ measurement of transmission voltage and power output of the wind power plant should be sent to the energy storage control system. The ESS should provide status information such as SOC and operation power to the superior control center.
3. Data and events. The system should record major operation data for a certain period, at least the transmission data by minute. The system should record peculiar events, such as changes in operation mode, acceptance instructions, faults, and shutdowns. All the data should be accessible on site and during remote downloading through computer interface or wireless operation.

5.2.1.6 Installation and maintenance

The system can be installed, operated, and maintained according to demands for power transformation and distribution and the requirements of the contractor. Sufficient supporting files and trainings should be provided. Standard electrical tools or specialized operation tools should also be provided.

The supplier should classify any change or removal of subsystems, such as power electronic module and storage unit, as faults. All the consumable and aged parts should be classified according to alternation cycle.

5.2.1.7 Safety

The system should be able to protect itself in case of any internal or grid faults. It should take into account public safety standards. The design for the installation, maintenance, and operation of the system should not cause personal nor public harm. Its environmental damage should be minimal too. The system's recovery process should be clear. It should be locked by disconnecting the switch.

5.2.2 Grid side

Economic growth and residents' improved standard of living has bridged the difference between the peak load in daytime and valley load at night as well as the difference between seasonal consumption. In summer, when the demand for power reaches its annual peak, there is a huge power supply gap, especially in some large- and medium-sized economically developed Chinese cities. However, the huge construction costs and the need for resource protection have made it increasingly difficult to meet the increasing demand for peak load by adding power generation, transmission, and distribution equipment. With the growing difference between the peak and valley loads, the load rate of the power generation equipment is declining, followed by efficiency of electrical equipment, idle capacity, and reduced economic efficiency of enterprises. The ESS, at the grid side, can acquire power from the grid at valley time and transmit power to the grid at peak time. This effectively relieves the imbalance between supply and demand and reduces the construction costs for scaling up the capacity of power plants. Take Shanghai as an example. From 2004 to 2006, the peak load duration each year was 183.25 h, which was met by investing more than RMB 20 billion each year on the grid side. Under such circumstances, the average annual utilization rate of the power transmission and distribution capacity was less than 2%. The investment can be reduced significantly by using large-capacity energy storage technologies, along with quality energy saving and emission reduction effects. A pumped storage power station is one of the most common energy storage facilities at grid side. However, the construction of such power stations is restricted by geographic conditions. Therefore, it is difficult for such facilities to be popularized. In contrast, battery-based energy storage boasts of small land occupancy, flexible establishment, and short establishment cycle.

The ESS at the grid side manages the load, regulates and controls regional grid frequency, expands power capacity, provides reactive support, and supplies power in case of any fault in the grid. It is usually installed near the transformer substation.

5.2.2.1 Technical requirements

1. Power level and discharge duration. The power level should reach about 1 MW for ESS used for low-voltage grid load management, and the discharge duration should be at a range of 2–6 h. The power level should reach about several MW for ESS used for medium-voltage grid load management, and the discharge duration should range between 2 and 6 h, while the storage duration should depend on the duration of regional peak load. Under the grid frequency regulation mode, the ESS can make active or reactive compensation to the grid, improving its power quality. In this case, the typical capacity of the ESS is tens of MW, with discharge duration of 15 min to 1 h.
2. Rated values. The net power level (kW or MW) of the system should be defined on the basis of constant discharge duration of 4 h, and other power levels, such as that of impulse power, should be specified. The net energy level (kWh or MWh) of the system should be defined based on continuous discharge underrated power. The reactive capacity of the system should be defined by kvar/Mvar or power factor. Under the island mode, the system should be able to run for 2–3 s underrated power in case of an impact load

of an overcurrent by 2.5–3 times while maintaining constant voltage as the user load changes.

3. Cycle life. The system should have a calendar life of 15 years or support 1500 full cycles. The system design should take into account the utilization conditions, such as the aging of the system over time and number of cycles, to guarantee the system's ability to meet energy demand till the end of its service life. If the product's service life is related to discharge depth, the design should take into account the capacity redundancy so the system is able to meet the energy demand at a given discharge depth till the end of its service life. The discharge depth shall be defined based on manufacturer's advice to meet the demand for cycle life.

5.2.2.2 Connection mode

Take the 2-MW/8-MWh ESS as an example. Two 1-MW/4-MWh energy storage circuits, each consisting of two 500-kW/2-MWh energy storage branches, together with 400-V breaker, 0.4/35-kV step-up transformer unit and grid-connected breaker should be used to connect with the 35-kV bus. See Fig. 5.14 for details.

The step-up transformer should be designed with a capacity of 1.2 MW, 1.2 times that of the energy storage converter, a normal capacity of 1250 kVA, and a transformation radio of 0.4/35 kV. The monitoring and protection systems (temperature control system) of the step-up transformer should be configured in the step-up transformer unit. A 400-V breaker and 35-kV grid-connected breaker should be equipped with control and communication units, and the monitoring and protection systems (TV, TA, and protectors) should be installed in the breaker cabinet.

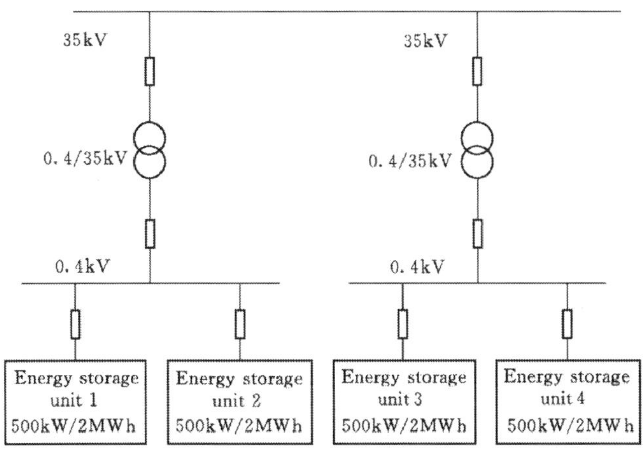

FIGURE 5.14 Connection mode for 2-MW/8-MWh ESS.

5.2.2.3 Operation mode

The operation mode of the ESS at grid side can be defined according to its usage, including the following:

1. Load tracking mode. The ESS discharges at different power levels depending on the control signals. It should be capable of responding to power demand based on remote threshold value.
2. Frequency regulation mode. The ESS should charge and discharge depending on the signals received each second. When supporting frequency, the ESS can request a long-term operation at given SOC value.
3. Constant power charging mode. The ESS charges under constant power.
4. Constant power discharging mode. The ESS discharges under constant power.
5. Reactive mode. The ESS can absorb or release reactive power as part of constant power charge and discharge.
6. Self-trained charging mode. The ESS can charge over a given period according to self-optimization solutions to achieve the expected state.
7. Self-maintenance mode. The ESS can conduct self-maintenance as required.
8. Standby mode. A stage without charge or discharge, and with auxiliary devices standby and contactor closed.
9. Shutdown mode. A state with the system's contractor disconnected from the grid.
10. Island mode. The system will shut down immediately as it perceives power loss on the large power grid. With preset mode, the system resumes power supply according to instructions from the grid.

5.2.2.4 Electrical interface

1. Interface standard. The system should keep to internationally acknowledged safety standards, electrical design, electrical connection, harmonics, DC injection, and isolation.
2. Breaker. The breaker should be able to cut off full-power operation, through manual or remote operation, and adapt to short-circuit current and normal pulse value.
3. Contactor. The system should include a contactor, which runs according to operation mode.

5.2.2.5 Communications and data management

1. Communications. The system should meet the requirements for multiple communication modes, such as Ethernet, Modbus, and CAN. The selection of communication modes is based on the power grid demand or requirement of the grid company. Typical control interface protocol is distributed network protocol, namely IEC 61850.

2. Integration interface. The system should be able to run under different modes and respond to load variation signals. It should be equipped with necessary communication and remote sensing devices, and it should support communication protocols and provide effective services.
3. Data and events. The system should record major operation data over a certain period, at least minute-by-minute transmission data. The system should record events such as change of operation mode, received instructions, faults, and shutdowns. All the data should be available for onsite and remote downloading through computer interface or wireless operation.

5.2.2.6 Installation and maintenance

The system can be installed, operated, and maintained according to demands for power transformation and distribution and the requirements of the contractor. Sufficient supporting files and trainings should be provided together with standard electrical tools or specialized operation.

The supplier should classify any change or removal of subsystems, such as power electronic module and storage unit, as faults. All the consumable and aged parts should be classified according to alternation cycle.

5.2.2.7 Safety

The system should be able to protect itself in case of any internal or grid faults. It should take into account public safety standards. The design for the installation, maintenance, and operation of the system should not cause personal and public harm. It should also be able to minimize environmental damage. The system's recovery process should be clear. It should be locked by disconnecting the switch.

5.2.3 User side

A reliable power supply grid is very important to the users, since blackouts may incur huge economic losses. For some high-tech enterprises, that results in a zero. A blackout may result in a great catastrophic equipment failure and economic losses amounting to more than RMB 10 million. Likewise, in public institutions, such as a hospital, outages may threaten public security and people's lives. In 2006, the user-side power supply reliability reached 99.973%, each user suffering blackout duration of about 2.324 h on average, based on a total installed capacity of 35,000 MVA, and the total power cut was about 4 million kWh, causing total GDP loss of about RMB 100 million (on basis of GDP/1 kWh of RMB 25). Therefore, for users that require a more reliable power supply and a big gap between peak and valley loads, the ESS can used as a uninterrupted power supply to start storage power supply within milliseconds to guarantee uninterrupted power supply. In addition, the ESS can reduce additional installations of power distribution systems, aimed at boosting power supply reliability, and therefore reducing investment in capacity storage. On the other hand, when two-part tariff is adopted, the total monthly expenditure for power generation will reduce, thus cutting down the power purchasing cost. One can therefore buy more low-priced power instead of high-priced power. The ESS at the user side can regulate peak load regulation, improve quality of power supply, and serve as a standby power supply. It can be installed in houses and buildings or at the low-voltage side of power distribution transformers.

5.2.3.1 Technical requirements

1. Power level and discharge duration. For ESSs at the user side, the power level should range from several to 200 kW, and the discharge duration should range between 2 and 4 h. ESSs for families should be able to support power supply for 2–4 h at a power level from 1 to 10 kW; the installed capacity of ESSs for communities is according to the capacity of the power distribution transformer, commonly supporting a power supply for 2–4 h at a power level from 25 to 200 kW; ESSs for large business users should be able to support a power supply for 2–4 h at a power level from 10 to over 100 kW.
2. Rated values. The net power level of the system should be defined on the basis of a constant discharge duration of 4 h (kW), and other power levels, such as that of impulse power, should be specified. The net energy level (kWh) of the system should be defined based on continuous discharge underrated power. The reactive capacity of the system should be defined by kvar/Mvar or power factor. Under the island mode, the system should support instantaneous overload operation, for example, operation for 2–3 s underrated power in case of an impact load of an overcurrent by 2.5–3 times while maintaining constant voltage.
3. Cycle life. The system should have a calendar life of 15 years. The design for the system should take into account the utilization conditions, such as the aging of the system over time and number of cycles, to guarantee that the system meets the following requirements: the system for peak load management should support at least 2250 cycles (full charge and discharge); if the product's service life is related to discharge depth, the design should take into account the capacity redundancy so the system is able to meet the demand for energy at given discharge depth till the end of its service life. The discharge depth shall be defined based on the advice of the manufacturer to meet the demand for cycle life.

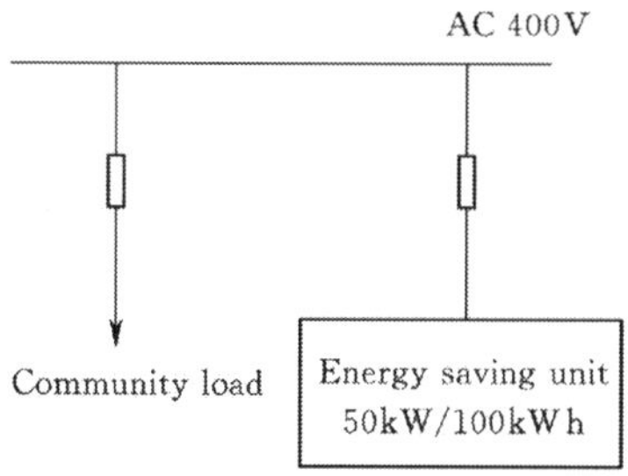

FIGURE 5.15 Connection mode for a 50-kW/100-kWh ESS.

5.2.3.2 Connection mode

An ESS for family users is usually connected with a 220-V single-phase line, while for communities with a 380-V three-phase line, and large industrial and commercial user energy storage system can be connected to single-phase 220V or three-phase 380V lines.

Take a 50-kW/100-kWh ESS for an example. The system should be connected to a 400-V low-voltage bus through a low-voltage switch and a 50-kW/100-kWh energy storage unit. See Fig. 5.15.

5.2.3.3 Operation mode

The ESS at user side can be operated under the following modes, each independent from others and not necessarily multifunctional.

1. Peak load management mode. The system may charge during power consumption valley and discharge during power consumption peak at adjustable power level based on user's setting.
2. Constant power charging mode. The ESS charges under designated power.
3. Constant power discharging mode. The ESS discharges under designated power.
4. Automatic charging mode. The ESS can charge over a given period according to self-optimization solutions to achieve the expected stable state.
5. Self-maintenance mode. The ESS can conduct self-maintenance as required.
6. Standby mode. ESS system neither charge nor discharge, with auxiliary devices on standby and contactor closed.
7. Shutdown mode. A state with the system's contractor disconnected from the grid.
8. Island mode. The system will enter into isolated operation by switching off the contactor as soon as it detects any abnormality in the public power grid (See IEEE P1547. 4). It can also detect the recovery of the public power grid and integrate into it at the same time.

9. Fast switch mode. The system should be able to respond quickly in case of any abnormalities (such as half cycle) and switch into island mode.
10. Delay switch mode. The system can delay switch to island mode (for minutes, for example) to allow other grid equipment to start.

5.2.3.4 Electrical interface

1. Interface standard. The system should keep to the internationally acknowledged standards on safety, electrical design, electrical connection, harmonics, DC injection, and isolation.
2. Breaker. The breaker should be able to cut off full-power operation based on manual operation. All operation and control privileges should be available at the grid side.

5.2.3.5 Communications and data management

1. Communications. The system should meet the requirements for multiple communication modes, such as CAN, MODBUS, and Ethernet. The selection of communication modes is based on the demand of the power grid or public grid regulations.
2. Integration interface. The system should maintain relevant standards on grid communication, ESS communication, and distributed power supply communication. The public grid should monitor the ESS through the operation center. The system should be able to run under different modes and reasonably respond to load signals. It should be equipped with necessary communication and remote sensing devices, support communication protocols, and provide effective services.
3. Data and events. The system should record major operation data over a certain period, at least minute-by-minute transmission data. The system should record events, such as change of operation mode, received instructions, faults, and shutdowns. All the data should be available by onsite and remote downloading through computer interface or wireless operation.

5.2.3.6 Installation and maintenance

The system can be installed, operated, and maintained according to power transformation and distribution demands and contractor's requirements. Sufficient supporting files and trainings should be provided together with standard electrical tools or specialized operation tools.

The supplier should classify any change or removal of subsystems, such as power electronic module and storage unit, as faults. All the consumable and aged parts should be classified according to alternation cycle.

5.2.3.7 *Safety*

The system should be able to protect itself in case of any internal or grid faults. It should take into account public safety standards. The design for the installation, maintenance, and operation of the system should avoid personal and public harm. It should also be able to minimize environmental damage. The system's recovery process should be clear. It should be locked by disconnecting the switch.

5.2.4 Microgrid side

A microgrid is a kind of small-scale power grid for both power generation and consumption. In a microgrid, power is consumed from the same source it was generated. This helps reduce losses caused by power transmission and distribution, improve energy utilization efficiency, reduce users' dependency on the grid, as well as enhance the reliability and safety of power use. Therefore, microgrids should be one of the most important components on the smart grid in future. However, as a major supplement, microgrids depend on the large grids. Essentially, microgrid users partly utilize the large grid. Usually, microgrids rely on a distributed gas-fueled triple-generation unit or a small-scale power generation unit such as wind and PV power generators for power supply. Currently, many power generation equipment, especially small-sized solar, wind, or biomass power generation equipment, do not need to be directly connected to large grids. Such power generation units are gaining in popularity. Considering the fluctuation and randomness of distributed power sources as well as the uncertainty of the load at the user side in the microgrid, an ESS installed in the microgrid can help stabilize fluctuation of the power supply.

5.2.4.1 *Technical requirements*

In a microgrid, ESS can stabilize fluctuation in PV and wind power generation as well as conduct emergency control of active power in the process of input and output of power supply to guarantee voltage stability of important load. Typically, the power level of the ESS in a microgrid ranges from tens to hundreds of kW, and the discharge duration ranges from 2 to 4 h. Considering the size and characteristics of a microgrid, ESSs based on battery, supercapacitor, and flywheel energy storage can be applied to a microgrid.

5.2.4.2 *Connection mode*

An ESS for microgrids is usually directly connected to a 400-V bus. Take a 100-kW/400-kWh ESS as an example. There is an energy storage circuit under the monitoring of a microgrid energy management system. See Fig. 5.16.

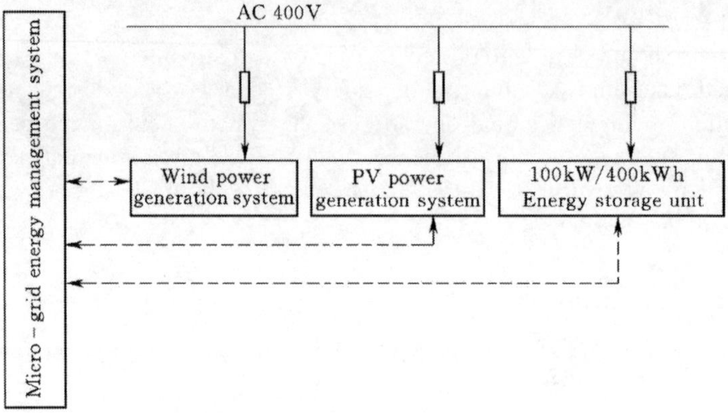

FIGURE 5.16 ESS connected with a microgrid.

5.2.4.3 Operation mode

ESS installed in a microgrid can be operated under the following modes, each independent from each other and not necessarily multifunctional:

1. Constant power charging mode. The ESS charges under designated constant power.
2. Constant power discharging mode. The ESS discharges under designated constant power.
3. Standby mode. A stage without charge or discharge, and with auxiliary devices on standby and contactor closed.
4. Island mode. The system will disconnect the contactor and get into isolated operation as soon as it detects any abnormality in the public grid. The system should be able to detect the recovery of the public grid and concurrently gear into the grid.
5. Fast switch mode. The system should be able to respond quickly in case of any abnormalities (such as half cycle) and switch into island mode.
6. Delay switch mode. The system can delay switch to island mode (for minutes, for example) to allow other grid equipment to start.
7. Frequency regulation. The ESS can charge or discharge as a response to the signal received each second. Under such a mode, the ESS can maintain a long-term operation at a set SOC value.

5.2.4.4 Electrical interface

1. Interface standard. The system should maintain safety standards, electrical design, electrical connection, harmonics, DC injection, and isolation.
2. Breaker. The breaker should be able to cut off full-power operation based on manual operation. All operation and control privileges should be available at the grid side.
3. Contactor. The system should include a contactor, which runs according to operation mode.

5.2.4.5 Communications and data management

1. Communications. The system should meet the requirements for multiple communication modes, such as CAN, MODBUS, and Ethernet. The selection of communication modes is based on the demand of the power grid or public grid regulations.
2. Integration interface. The system should stick to relevant standards on grid communication, ESS communication, and distributed power supply communication. The public grid should monitor the ESS through the operation center. The system should be able to run under different modes and respond to the load signals reasonably. It should be equipped with necessary communication and remote sensing devices and support communication protocols and provide effective services.
3. Data and events. The system should record major operation data over a certain period, at least the minute-by-minute transmission data. The system should record events, such as change of operation mode, received instructions, faults, and shutdowns. All the data should be available by onsite and remote downloading through computer interface or wireless operation.

5.2.4.6 Installation and maintenance

The system can be installed, operated, and maintained according to power transformation and distribution demands and the requirements of the contractor. Sufficient supporting files and trainings should be provided together with standard electrical tools or specialized operation tools.

The supplier should classify any change or removal of subsystems, such as power electronic module and storage unit, as faults. All the consumable and aged parts should be classified according to alternation cycle.

5.2.4.7 Safety

The system should be able to protect itself in case of any internal or grid faults. It should take into account public safety standards. The design for the installation, maintenance, and operation of the system should be safe against personal and public harm. It should also minimize environmental damage. The system recovery process should be clear. It should be locked by disconnecting the switch.

5.3 Analysis of economic efficiency of ESS in typical application scenarios

5.3.1 Types of application scenarios

The existing ESSs are costly. Therefore, the analysis of ESSs' economic efficiency in various application scenarios has attracted great attention both at home and abroad. It has become a major influential factor for advancements in energy storage technologies. Sandia National Laboratories (SNL) in the

United States has analyzed costs of different energy storage technologies, as well as the relations between energy storage costs and discharge duration and power quality of large-capacity system and distributed energy storage units in specific application scenarios. It classified the application scenarios and estimated the benefits in each application scenario, such as low-storage and high-generation arbitrage, frequency regulation benefits, reserved benefits, reduction of transmission congestion, and delayed investment in construction of power transmission and distribution facilities. SNL divided the benefits of ESSs as follows:

1. Energy exchange. The ESSs charge power in case of a low demand or low power price; it sells power in case of high demand or high power price.
2. Concentrated power generation. The ESSs help reduce construction costs of new power plants and cut budgets of buying new power capacity or leasing capacity from power markets.
3. Support services. The ESSs provide a multitude of support services, including early arrangement for power generation and trading as well as real-time regulation for balance of power generation and load; reactive support/voltage control to maintain power transmission system's voltage falls into allowed range; minute-by-minute regulation to ensure power supply/load balance based on the standard requirements of public grid; online standby for power compensation or temporary supply in case of breakdown of power transmission, including online response within 10 s to maintain system frequency and offline response within 10 min for power compensation or temporary supply; correction of difference between actual and planned energy trades by hour; load tracking to meet load fluctuation by hour and day; standby power source that can supply power for up to 1 h or for market trading; compensation for power transmission and distribution systems losses; dynamic regulation, namely real-time regulation of power generation and output or load; black start, meaning to start part of the power grid without outer assistance in case of breakdown of large power grid; maintenance of grid stability, required to give real-time response to system disturbances for the purpose of maintaining system stability and security.
4. Power transmission support. Improving dynamic stability and enhancing bearing ability; providing active and reactive compensation and improving line capacity in case of subsynchronous resonance; reducing instantaneous voltage drop and improving dynamic voltage stability; reducing load switch-off under low frequency in case of disturbances to the large grid.
5. Reducing demand for expansion of power transmission capacity.
6. Reducing power transmission congestion. Reducing costs and expenses caused by power transmission congestion during peak hours.
7. Delaying investment in upgrading and renovation of power transmission and distribution facilities. Reducing costs to local taxpayers, improving

utilization efficiency of public assets, saving money for construction of more important projects as well as reducing risks of low-efficiency utilization of equipment after investment.

8. Power cost management. It means charge at low demand and price and discharge at high demand and price based on the user pricing mechanism.
9. Capacity charge management. Avoiding spike in monthly or annual prices caused by power consumption during peak hours.
10. Reliability of power consumption service. For orderly closedown, discharge duration should at least last for an hour; for transmission to local power plant, discharge duration should be in specific hours.
11. Power quality. The ESSs should be able to respond to problems in power quality, such as voltage fluctuation, frequency fluctuation, low power factor, harmonics, and service interruption.
12. Guaranteeing fulfilment of contract on renewable energy power generation and supply. Storing renewable energy power at low power demand and discharge power at high power demand to make sure the contract on renewable energy power generation and supply is fulfilled.

5.3.2 Analytical methods for costs/benefits of ESSs

ESSs have been widely applied in Japan, the United States, and some European countries. China has also launched a number of pilot programs for use of ESSs. Therefore, there have been all-around and specific studies on the value of ESSs at home and abroad.

Currently, evaluation of ESSs' value is based mainly on various application scenarios, usually including power grid side, user side, and new energy power generation side. ESSs can bring benefits by delaying investment in grid upgrading, reducing transmission congestion, providing support services, and improving power supply reliability. Such systems can also support arbitrage through power storage at low price and discharge at high price. Investors are not the only ones benefiting from the ESSs. The benefits should be analyzed according to the functions of the ESSs.

There are numerous documents about modeling and analysis methods for estimating the benefits generated by using ESSs at the power transmission and distribution side. Some studies have recorded benefits generated by vanadium BESSs, as the systems are used to delay investment in grid upgrading, reduce congestion cost and arbitrage, targeting at optimizing the benefits through genetic algorithm and linear programming. Some documents have given full consideration of ESSs' value in delaying investment in grid upgrading, providing support services, improving utilization efficiency of power transmission equipment, and optimizing load and power management to acquire a multipurpose model with the biggest net present value and lowest cost. Some documents have given comparison of benefits and costs as well as numerical simulation models of battery-based ESSs for primary frequency regulation after taking into account the frequency fluctuation curve and battery features in charge and discharge, with the purpose of acquiring the best capacity and maximizing the annual benefits generated by the systems. Some documents have shown comparative analysis of optimization solutions using batteries, fuel cells, diesel generators, and PV generators to delay grid upgrading for the purpose of maximizing the benefits brought by ESSs. In these documents, dynamic planning has been applied to get the most economically feasible size and unit cost in each solution.

Some studies have targeted dispatching and operation strategies designed to optimize the economic efficiency of ESSs with defined power and capacity. Some documents have recorded a mixed integer nonlinear programming model for optimizing dispatching and operation of a 10-MW/70-MWh vanadium and sodium–sulfur battery ESS based on consideration of the systems' arbitrage, standby, and frequency regulation earnings, getting solutions through General algebraic model system. Some documents have shown net revenue generated by 1-MW/10-MWh sodium–sulfur and 1-MW/0.25-MWh flywheel ESSs under different solutions, taking into account arbitrage and frequency regulation earnings. These documents also include evaluation of the economic efficiency of the systems. Some documents have demonstrated modeling calculation taking into account ESSs' functions in delaying investment in grid upgrading, arbitrage, and frequency regulation earnings as well as analysis and comparison of the economic efficiencies of different battery-based energy storage technologies (mainly including lead–acid, nickel–cadmium, sodium–sulfur, and vanadium batteries). There are also other documents that have given comparisons of economic efficiencies of different sizes of battery-based ESSs as used in local grids as standby, load regulation, and frequency control. These documents also include analysis of the ripple economic benefits generated by the systems following the growth of their sizes.

ESSs installed at the new energy power generation side generate benefits by stabilizing power output fluctuation and time shift of power generation. In addition to the arbitrage expected by the power producer, the ESSs also generate huge social benefits by reducing the costs spent in buying reserve capacity. Many documents reveal studies and reports on the installation and operation of ESSs at new energy power generation side. Calculations recorded in these documents are based on PV battery united ESSs installed at factories, commercial buildings, and shipping centers. Analysis of best sizes and economic efficiency was made on basis of total costs and reduction of CO_2 emissions. Best economic efficiency was arrived at based on calculation and comparison of unit power generation costs using different energy storage technologies (pumped storage, compressed air energy storage, and battery energy storage) in a wind–PV storage united power generation and storage system and independent power supply systems such as diesel engine generator sets. A dynamic economic scheduling model for minimizing the total operation of a microgrid was conducted, and dynamic programming algorithms were applied to get the best storage size and economic efficiency of ESSs in microgrids with wind power generators, PV power generators, fuel cells, diesel engine sets, and battery storage systems.

ESSs at the user side are mainly used to regulate load to save power charge and guarantee uninterrupted power supply. Some documents focus on return on investment, including reduction of users' expenditure in power consumption and capacity charge, seeking best energy storage size and contracted capacity through a programming model based on multistep dynamic programming and rules from an expert knowledge base. Some documents aim to maximize the net revenue generated by saved power charge due to the load regulation of ESSs. Simulating optimization of storage size and operation strategies was done through dynamic programming. Current studies mainly focus on ESSs' function in reducing power charge, rarely paying attention to the value of ESSs in reducing users' outage cost by guaranteeing uninterrupted power supply. Therefore, these studies have not fully revealed the actual value of ESSs at the user side.

5.3.2.1 Economic terms

5.3.2.1.1 Discounted present value

Discounted present value is the current worth of a future sum of money or stream of cash flow given a specified interest rate. Compound interest is calculated on basis of a principal that equals the sum of the previous year's principal and interests. The sum of the principal and interest in the nth year should be the following:

$$A = A_0(1 + r)^n \tag{5.11}$$

Where A_0 is the principal; r is the annual rate of interest; n is the year; A is the sum of principal and interests.

Therefore, we have this:

$$A_0 = \frac{A}{(1 + r)^n} \tag{5.12}$$

Where A_0 refers to discounted present value; A refers to earning in a future year; n refers to year; r refers to discount rate; $(1 + r)^n$ refers to compound discount rate.

The formula in Eq. (5.11) is used to calculate the discounted present value as future returns on the investment by the investor.

5.3.2.1.2 Equivalent annual value

Generally, the return on investment does not only show up once a year. Instead, it lasts for a number of years, with a certain amount of revenue received each year. This is called payback flow, which should be calculated as follows:

$$A_0 = \frac{A_1}{(1+r)} + \frac{A_2}{(1+r)^2} + \cdots + \frac{A_n}{(1+r)^n} \tag{5.13}$$

Where $\frac{A_i}{(1+r)^i}$ is the discounted present value in the ith year; $(i = 1, \quad 2 \cdots n)$; A_0 is the present value of payback flow.

If $A_1 = A_2 = \cdots A_n = A$, this is called equivalent annual value, which should be calculated as follows:

$$A_0 = \frac{A}{r}\left[1 - \left(\frac{1}{1+r}\right)^n\right] \tag{5.14}$$

Where A_0 is the equivalent annual value; A is the annual earnings; r is the rate of interest for discounted prevent value; n is the year;

$\frac{1}{r}\left[1 - \left(\frac{1}{1+r}\right)^n\right]$ is the present value factor for equivalent annual value.

5.3.2.1.3 Break-even point

Break-even point, also called zero profit point, is the point of balance between making either a profit or a loss. It can be represented by total earnings or unit earnings.

5.3.2.2 ESS cost calculation methods

5.3.2.2.1 Total investment cost

The cost of an ESS consists of two parts: (1) cost of fixed equipment, including energy storage equipment, PCS, and auxiliary facilities; (2) operation and maintenance cost.

The cost of fixed equipment and auxiliary facilities is related to energy storage, while the cost of PCS is related to the peak power value for discharge. The total investment cost should be calculated as follows:

$$C_{\text{system}} = \frac{C_{\text{energy}} E_{\text{system}}}{\eta} + C_{\text{power}} P_{\text{system}} \tag{5.15}$$

Where C_{system} is the unit energy storage cost of the system, RMB/kWh; E_{system} is the total energy storage of the system, kWh; η is the charge and discharge efficiency of the system; C_{power} is the cost of power of energy conversion system, RMB/kW; P_{system} is the power of the system, kW.

Established energy storage power stations should be overhauled and maintained regularly to reduce equipment faults and guarantee operation of the battery packs. The operation and maintenance costs are related to the size of the power station, the type of batteries, the charge and discharge operations, and the battery aging, among other factors. Here, the formula for calculating the costs is simplified as follows:

$$C_{\mathrm{OM,s}} = C_{\mathrm{o}} P_{\mathrm{system}} \left| C_{\mathrm{m}} P_{\mathrm{system}} \right. \tag{5.16}$$

Where C_{o} is the annual unit power operation cost of the system, RMB/(kW·year); C_{m} is the annual unit power maintenance cost of the system, RMB/(kW·year).

The costs of ESSs are related to application scenarios. Electric Power Research Institute has carried out surveys and studies on the costs of existing ESSs in three application scenarios, respectively, for large-capacity energy storage, distributed energy storage, and improvement of power quality. Please refer to Tables 5.3–5.5 for the results.

5.3.2.2.2 Annual average cost in life cycle

Equivalent annual cost present value should be calculated as follows:

$$A = A_0 \left[r \frac{(1+r)^n}{(1+r)^n - 1} \right] \tag{5.17}$$

Where A_0 is the total investment cost in life cycle; r is the discount rate; n is the service year.

TABLE 5.3 Cost of large-capacity ESSs.

Type	Energy cost ($/kWh)	Power cost ($/kW)	Supporting facilities ($/kWh)	Efficiency % (AC/AC)	Cost of regular operation and maintenance [S/(kW·year)]
Flooded lead–acid battery	150	125	150	0.75	15
VRLA battery	200	125	150	0.75	5
Sodium–sulfur battery	250	150	50	0.7	20
Compressed -air energy storage	3	425	50	0.73	2.5
Pumped storage	10	1000	4	0.75	2.5

TABLE 5.4 Cost of distributed ESSs.

Type	Energy cost ($/kWh)	Power cost ($/kW)	Supporting facilities ($/kWh)	Efficiency% (AC/AC)	Cost of regular operation and maintenance ($/kWh)
Flooded lead–acid battery	150	175	50	0.75	15
VRLA battery	200	175	50	0.75	5
Zinc–bromine battery	400	175	0	0.60	20
Sodium–sulfur battery	250	150	0	0.7	20
Lithium-ion battery	500	175	0	0.85	25
All vanadium redox flow battery	600	175	30	0. 7	20
Ground compressed air energy storage	120	550	50	0.79	10
High-speed flywheel	1000	300	0	0. 95	1000

5.3.2.3 *Methods for calculating economic efficiency of ESSs for a single purpose*

5.3.2.3.1 Arbitrage

Both users and power generation companies can benefit from the ESSs as they function to purchase power at load valley and low price and discharge power at load peak and high price, helping to reduce consumption of high-priced power. Corresponding annual revenue can be calculated as follows:

$$R_1 = n\sum_{i=1}^{24}\left(P_{\text{discharge},i} - P_{\text{charge},i}/\eta\right)e_i \tag{5.18}$$

Where $P_{\text{discharge},i}$ is the discharge power of ESS during i period; η is the efficiency of ESS; $P_{\text{charge},i}$ is the charge power of ESS during i period; e_i is the power price during i period; n is the ESS operation days.

TABLE 5.5 Costs of ESSs for power quality improvement.

Types	Energy cost ($/kWh)	Power cost ($/kW)	Efficiency % (AC/AC)	Cost of regular operation and maintenance ($/kWh)
Lead–acid battery	300	250	0.75	10
Lithium-ion battery	500	200	0.85	10
Superconducting magnetic energy storage	50,000	200	0.95	10
High-speed flywheel (150 kW × 15 min)	1000	300	0.95	5
High-speed flywheel (120 kW × 20 s)	24,000	333	0.95	5
Low-speed flywheel	50,000	300	0.9	5
Supercapacitor	30,000	300	0.95	5

5.3.2.3.2 Saving cost of new power plants

ESSs can reduce the need for new plants and lease of power capacity and therefore reduce related costs. Annual revenue thereof can be calculated as follows:

$$R_2 = C_{\text{margin}} P_E \tag{5.19}$$

Where C_{margin} is the annual present value of marginal cost of a new power plant, RMB/(kW·year); P_E is the demanded power of a new plant.

5.3.2.3.3 Auxiliary services

ESSs can provide a multitude of auxiliary services, including serving as on-line/offline standby power source, power source for accidents and reactive power support or for frequency regulation and black start. They can also be used to ensure the implementation of power generation plans and guarantee the fulfilment of power trading contracts. Revenue generated from ESSs' auxiliary services is closely related to the service type, discharge duration, power price, service duration and seasonal factors. Herein, they can be calculated as follows:

$$R_3 = C_{\text{service}} P_E \tag{5.20}$$

Where C_{service} is the annual present value of auxiliary Services, RMB/(kW·year); P_E is the demanded power for auxiliary services.

TABLE 5.6 Revenue of ESSs as power transmission and distribution support.

Type of revenue	Annual present value of revenue [\$/(kW·year)]
Expansion of power transmission capacity	13
Voltage support (cost of capacitor)	—
Low-frequency load switch-off (3 times/year)	11

5.3.2.3.4 Power transmission support

ESSs installed at the power transmission side can help reduce investment in traditional voltage support and equipment for controlling subsynchronous oscillation as well as improve voltage stability and increase revenue generated following the reduction of low-frequency load switch-off. Please refer to Table 5.6 for the annual present value of the revenue. The revenue should be calculated as follows:

$$R_4 = \sum_i C_{\mathrm{trans},i} P_E \tag{5.21}$$

Where $C_{\mathrm{trans},i}$ is the annual present value of revenue generated by ESS for single purpose, RMB/(kW·year); P_E is the demanded power for serving as added power transmission support.

5.3.2.3.5 Reducing demand for power transmission capacity

The ESSs can help reduce cost of power generation companies in power transmission services, including both operation and maintenance fees paid to power transmission companies. Annual revenue thereof, namely reduced cost, should be calculated as follows:

$$R_5 = \left(C_{\mathrm{trans,o}} + C_{\mathrm{trans,m}}\right) P_E \tag{5.22}$$

Where $C_{\mathrm{trans,o}}$ is the unit operation cost for added power transmission capacity, RMB/(MW·year); $C_{\mathrm{trans,m}}$ is the unit maintenance cost for added power transmission capacity, RMB/(MW·year); P_E is the maximum power for continuous operation of ESSs, MW.

5.3.2.3.6 Reduction of power transmission congestion

Cost of power transmission congestion can be calculated based on regional power grid. Annual revenue thereof should be calculated as follows:

$$R_6 = C_{\mathrm{trans,b}} P_E \tag{5.23}$$

Where $C_{\mathrm{trans,b}}$ is the unit cost power transmission congestion, RMB/(MW·year).

5.3.2.3.7 Delaying upgrading and renovation of power transmission and distribution facilities

An ESS installed in a power distribution network can serve as a power source for peak regulation. If the charge and discharge power of the ESS is larger than the difference between peak load and average load, and the ESS is able to level peak load during peak hours, its economic efficiency in reducing the cost of upgrading and renovation of power transmission and distribution facilities should be calculated as follows:

$$R_7 = dC_{\mathrm{d}}\eta(P_{\max} - P_{\mathrm{a}}) \tag{5.24}$$

Where d is the rate of depreciation of fixed assets for power distribution to users; C_{d} is the unit cost of power distribution system for users, RMB/MW; η is the efficiency of energy storage devices; $P_{\max}$ is the maximum load per day, MW; P_{a} is the average load power per day, MW.

5.3.2.3.8 Reducing end users' power consumption costs

ESSs installed at the user side can help reduce consumption of high-priced power by purchasing power at load valley and low price and discharging power at load peak and high price. Corresponding annual revenue should be calculated as follows:

$$R_8 = n\sum_{i=1}^{24}\left(P_{\mathrm{discharge},i} - P_{\mathrm{charge},i}/\eta\right)e_i \tag{5.25}$$

Where n is the ESS operation days per year.

5.3.2.3.9 Reducing end users' capacity cost

ESSs installed for large- and medium-sized users can help shift peak load, reduce users' power consumption load during peak hours, regulate reactive power, and reduce demand by power distribution stations for power distribution and transformation capacity. This leads to a reduction of monthly capacity cost for the users. Annual revenue thereof should be calculated as follows:

$$R_9 = \sum_{i=1}^{12} e_{\mathrm{r},i}\eta(P_{\max} - P_{\mathrm{a}}) \tag{5.26}$$

Where $e_{\mathrm{r},i}$ is the user's basic power cost on basis of maximum capacity, RMB (in ten thousand)/(MW·month).

5.3.2.3.10 Cut down losses related to reliability of power supply

Benefits thereof can be estimated based on two standards: (1) Annual power outage time; (2) value of power not in service (RMB/kWh). Economic efficiency thereof should be calculated as follows:

$$R_{10} = C_{\mathrm{interr}}tP_E \tag{5.27}$$

Where C_{interr} is the unit electrical energy losses caused by power outage; t is the annual power outage time; P_E is the power of ESS.

5.3.2.3.11 Reducing losses caused by power quality

It is hard to estimate the economic efficiency related to power quality. Generally, such economic analysis is made based on power quality events. For example, if a commercial or industrial user sees N times of power quality events, and the value generated by each event is C_{qual} (RMB/kW), the economic efficiency thereof should be calculated as follows:

$$R_{11} = C_{\text{qual}} N P_E \tag{5.28}$$

5.3.2.3.12 Revenue from time shifting of renewable energy power supply at on-grid prices during peak and valley hours

If the wind power is priced based on the power prices during peak and valley hours at the user side, the economic efficiency of the ESS thereof should be calculated as follows:

$$R_{12} = e_i\left(P_{\text{discharge},i} - P_{\text{charge},i}/\eta\right)\Delta t \tag{5.29}$$

Where e_i is the on-grid power price during i period; Δt is the periods of a day divided based on power price mechanism.

5.4 Analysis of energy storage efficiency in Chinese power market

5.4.1 Comprehensive economic efficiency of ESS at new energy power side

5.4.1.1 Reducing demand for standby capacity

Take a wind power plant as an example. When wind power increases its proportion in the power generation system, the impact of forecast error on grid operation will be expanded. In that case, positive and negative standby power should be added to the system. However, an ESS installed in the wind power plant can help regulate the output of the plant, shrink forecast error, and reduce the demand for standby capacity. Suppose the ESS can meet every demand for standby capacity at any time, for example, every 5 min (288 periods per day); its efficiency in reducing demand for standby capacity during the ith period should be calculated as follows:

$$R_{1,i} = e_i(1 - \sigma)P_{w,i} \tag{5.30}$$

Where e_i is the price of standby capacity during i period, RMB (in 10,000)/MW; σ is the reliability of forecast technology at wind power plant; $P_{w,i}$ is the output of wind power plant during i period.

5.4.1.2 Reducing capacity demand for renewable energy power transmission

In China, the distance between areas with rich wind resources and areas with high demand for power supply are wide. Large wind power plants require high-voltage channels for linking into the grid. This leads to high cost of power transmission channels that accounts for a large proportion of total investment in a wind power plant. The optical power transmission capacity of a wind power plant cluster is usually smaller than its installed capacity.

If the grid connection channel of a wind power plant is planned based on typical maximum daily output, an ESS can help control output fluctuation and briefly reduce the peak power, thus reducing demand for channel capacity and therefore saving related costs. The saved costs should be calculated as follows:

$$R_2 = K\left[\max\left(P_{w,i}\right) - \max\left(P_{w,i} + P_{\text{discharge},i} - P_{\text{charge},i}\right)\right] \tag{5.31}$$

Where K is the unit channel construction cost, RMB (in 10,000)/MW; $P_{w,i}$ is the typical maximum daily output of wind power plant.

5.4.1.3 Arbitrage through peak and valley prices

Wind power generation systems' discordancy with the grid during peak and valley hours will result in shutdown and low-output operation of thermal power generating systems during these hours and thus lead to increase of coal consumption and significant reduction of economic efficiency. Moreover, the connection of large-scale wind power with the grid has an impact on the grid's security. Therefore, connection between wind power and the grid is restricted.

When fixed on-grid price mechanism is applied, ESSs, when used to regulate power supply in peak and valley hours, cannot bring benefits to the wind power plant operators or encourage power generation companies to take technical measures to improve power quality. If reasonable adjustable peak and valley prices can be applied, wind power plants can be guided to partake in the peak regulation that will alleviate the pressure for peak regulation at the power source and grid sides and beget optimization of the power generation systems in the grid. On the other hand, impact of wind power on the grid can be reduced and security of the grid can be better ensured. This will increase the uptake of wind power by the grid. If wind power is priced based on the power prices during peak and valley hours at the user side, the economic efficiency of the ESS thereof should be calculated as follows:

$$R_3 = e_i\left(P_{\text{discharge},i} - P_{\text{charge},i}/\eta\right)\Delta t \tag{5.32}$$

Where e_i is the on-grid power price during i period; Δt is the periods of a day divided based on power price mechanism.

5.4.2 Comprehensive economic efficiency of ESSs at power distribution network side

5.4.2.1 Reducing demand for capacity expansion of the grid

When the load in a power distribution network exceeds its capacity, the transmission network needs to be upgraded or expanded, traditionally by upgrading or expanding the transformer at the substation and the power transmission and distribution lines. An ESS installed at the spot where load exceeds capacity can delay such upgrading and distribution and thus save costs. It helps the grid operator reduce total investment as well as improve equipment utilization efficiency and lower investment risks. Usually, ESSs can be applied under the following circumstances:

1. When slight and short-term overload appears;
2. Load increases slowly;
3. It is expensive to upgrade the power distribution network, but small-capacity ESS can help delay the investment;
4. Application of conventional renovation methods is restricted, for example, no circuit corridor.

Power transmission and distribution capacity of a power grid is generally set based on maximum local load demand. So, load power may be greatly smaller than the grid capacity during valley hours and approach the rated capacity of the grid during peak hours, and an overload may appear at the transformer substations and on some transmission and distribution lines. The ESS installed at the power distribution network serves as a power source for peak regulation. Its regulation ability is determined by its power and energy. If the charge and discharge power of the ESS is larger than the difference between peak load and average load, and the ESS is able to level peak load during peak hours, its economic efficiency in reducing the cost of grid expansion should be calculated as follows:

$$R_1 = dC_d\eta(P_{max} - P_a) \tag{5.33}$$

Where d is the rate of depreciation of fixed assets for power distribution to users; C_d is the unit cost of power distribution system for users, RMB/MW; n is the efficiency of energy storage devices; P_{max} is the maximum load per day, MW; P_a is the average load power per day, MW.

5.4.2.2 Reducing total grid loss

The ESS charges power during the load valley hours and discharges during the load peak hours during the day. In charging, the system's load increases, followed by an increase in grid loss. In discharging, the current through the main distribution line reduces, followed by lowered lines and transformer losses, and thus a minimal grid loss. Considering that electric energy loss is in direct proportion to the squared current value, within certain capacity range, the reduction of grid loss caused by the ESS is bigger than the increase of grid loss caused by it. Moreover, the high power price during peak hours leads to high unit cost of grid loss. Therefore, ESS can play a considerable role in reducing grid loss and saving related costs. Corresponding revenue should be calculated as follows:

$$R_2 = n\sum_{i=1}^{24}\frac{\left[P_i^2 + Q_i^2 - \left(P_i + P_{discharge,i} - P_{charge,i}\right)^2 - \left(Q_i + Q_{discharge,i} - Q_{charge,i}\right)^2\right]Re_i}{U^2} \tag{5.34}$$

Where P_i, Q_i is the active and reactive power of load during i period; $P_{charge,i}, Q_{charge,i}$ is the active and reactive power absorbed by the ESS during i

period; $P_{\text{discharge},i}$, $Q_{\text{discharge},i}$ is the active and reactive power provided by the ESS during i period; e_i is the power price during i period; n is the equivalent resistance from superior transformer substation to ESS location.u is the voltage at access of ESS.

5.4.2.3 Arbitrage

With increasingly enlarged difference between system load peak and valley, to some extent, the power prices during peak and valley hours reflect the difference between power generation costs. However, ESSs can be used during these hours to guarantee economically efficient operation of power generation units with higher economic benefits. ESSs of certain size can help improve the efficiency of power generation units during these hours. For example, power generated using each ton of coal can rise by 50–150 kWh, even considering the cycle efficiency of ESSs. Since ESSs charges power at load valley and low power price and discharge at load peak and high price, grid companies can gain economic benefits by buying low-price power and selling it at a higher price. However, in China, such economic benefits are not available to the grid companies, because there are no peak and valley prices. Economic benefits of this kind can only be evaluated based on the peak and valley price system for user price. Moreover, for power stations, in addition to economic benefits, wide application of ESSs can also help power stations reduce emission and improve efficiency without causing any change in power generation capacity. This perfectly matches with the country's energy policies and can generate remarkable social benefits. Annual revenue thereof should be calculated as follows:

$$R_3 = n\sum_{i=1}^{24}\left(P_{\text{discharge},i} - P_{\text{charge},i}/\eta\right)e_i \tag{5.35}$$

5.4.2.4 Reducing standby capacity

1. Standby capacity for distributed power generation. Following the increasing attention paid to energy saving and emission reduction, distributed new energy power generation (mainly including wind and solar power generation) has witnessed rapid development. Distributed power generation units are usually small-capacity units that are deployed in the power grid or near the load, designed to meet the load demand of specialized users or complement the grid for economic efficiency. Among them are micro gas turbines, fuel cells, and power generation units driven by renewable energy such as PV and wind resources. Power generation units using renewable energy feature obvious intermittency and fluctuation, with low power supply reliability in case of independent operation. However, when the units are incorporated into the grid, they play an important role in regulating the grid capacity. Even so, their fluctuation

may still cause disturbance to the power grid, especially when they are widely used. In that case, standby power sources are needed for fast capacity regulation. ESSs installed in the grid can serve as standby capacity as well as improving the power system safety. If ESSs can charge and discharge evenly, their expected capacity for power regulation is 0.5 P_{max}T, and revenue thereof should be calculated as follows:

$$R_4 = 0.5P_{max}Te_s \tag{5.36}$$

Where P_{max} is the maximum power of ESS for continuous discharge; T is the duration of continuous discharge of ESS; e_s is the price of standby capacity, RMB (in 10,000)/(MW·year).

2. Standby capacity to guarantee reliability of power grid. Cost for guaranteeing the reliability of the power grid refers to extra investment made to guarantee a certain amount of power supply and related operational costs. Revenue from power grid reliability can be defined as revenue for users when power supply reaches a certain level. Herein such revenue is calculated based on user's economic losses caused by insufficient power supply, shortage of power due to power supply interruptions, and blackouts. At constant unit power shortage cost, the lower the cost is, the higher the revenue is. ESSs installed in the distribution network can help improve reliability of power supply and therefore save related investment. It should be calculated as follows:

$$R_5 = (E_1t + E_2N)P_E + E_3 \tag{5.37}$$

Where E_1 is the loss per unit period caused by power supply interruption, RMB/kWh; t is the annual interruption duration, h; E_2 is the loss caused by each power quality accident, RMB/(kW·time); N is the number of power quality accidents each year, time/yr; E_3 is the present value of annual cost of standby power source construction; P_E is the ESS's power, kW.

5.4.3 Comprehensive economic efficiency of ESSs at user side

5.4.3.1 *Reducing the need to construct new distribution stations for users*

The capacity of distribution systems is usually determined by maximum load. In addition, for users that require higher power supply reliability, the distribution systems' redundancy should be increased. However, ESSs installed at the low-voltage side of the distribution systems can help fulfill power supply demand during peak hours and reduce the required capacity of the distribution systems and therefore save related investment.

Equivalent annual revenue present value thereof should be calculated as follows:

$$R_1 = dC_d\eta(P_{max} - P_a) \tag{5.38}$$

Where d is the rate of depreciation of fixed assets for power distribution to users; C_d is the unit cost of power distribution system for users, RMB/MW; η is the efficiency of energy storage devices; P_{max} is the maximum load per day, MW; P_a is the average load power per day, MW.

5.4.3.2 Reducing cost of power distribution and transformation losses

ESSs charge power during valley hours, increasing the load rate of power distribution and transformation as well as the users' power distribution and transformation losses during valley hours; when the ESSs discharge power, the load rate of power distribution and transformation is lowered, so are the users' power distribution and transformation losses during peak hours. Annual revenue thereof should be calculated as follows:

$$R_2 = n\sum_{i=1}^{24}\frac{\left[P_i^2 - \left(P_i + P_{discharge,i} - P_{charge,i}\right)^2\right]P_k e_i}{\left(S_d \cos\varphi\right)^2} \tag{5.39}$$

Where P_i is the load power during i period; $P_{charge,i}$ is the ESS' charging power during i period; $P_{discharge,i}$ is the ESS' discharging power during i period; e_i is the power price during i period; P_k is the power distribution and transformation losses due to short circuit; S_d is the power distribution and transformation capacity; u is the voltage at access of ESS; cos φ is the power factor at load side of the transformer.

5.4.3.3 Reducing users' power consumption

The bigger the difference between the peak and valley prices is, the bigger the difference between the users' expenses in power during the peak and valley hours. Therefore, with the ESSs, the users can buy lower price power during valley hours and use it during peak hours. This will assist the users to reduce power consumption costs. Annual revenue thereof should be calculated as follows:

$$R_3 = n\sum_{i=1}^{24}\left(P_{discharge,i} - P_{charge,i}/\eta\right)e_i \tag{5.40}$$

Where n is the ESS annual operation days.

5.4.3.4 Reducing users' basic power expense

China currently adopts two power price mechanisms, namely, single price mechanism and two-part price mechanism. The latter is mainly applied to large industrial users, consisting of a basic power price, consumption price, and regulated price based on power factor. Basic power price is fixed based on user equipment capacity (kVA) or user's maximum demanded capacity (kW). It is only related to capacity, having nothing to do with actual consumption.

Large- and medium-sized users who have applied for special power distribution and transformation should pay a basic power charge according to the maximum demanded capacity. ESSs can help reduce users' power consumption during peak hours, regulate reactive power of load, as well as reduce demanded distribution and transformation capacity. This leads to a reduction in the monthly payment of the users for power capacity. Revenue thereof should be calculated as follows:

$$R_4 = \sum_{i=1}^{12} e_{r,i}\eta(P_{max} - P_a) \tag{5.41}$$

Where $e_{r,i}$ is the basic power expense of users according to maximum demanded capacity, RMB (in 10,000)/(MW·month).

5.4.3.5 Reducing losses caused by power supply reliability and power quality accidents

Users' losses due to blackouts include direct and indirect losses. Direct losses refer to actual losses caused by blackouts, while indirect losses refer to extra expenses spent by the users in reducing or avoiding occurrences of such losses, for example, installation of standby power source. Revenue thereof should be calculated as follows:

$$R_5 = C_1 t P_E + C_2 P_E N + E_3 \tag{5.42}$$

Where C_1 is the unit loss caused by power supply interruption, RMB/kWh; t is the annual power supply interruption duration; C_2 is the unit loss caused by each power quality accident, RMB/(kW·time); N is the times of power quality accidents each year, times/yr; E_3 is the present value of annual cost of standby power source installation.

The aforementioned calculation methods cover all potential comprehensive economic benefits that can be generated by ESSs. However, specific evaluation should be made according to the operation modes of the ESSs.

References

[1] Xue J, Ye J, Wu F, et al. Application and development of energy storage monitoring system in smart grid. Electrotechnical Application 2012;21:53–9.

[2] Xue J, Ye J, Zhang Y, et al. Battery Module Technology and Application in ESS. 11. 2013. p. 1944–7.

[3] Wu F, Yang B, Ye J, et al. Application and typical design of large - capacity battery ESSs. In: The 13th annual meeting of China association for science and technology; 2011.

[4] Ju J, Wu F, He W, et al. Structural design of energy storage monitoring system. In: The 13th annual meeting of China association for science and technology; 2011.

[5] Che Z. Analysis and research of battery pack connection modes. Mechanical and Electrical Technology 2013;(2):109–11.
[6] Lei J, Jiang X, Xie J. Current development of equalizer circuit in Li - ion battery packs. Battery 2007;(1):62–3.
[7] Li S. Research of equalization technology for power lithium battery packs. Beijing Jiaotong University; 2011.
[8] Wang Z, Sun F, Cheng L. Analysis on impact of inconsistency on power battery packs. Transaction of Beijing Institute of Technology 2006;(7):577–80.
[9] Qiu M. Study on several issues concerning current collection systems in power stations with energy storage system. Hangzhou: Zhejiang University; 2012.
[10] Wang Zhao'an, Huang J. Power electronics technology (4th edition). Beijing: China Mechanical Industry Press; 2000. p. 100–65.
[11] Zhang F. Study on Bi - directional DC - DC converters. Nanjing: Nanjing University Aeronautics and Astronautics; 2004. p. 5–10.
[12] Tong Y, Wu T, Jin X, et al. Topology research of Bi - directional DC - DC converters. Proceedings of the Chinese Society for Electrical Engineering 2007;27(13):81–6.
[13] Mears LD, Cotschau HL, key T, et al. America; electric power research Institute and department of energy. Energy Storage for grid connected wind generation applications; 2004.
[14] Yan Z. Estimation and research of value of battery energy storage systems in smart power grids. Shanghai Jiaotong University; 2012.
[15] Bi D, Ge B, Wang W, et al. Power control for grid - connected wind power plants with vanadium - battery-based energy storage systems. Power System Automation 2010;34(13):72–8.
[16] Hadi B, Luo C, Boon TO. Impacts of wind power minute - to - minute variations on power system operation. IEEE Transactions on Power Systems 2008;23(1):150–60.
[17] Yuan T, Qin C, Li Y, et al. Research of economic dispatch opti- mization models based on wind power penetration limits. Power System Protection and Control 2011;39(1):15–22.
[18] Zhao S, Zhang D, Yin Y, et al. Policies on wind power price and risk control strategies. Power System Technology 2011;35(5):142–5.
[19] Li H, Duan J, Wang C. Battery-based energy storage technology for smart grid and its value evaluation. China Machine Press; 2012.

Chapter 6

Application of energy storage technology in grid-connected new energy power generation

Chapter outline

Grid-scale Energy Storage Systems and Applications. https://doi.org/10.1016/B978-0-12-815292-8.00006-X

Due to weather and geographic influence, the power output of new energy power generation is characterized by great volatility and unpredictability, exerting heavy influence on the power system's stability and quality during operation. With the rapid growth of grids connected to new energy power generation domestically that accounts for a larger percentage of the total capacity, it is necessary, therefore, to effectively find solutions to challenges experienced by large-scale grids that are connected to new energy power generation. It is conducive not only in boosting stability of the power system, but also improving the utilization of new energy power generation and the overall energy structure adjustment.

The energy storage devices play multiple roles in the new energy subsector. For example, they can restrict power fluctuation while meeting the requirements of the grid-connected new energy; they also improve quality of power generated and enhance reliability during operation. The controllability and adjustability of new energy power generation can be improved to provide stable power output, so it is of great importance to the dispatch control while making the grid safe and economical to operate.

6.1 Impact of energy storage system on grid-connected new energy power generation

6.1.1 Smooth power fluctuation

The increase of large-scale wind power and photovoltaic (PV) grid-connected generation is compromising the reliability and stability of the traditional power grid. The application of large-scale grid-connected new energy system has seen traditional unit's ramp rate unable to adjust to the sharp and short-term power fluctuations, hence limiting the new energy's output to the grid. However, the advancement of large-scale energy storage technology has enabled smoother power fluctuation, effectively alleviating pressure on the grid to regulate frequency in the peak, improving controllability, continuity, and stability of the new energy power generation, as well as improving the grid's compatibility with the new energy power generation system.

The charge/discharge of the energy storage system can control the output power of the new energy power generation. The energy storage device absorbs surplus output power through controlling the converter when the output power of the new energy soars. On the contrary, when the output power slumps, the energy storage device releases new power by controlling the converter. The principle of energy storage devices suppressing power fluctuation is similar to filters processing the signal. The low-flux filter processes the amplitude of input signal and makes it smooth. The coordinated operation between the storage system and power generation can easily identify change in the new energy power generation and adjust its power output to smooth the power fluctuation and ensure the system is operating safely and stably.

6.1.2 Reduce power system's demand for peak regulation capacity

The peak regulation is related to power structure and cost effective operation, heavily influencing the power system. It is the main factor that controls the compatibility of a power system to the new energy. The peak regulation, in essence, is for striking a balance between power generation and load demand in the power system. If the power system cannot effectively control the balance of supply and demand, it will affect the reliability of the electricity to the users, and it may result in a wide range of power accidents. Generally, the power system's peak regulation capacity must be greater than the peak—valley difference of the equivalent load.

The output power fluctuation of the new energy is often inconsistent with the system's load fluctuation, generally more violent. In addition, wind energy features the antipeak regulation. During the process of connecting large-scale grid to new energy power generation, the power system not only faces the risk of insufficient peak regulation capacity, but also a big challenge in its peak regulation ability. In this case, the system's demand for peak regulation capacity will be increased and the sufficient peak regulation units will be provided. If so, it will greatly increase investment in the construction of a generator.

Storage energy devices can store excessive new energy when the power system is at low load; on the contrary, it can release new energy when the power system is at high load. Therefore, storage energy devices are needed to reduce demand for spare peak regulation units and then avoid the difficulties caused by these units.

6.1.3 Trace new energy power schedule output

The new energy power plant that adopts the prediction technology as the output power is one of the important means of reducing difficulties experienced in the grid dispatch. The prediction for the power generated by new energy can provide data support for the power grid dispatch, effectively reducing the unpredictable impact of the grid-connected new energy and the operation cost of power system. However, predictive error is inevitable due to the influence of various factors such as weather, climate, and prediction technology. To boost safe and stable operation, it is necessary that the energy storage technology be employed to reduce prediction error and improve reliability of new energy output.

The energy storage system can trace the new energy power output schedule. It controls the input/output power of an energy storage system to make the output of new energy and the energy storage system close to the predictive curve for the purpose of improving the dispatch ability and credibility of new energy output power. When the difference between the real-time new energy power and schedule output power (predictive power) exceeds the permissible error, the energy storage system can absorb or release power to bridge the difference and maximize its ability to trace new energy schedule output. In brief, under a certain storage capacity, the hybrid application of storage energy system and new energy power generation can make the latter controllable and adjustable, and ensure the new energy output curve is close to the demand curve. In addition, it can compensate for predictive inaccuracy of new energy power generation without an energy storage system and reduce the risk of exceeding the limit of prediction error in extreme cases.

6.1.4 Regulate the power system's frequency and voltage

The power system requires power generation and consumption to always be balanced, and to maintain this, the ramp rate must be greater than the load rate every time. However, the load rate keeps changing, especially during the peak times. If the power system's generator unit cannot adapt to the rapid load change, it would lead to mismatch between load and power generation, affecting the quality of power supply. Meanwhile, the new energy output power is also random and intermittent, making it difficult to maintain system stability.

The energy storage system maintains the stability of a new energy generation system by improving the balance in the power grid frequency support, damping oscillation, inertia, voltage support, and other aspects. The energy storage system can quickly absorb or release active and reactive power to enhance stability of the power system. The energy storage system can be considered a spare unit. When the output of a new energy power generation system changes suddenly, the energy storage device absorbs or releases energy till the conventional power generation mode adjusts. In this case, the energy storage system only provides for a few minutes of energy, but it needs to respond readily to the change. Therefore, only the battery energy storage and the flywheel energy storage with rapid response would satisfy the requirements. In addition, the energy storage system can deal with a variety of changes in the operation of power grid, effectively replacing other devices used to adjust power. In this case, the storage energy system features larger capacity and longer cycle life. Therefore, the large compressed air storage, the pumped storage, and the large-capacity battery storage can meet requirements.

Currently, most of the new energy power plants in our country are built at the end of the power grid; that is to say, they are vulnerable. The generator unit not only generates active power but also absorbs the grid's reactive power. However, fluctuation in output power can cause a similar reaction in input reactive power. With the grid voltage declining, the grid will reduce its compensation for the reactive power, and the capacitor installed at the end of the generator unit will reduce its compensation for reactive power. As a result, the increasing demand for the grid's reactive power will lead to grid voltage fluctuation, flicker, and unbalance, and even collapse. Finally the generator unit will be forced to stop operation because of low voltage protection. The large-scale grid-connected new energy will have an adverse effect on the voltage quality and stability of the local power system. In the case of dipping grid voltage, the energy storage system can come in to provide power to the grid. On the contrary, in the case of power surge, the grid will absorb some reactive power, enabling the voltage to quickly recover to normal levels.

6.2 Design of an energy storage system in a new energy grid-connected power generation system

6.2.1 Storage energy system's configuration method

The configuration of the storage system in the new energy grid is divided into two modes: distributed and centralized configuration. The configuration methods are widely applied in wind farms. The distributed configuration is applied on the excitation DC link of a wind turbine or on the output terminal of each wind turbine. The centralized configuration is applied on the outlet of the wind farm. Figs. 6.1 and 6.2 show these two configuration methods.

As shown in Fig. 6.1, the distributed configuration means an energy storage system is added to the DC link of the wind turbine. The configuration suppresses the output power of each wind turbine to achieve the control of all the wind turbines. There are two configuration methods. One method is to take advantage of the original grid-side converter. Another is to use an AC/DC converter to configure an energy storage system on the output terminal of a wind turbine, achieving the power decoupling control for the AC/DC converter and the smooth control of output power from the wind turbines. The former does not need the power converter configured, but it changes the original control mode of the grid-side converter, increasing the control complexity of the wind turbine. In the latter method, an additional AC/DC power converter is installed, thus adopting an independent energy storage control system without changing the structure and the control mode of the wind turbine. It is more convenient and flexible.

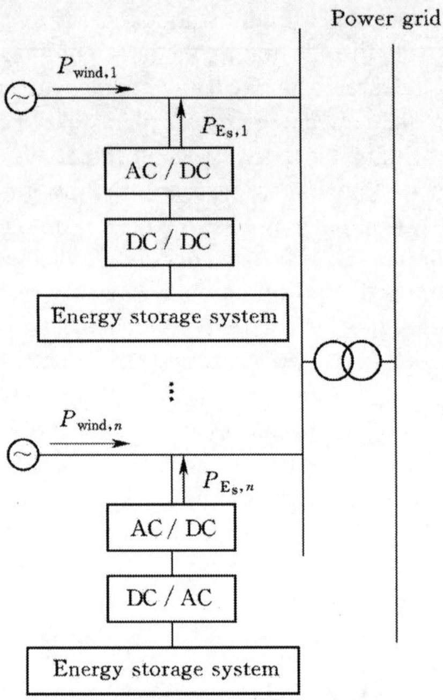

FIGURE 6.1 Distributed configuration.

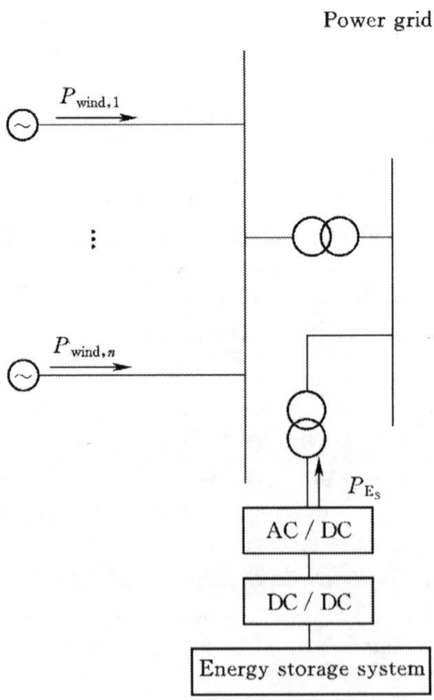

FIGURE 6.2 Centralized configuration.

As shown in Fig. 6.2, the centralized configuration means that an independent energy storage system is configured at the exit bus of the wind farm, regulating and controlling the output power of the whole farm. The method is to control and adjust the output power from a macro perspective.

In theory, the listed two configurations aim at suppressing and controlling the grid-connected power fluctuation of the wind farm. Their advantages and disadvantages are as follows:

First, the distributed configuration programs the energy storage system on the original excitation converter's DC link to exploit existing resources. It saves the investment cost without purchasing an additional converter, but it changes the original excitation control scheme of the wind turbine and needs manufacturers to provide technical support practically. The centralized configuration does not change the existing control structure of the wind turbine, and it only needs to be relatively independent of the energy storage system that is connected to the exit bus of wind farm through an AC/DC converter to control the wind grid-connected power fluctuation.

Second, the distributed configuration is aimed at adjusting and controlling power of each wind turbine, so power and capacity of each storage system is small. The centralized configuration aims at adjusting and controlling the power of the farms, so the energy storage system boasts of larger power and capacity. So far, in addition to pumped storage hydro technology, other larg-scale energy storage technologies that are expensive are yet to be mature.

Third, large wind farms have several sets of wind turbines that occupy a large area and are placed at different locations. In addition, under the influence of geographic conditions and internal wake effect of the wind farm, the output power fluctuation of each wind turbine is different, so there is a certain harmonization between each wind turbine. If so, to a certain extent, the output power fluctuation of the farm can be decreased, meaning the smooth characteristic of several sets of wind turbines can decrease the power output fluctuations and reduce the demand for the energy storage system's capacity. Therefore, with the scaling up of wind farms, the capacity ratio of the energy storage system and the farm is on the decrease. If energy storage technology and wind farm's scale meets requirements, the required capacity of centralized configuration is less than that of the distributed configuration.

Fourth, in terms of operation and maintenance, if the distributed configuration was employed, the fault or overhaul of wind turbines will affect the operation of the storage system. This is contrary to what will happen when the centralized configuration is adopted. In addition, the distributed configuration needs an energy storage system installed on each wind turbine, greatly increasing the workload of the system in operation and maintenance, but it reduces the reliability and cost effectiveness of the whole system.

In brief, with the development of power electronic devices, high-power converters and large-scale energy storage technology are becoming mature, so the application of the latter, based on the centralized configuration, is more advantageous in the grid-connected new energy power generation.

6.2.2 Technical/economic analysis of energy storage system

Without considering the application of pumped storage technology and compressed air energy storage technology in new energy grid-connected power generation, the section mainly analyses the technologies and economy of the battery flywheel energy storage system. The system with a short construction period is flexible and easy to be installed.

Section 5.2.1 gives a detailed introduction of energy storage technology parameters in the application of new energy side from different aspects, such as power level, continuous discharge time, response time, and cycle life. In addition, the costs and benefits of an energy storage life cycle should be analyzed, and suitability should be chosen depending on both technical and economic aspects.

6.2.2.1 Energy storage technologies applied for new energy grid-connected power generation

Presently, the main energy storage technologies used in wind farms include lead–acid, lithium-ion, all vanadium liquid, and sodium–sulfur battery, together with flywheel energy storage technologies.

The lead–acid battery with moderate specific energy, high reliability, mature technology, and less costly is widely used in wind farms. Compared with the lead–acid battery, the lithium-ion battery boasts of higher energy density, more cycle times, and higher energy conversion efficiency. However, its costs are higher, so if it is widely used in wind farms, its technology/economy is still needed for further testing. The vanadium redox flow battery with high response ability and high cycle times can flexibly expand its capacity and deeply charge/discharge, having great advantages in controlling the wind farm power fluctuations. Currently, dozens of sets of the vanadium redox flow battery energy storage system in Japan have been operating under the demonstration operation stage. For example, in Hokkaido, the 6-MV vanadium redox flow battery energy storage system has been configured to the 30-MW wind farm for peak frequency regulation. The sodium–sulfur battery energy storage technology has a long, continuous discharge and high cycle time, shifting energy in time and trace wind power schedule output. The MMW sodium–sulfur battery energy storage system has advanced into the demonstration application stage. The sodium–sulfur battery, the world's largest energy storage system, has been deployed to a wind farm in northeastern Japan. The system uses 34-MW sodium–sulfur battery to mesh with a 51-MW wind farm, and the power that stabilizes the system to operate is 42 MW. Flywheel energy storage technology boasts of high conversion efficiency, rapid response speed, high power density, and long cycle life. In addition, its storage life cannot be affected by the degree of discharge, which is not long, and is more suitable for controlling the frequent wind power fluctuations. When the flywheel energy storage system is applied to a wind power generation system that is operating independently, in the case of wind power fluctuations and load disturbance, it can quickly deliver or absorb active power to balance power generation and consumption and significantly improve the efficient utilization and the quality of the power supply.

6.2.2.2 Battery energy storage technological/economic analysis case

The following aspects should be taken into consideration during the design and application process of a large-scale sodium—sulfur battery energy storage system.

1. The selection of the module size and the charge/discharge pulse coefficient of the sodium—sulfur battery should match the minimum discharge voltage of the battery.
2. The state of charge (SOC) management makes certain that the battery can acquire power and capacity in real time, and it should be charged at the best state. The minimum value of a battery's SOC should be limited to 30%, ensuring the battery performs optimally in the expected life cycle within many days.
3. The thermal management system should regulate the core temperature of the sodium—sulfur battery within the allowable range, and the thermal loss rate should meet the application requirements.
4. The life cycle of the device should be within the service life. To meet various functions, long cycle life is important. The application of energy storage system not only meets the load transfer, but also power regulation.

Take the application of the 10-MW/65-MWh sodium—sulfur battery energy storage system to reduce the gap between wind power prediction and actual output as an example. Its performance parameters' selection and economic analysis results are introduced as follows.

The typical technical parameters of a10-MW sodium—sulfur battery energy storage system are shown in Table 6.1. According to current production and standard of the sodium—sulfur battery, 200 battery modules with 50 KW can be integrated. The charge/discharge voltage range of the sodium—sulfur battery energy storage system is from 930 to 1550 V, and the system's net efficiency, battery's standby efficiency, and power control systems (PCS) efficiency are 89.1%, 96.1%, and 98.8%, respectively, and the maximum discharge depth of the sodium—sulfur battery is 90%.

TABLE 6.1 The typical parameters of the sodium–sulfur battery energy storage system.

Applications	Single function: reduce the difference between predictive output and the actual output of wind power
Type	
Battery module type	50 kW/65 kWh
Module number	200
Pulse coefficient	1.0
Maximum charge voltage (V)	1550
Minimum discharge voltage (V)	930
Maximum discharge depth	90%
Life cycle (year)	>15
System data	
Net efficiency (%)	89.1
NaS battery system's standby efficiency (%)	96.1
PCS standby efficiency (%)	98.8

The cost-benefit analyses of a 10-MW/65-MWh sodium–sulfur battery energy storage system in the whole life cycle are introduced.

According to Table 6.2, the costs of a 10-MW/65-MWh sodium–sulfur battery energy storage system include initial cost of PCS, battery and balance of plant, operation and maintenance costs, and postprocessing costs. There are two methods of calculating the total costs. One way is to calculate the power costs of PCS, balance of plant (BOP), and battery; the other way is to calculate the various power costs of PCS and BOP as well as the energy cost of the battery.

In calculating the cost-benefit of the life cycle, the market environment should be considered to determine the required economic parameters. Suppose the energy storage system's life cycle is 15 years, and its discount rate is 7.5%. Its asset tax and insurance premiums are 2% each year, and its annual fixed rate is 9.8%. According to the calculation method of present-worth factor (PWF), its PWF is 8.99.

The energy storage system used in reducing the difference in wind power between predictive and actual output is in essence reducing penalties in the case

TABLE 6.2 The costs of 10 MW/65 MWh of the sodium–sulfur battery energy storage system.

Application	Single function: reduce the difference between predictive output and actual output of the wind power
Battery capacity (MWh)	65
PCS initial cost ($/kW)	239
Balance of plant (BOP) initial cost ($/kW)	100
Battery initial cost ($/kW)	1382
Battery initial cost ($/kWh)	213
Total cost (M$)	17.2
Fixed operation and maintenance cost [$/(kW·year)/	39.4
Variable operation and maintenance cost [$/(kW·year)/	16.9
Net PWF of processing cost ($/kW)	25.4

of inaccurately predicted output. Based on engineering experience, assume the difference is $400 (kWh·year), the following value can be calculated:

Total costs of energy storage system

= initial costs + operation and maintenance costs + postprocessing cost = (battery initial cost + PCS initial cost + BOP initial cost) * total power + (fixed operation and maintenance cost + variable operation and maintenance cost) * total power * present-worth factor + (net present-worth factor of postprocessing cost) * total power

The total benefits of energy storage system

= annual value of unity power * total power * present-worth factor in calculation

The total costs and total benefits of a 10-MW sodium–sulfur battery energy storage system used in bridging the difference between wind power prediction and the actual output are $24.8 million and $23.3 million. The net present-worth factor is $1.5 million, and the ratio between benefits and costs is 0.94.

The preceding assessment is based on the biding market price and penalties in the power market of wind farms. Notably, the net present-worth factor is significantly affected by the cost of an energy storage system. When the price of a 50-kW sodium—sulfur battery module falls from $68,000 to $62,000, the total costs and benefits are balanced.

6.2.3 Configuration of energy storage system capacity

The National Development and Reform Commission of the People's Republic of China in 2011 pointed out that, by 2020, China's installed wind and solar power capacities will reach 200 and 50 GW. Using the 20% of installed capacities to configure energy storage capacity, renewable energy power generation has 50 GW demand in the energy storage market. Presently, a high cost is one of the main factors restricting the large-scale application of energy storage systems. The optimal capacity configuration of the storage system needs to satisfy the safe and stable operation of wind power grid-connected generation and the optimization operation of the energy storage system.

6.2.3.1 Factors that affect capacity configuration of an energy storage system

In the new energy generation system, the factors affecting configuration of storage system capacity include the characteristics of new energy and the energy storage system's access mode, as well as technical characteristics such as efficiency and allowable charging/discharging depth. This section focuses on the external influence factors such as new energy's characteristics and evaluation methods.

1. The characteristics of new energy power generation. The new energy power generation includes PV, wind, biomass, and geothermal power generation. Each system has different characteristics. For example, PV power generation only generates power during the day, and compared with wind power generation, most of its power fluctuations are at low frequency; Wind power generation can generate power day and night, but the difference between day and night is very large, and the frequency fluctuations are high. For the wind/light integrated power generation system, due to its own dynamic complementary balance characteristics, its power fluctuations are less than that of a single power source. Therefore, the type of energy storage system should be selected according to the fluctuation characteristics of different new energy as well as local geographic and climatic conditions.
2. New energy generation data. The larger in seasonal time span and the more extensive in time distribution the new energy generation data, the better the authenticity and the stronger the representativeness are. In addition, predictive accuracy of new energy power generation also directly influences the power/energy configuration of energy storage system.

3. Access of the energy storage system. According to researches, when the scale of wind power generation gradually increases from single wind turbine to wind farm and then clusters, the mutual effects among wind turbines are obvious. The bulk of output powers are at low frequency and smoother. Therefore, the power/capacity configuration of the energy storage system can be optimized by the energy storage system based on the centralized configuration in the wind farm (wind cluster).
4. Target optimization. In different applications, the access feature of the energy storage system aims at optimizing the different objectives of new energy generation. Presently, the main optimization objectives are to meet the requirements of the grid-connected access technology, minimum power output fluctuations, stabilize output according to smooth/predictive curve, and improve the utilization rate of new energy.
5. Evaluation standard. When targeting the same optimization target, there are various methods of configuring the power/capacity of the energy storage system, so the quantitative evolution on system's configuration effects are needed. The common evaluation standards include fluctuation rate, standard variances, symetric distribution characteristics of energy storage power, and minimal investment costs.

The different methods of selecting the energy storage system capacity are described in detail, based on the applications of an energy storage system on controlling the output power fluctuations of a wind farm and tracing the power schedule output.

6.2.3.2 The different configuration methods for smoothing new energy output power fluctuation

The first step is to combine the spectral analysis and low-flux filter. According to the spectrum analysis results for new energy power, the optimal cutoff frequency of the first-order low-flux filter within the range of frequency fluctuations is determined; namely, the target power that is capable of smoothing the fluctuation of the grid-connected generation system is gained. Later, the optimal power and capacity of the energy storage system can be calculated.

1. Optimal objective

The rules for maximum wind power rate that Technical Rule for Connecting Wind Farm to Power System (GB/T 19963− 2011) mentioned are as follows: the wind farm should limit the change rate of the output power. The maximum power change rates at 1 and 10 min are shown in Table 6.3.

Therefore, the optimal target for smoothing the power fluctuation of wind farm can be gained:

$$\Delta P_t \leqslant \Delta P_t^{up} \text{ or } F \leqslant F_t^{up} \tag{6.1}$$

TABLE 6.3 The maximum limit of active power change in a wind farm under normal operation.

Installed capacity of wind farm (MW)	Maximum limit of active power change at 10 min (MW)	Maximum limit of active power change at 1 min (MW)
<30	10	3
30–150	Installed capacity/3	Installed capacity/10
>150	50	15

Where F_t is the power fluctuation rate within a time range; F_t^{up} is the maximum limit of fluctuation rate; ΔP_t is the maximum change of power within a time range; ΔP_t^{up} is the maximum limit of power change within a time range.

2. Original data's sampling and analysis

The original data refers to the historical power data or predicted power data sampled by monitoring system samples or recorded by a forecasting department within the typical cycle of smoothing new energy fluctuations. The typical cycle can be chosen by year/month/day, based on the actual situation of wind farms or PV power stations.

According to the sampling theorem, the highest resolution frequency of the spectrum analysis is half the sampling frequency, namely, fs/2. Therefore, the sampling frequency must be at least two times the maximum frequency of the signal to avoid frequency domain aliasing. The smaller the sampling period, the more accurate the date and the wider the scope of the spectrum analysis are.

Based on fluctuation constraints, it can be determined whether each successive time period meets the requirement. If the power fluctuations meet the requirement, there is no need to configure the energy storage system; otherwise, it should be configured as follows.

3. Spectral analysis

Amplitude frequency characteristics can be gained by using discrete fourier transform (DFT) for original data.

$$P(k) = \sum_{n=0}^{N-1} P(n)e^{-j\frac{2\pi}{N}kn}, k = 0, 1, \cdots, N-1 \tag{6.2}$$

Based on amplitude frequency characteristics, the main frequency range of power fluctuations can be calculated.

4. Determining the rated power of an energy storage system

According to preceding frequency range of power fluctuations, the cutoff frequency is selected to perform first-order low-flux filter on the original power. Following that, the tie-line power fluctuation rate outputted by low-flux filter is calculated, and based on the constraint condition, the cutoff frequency that not only meets the requirement of fluctuation but also makes the configuration of the energy storage capacity as small as possible is gained. Finally, the corresponding tie-line power and the energy storage system's reference power are obtained.

The principle of using first-order low-flux filter is shown in Fig. 6.3.

In the energy storage system, the discharge power is positive, and the charge power is negative. Their variables in Fig. 6.3 are as follows:

$$P_{\text{line}}(s) = \frac{\omega_c}{\omega_c + s} P_v(s) \tag{6.3}$$

$$P_E(s) = P_{\text{line}}(s) - P_v(s) = \frac{s}{\omega_c + s} P_v(s) \tag{6.4}$$

Where P_v is the output power of new energy; P_{line} is the tie-line power of grid-connected new energy system; P_E is the compensation power provided by the energy storage system; ω_c is the cutoff fluency of filter.

According to $s = j\omega$ and the formula of Fig. 6.4, the amplitude frequency characteristic can be calculated:

$$P_{\text{line}}(\omega) = \frac{1}{\sqrt{1 + (\omega/\omega_c)^2}} P_v(\omega) \tag{6.5}$$

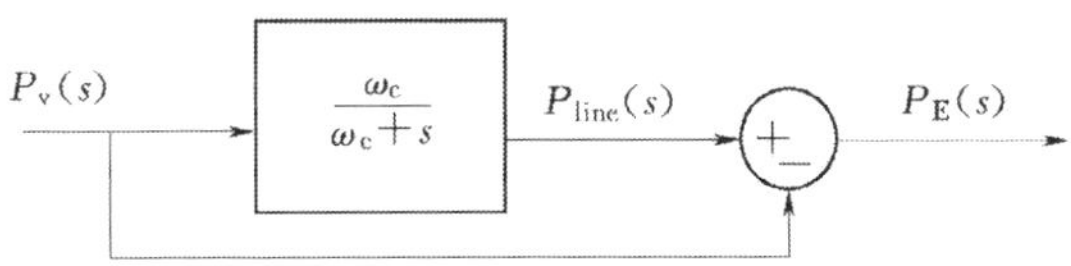

FIGURE 6.3 The principle of first-order low-flux filter.

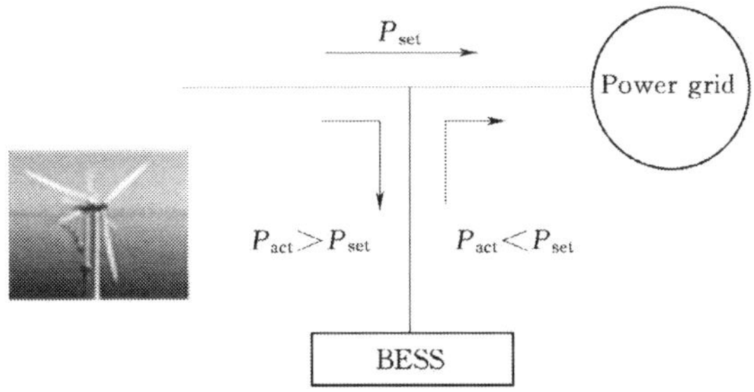

FIGURE 6.4 Schedule output of the charge/discharge power of the battery energy storage system.

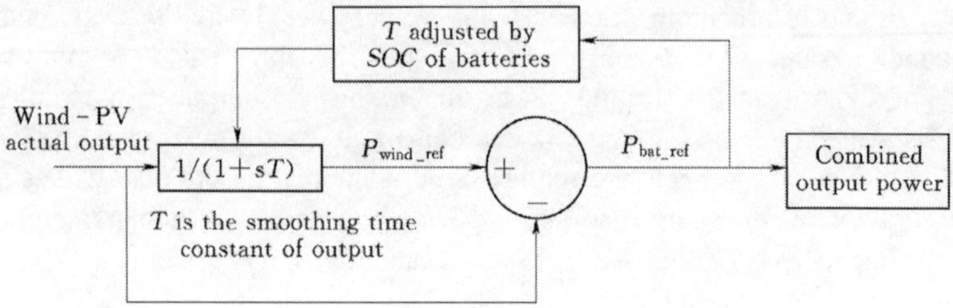

FIGURE 6.5 Control strategy of smoothing the power fluctuations.

As shown in Fig. 6.5, the amplitude frequency function of the first-order low-flux filter is a monotonically decreasing function. When $\omega = 0$, the amplitude reaches the maximum, up to 1; When $\omega = \omega_c$ (the cutoff frequency), the amplitude is 0.707. With the increase of amplitude, the frequency response of the system gradually declines to 0. In other words, when the appropriate filter cutoff frequency is taken, the amplitude of the fluctuation component, which is higher than cutoff frequency, can be attenuated to 0.

The compensation capacity provided by energy storage system is directly related to the compensation frequency band provided by the storage system. The smaller the cutoff frequency, the greater the compensation frequency ranges. In this case, although the overall smoothness effect is better, the energy storage capacity required is larger. The system's cutoff frequency is determined by testing different frequencies starting from low frequency to high frequency. If the filtered tie-line power fluctuation rate is far less than the constraint condition, the required energy storage capacity is larger, so the cutoff frequency should be disconnected from the high frequency. If the tie-line power fluctuation rate is greater than the constrain condition, the cutoff frequency should be taken from low frequency. If the tie-line power fluctuation rate is less than or close to the constraint condition, the corresponding cutoff frequency is ideal.

When it comes to determining the reference power that should be compensated by the energy storage system, it is necessary to consider the energy loss in the charge/discharge process, namely charge/discharge efficiency. In the discharge process, the actual discharge power in the storage energy system is equivalent to the reference discharge power divided by efficiency. In the charge process, the actual charge power of the energy storage system is that the reference charge power times the charge efficiency. Given the charge/discharge efficiency, the power of energy storage system can be calculated following this formula;

$$P_E[n] = \begin{cases} \dfrac{P_{E0}[n]}{\eta_d} & P_{E0}[n] \geq 0 \\ P_{E0}[n] \cdot \eta_c & P_{E0}[n] < 0 \end{cases} \quad n = 1, 2, \cdots N \tag{6.6}$$

Where P_{E} $[n]$ is the actual charge/discharge power of the energy storage system; P_{E0} $[n]$ is the reference charge/discharge power of the energy storage system after low-flux filter; η_{d} is the discharge efficiency; η_{c} is the charge efficiency; N is the number of sampling data.

The charge/discharge power of an energy storage system must ensure continuous, steady operation, meaning the shortage/surplus in energy storage will never appear in the entire operation cycle. Therefore, the net charge/discharge is zero. Given similar corresponding time of each power's sampling value, it can be seen how energy balance embodies the power balance according to the formula of $E = PT$. The compensation power provided by energy storage system can conduct translational motion along the y-coordinate, while the tie-line output power can do translational motion toward the opposite direction. The formula is as follows:

$$\Delta P = \frac{1}{N}\sum_{1}^{N} P_{\mathrm{E}}[n], n = 1, 2, \cdots N \tag{6.7}$$

$$P'_{\mathrm{E}}[n] = P_{\mathrm{E}}[n] - \Delta P, n = 1, 2, \cdots N \tag{6.8}$$

$$P'_{\mathrm{line}}[n] = P_{\mathrm{line}}[n] - \Delta P, n = 1, 2, \cdots N \tag{6.9}$$

Where ΔP is the translation amount; $P'_{\mathrm{E}}[n], P'_{\mathrm{line}}[n]$ are the complementation and tie-line power, respectively, after translation motion.

It is easy to see that the tie-line fluctuation rate has not changed after translation.

In the entire operation cycle, the absolute maximum value of the complementary power provided by the energy storage system is in essence the maximum charge/discharge power of the energy storage system, known as rated power.

$$P_{\mathrm{EN}} = \max\left\{\left|P'_{\mathrm{E}}[n]\right|\right\} \tag{6.10}$$

5. The determination of an energy storage system's capacity and initial state

Based on the complementary power curve, the charge/discharge of the energy storage system can be calculated:

$$E[n] = \sum_{0}^{N} P'_{\mathrm{E}}[n]\frac{T_{\mathrm{s}}}{3600}, n = 0, 1, 2\cdots N \tag{6.11}$$

The unit of E needs to be converted into kWh.

The real-time charge state of the energy storage system is shown in Formula (6.12).

$$SOC = SOC_0 - \frac{E[n]}{E_{\mathrm{N}}} \tag{6.12}$$

Where SOC_0 is the initial charge state.

In the charge process, when $E[n]$ is negative, the residual capacity and SOC increases; in the discharge process, when $E[n]$ is positive, the residual capacity and SOC decreases.

The initial SOC and energy of the capacity storage system should meet the following demands: when the discharge is at its maximum, SOC cannot be lower than its minimum limit; when the charge is at its maximum, SOC cannot be higher than its maximum limit. Based on Formula (6.12), the maximum discharge max $\{E\,[n]\}$ and the maximum charge min$\{E\,[n]\}$ can be calculated.

Assume the rated capacity of an energy storage system is E_N, and the initial SOC is SOC_0, and the maximum and minimum values of SOC are SOC_{up}和 SOC_{low}, and the following can be concluded:

$$\begin{cases} SOC_0 - SOC_{low} \geq \dfrac{\max\{E[n]\}}{E_N} \\ SOC_{up} - SOC_0 \leq \dfrac{\min\{E[n]\}}{E_N} \end{cases} \tag{6.13}$$

the minimum E_N that satisfies the condition.

$$\begin{cases} E_N = \dfrac{\max\{E[n]\}}{SOC_0 - SOC_{low}} \\ E_N = \dfrac{|\min\{E[n]\}|}{SOC_{up} - SOC_0} \end{cases} \tag{6.14}$$

$$E_N = \frac{\max\{E[n]\} + |\min\{E[n]\}|}{SOC_{up} - SOC_{low}} \tag{6.15}$$

Based on Formula (6.12), the initial SOC can be calculated:

$$SOC_0 = \frac{\max\{E[n]\}SOC_{up} - \min\{E[n]\}SOC_{low}}{\max\{E[n]\} - \min\{E[n]\}} \tag{6.16}$$

Later, the rated capacity of the energy storage system can also be gained by Formula (6.14).

Therefore, based on the complementary power provided by the energy storage system, the maximum discharge and charge can be calculated. After the range of SOC is defined, the minimum rated capacity and initial state of the energy storage system can be calculated. It is only in the initial state that the energy storage system can meet the conditions of charge/discharge in the entire operation cycle.

6.2.3.3 Configuration methods of energy storage capacity in tracing the new energy schedule output

One of the reasons an energy storage system is configured on the wind farm is to meet the power grid dispatch schedule curve or constant power output.

In this case, the main task is to bridge the difference between the dispatch output of the power grid and the actual output of the wind farm. As shown in Fig. 6.4, during a certain period of time, the energy storage system is discharged when the actual output of the wind farm is less than the power dispatch value; the storage system is changed when the actual output is greater than the power dispatch value that is determined by the grid company and wind farm based on the forecast of the hourly wind power output. From the power grid, the hourly steady output of the wind farm can be achieved. For the wind farm, the schedule dispatch of the power grid can be satisfied.

According to the constant power value given by grid dispatch and new energy output predictive curve, the energy storage system compensates the power higher or lower than the constant value or the curve. Hence, the difference of actual output and predictive power is the reference power value of the energy storage system. Once the power value is determined, it need not be corrected, and the complementary capacity of the energy storage system can be calculated based on the content of Section 6.2.3.2. If the charge power of the energy storage system is negative and the discharge power positive, the reference power value can be calculated based on Formula (6.17).

$$P_E = P_{set} - P_{act} \tag{6.17}$$

Where P_{set} is the fixed power value; When P_{set} is the P_{act}, the energy storage system is discharged to deliver power to the power grid; When $P_{set} < P_{act}$, the energy storage system is charged to absorb the residual power.

When the power curve of the energy storage system is gained, its capacity and initial state can also be determined according to Section 6.2.3.2.

6.3 Technology that controls the operation of new hybrid integrated energy storage generation

Based on active power regulation and control technology of the power grid, the automatic generation control of traditional hydropower and thermal power plants targets a single power plant or a single generator unit. The dispatch end can combine its own output and load characteristics as well as the specific control mode to select the control power plant or unit. Based on the operation characteristics of hybrid new integrated energy storage generation, the dispatch end generally takes the whole power station as the smallest unit of power control.

The new integrated energy storage automatic generation control systems consists of a wind turbine, PV PCS, energy storage PCS, hybrid power generation monitoring systems, and remote-control signal receiving devices (or communication work stations). It can be divided into two layer control modes. The dispatch end issues specific instructions to the farm end of the hybrid generation control system, and then based on the generation, the control system automatically searches for a strategy, close to the dispatch construction, to control the corresponding output of the wind turbine, PV, and energy storage system, tracing the dispatch schedule.

The system can control the energy storage and intermittent power optimization operation according to specific targets in different applications. As research shows, when SOC estimate and first-order filter are introduced into the energy storage system operation strategy, the energy storage system can smooth the wind power fluctuation and improve its operation life. When an energy storage system combines with wind power prediction technology, it can improve the predictive accuracy and reduce the number of charge/discharge energy storage batteries to greatly improve the overall efficiency of the energy storage battery. The combination of wind, PV power, and energy storage can effectively reduce the active output power fluctuations of the hybrid power generation system to protect the power grid. Through the establishment of a hybrid wind−PV storage power generation system model, the wind−PV power prediction, the combined smart dispatch, the energy storage system control strategy, and reactive voltage control are combined to optimize operation of the new-generation integrated energy storage.

6.3.1 Smooth the power fluctuation

6.3.1.1 Operation mode

The operation mode of maximum power output of the new energy refers to the wind and PV power station output that is not restricted and boasts their maximums. The schedule output of the wind farm and PV power station issued by the dispatch end refers to the installed capacity that can ensure the grid-connected wind−PV generation is at its maximum. At this point, the energy storage battery adopts the smooth power fluctuation control strategy to make real-time output compensation, and smooths the output power fluctuation of the system, ensuring the total output power change rate of the new energy generation meets the active power change rate of the connected grid and realizing the maximum power output of wind and PV power generation.

Let us take the operation data of national wind−PV energy storage demonstration projects as an example. In the operation of the hybrid wind−PV energy storage power generation system, it can fully utilize its regulation ability to remove the restrictions on new energy power outputs in addition to utilizing the wind power and PV resources and controlling the active power fluctuations of the grid-connected generation and improving the power quality and the adaptability of the wind−PV energy storage grid-connected generation.

6.3.1.2 Control strategy

For wind power generation, the fluctuation is caused by random variation of wind speed, and it is widely distributed in various bands in the frequency domain. The power fluctuation of different frequencies will lead to varied degrees of damage to the grid-connected power system. Therefore, it is necessary to first analyze the frequency characteristics of the power fluctuation and identify the frequency that causes tremendous harm, then determine the control objective to smooth the fluctuations. The DFT can be used to decompose the wind power fluctuation into three sections based on varying frequencies: low frequency zone (0.01 Hz and below), high frequency zone (1Hz and above), and intermediate frequency zone (0.01−1 Hz). The high frequency zone component can be absorbed by the inertia of the wind turbine rotor. The intermediate frequency zone component with its power varying rapidly can cause serious impact to the power grid within a short period of time and threaten the safety of the power system. When the low frequency zone component with slow fluctuation and power change rate is connected into the power grid, the power system's automatic generation control (AGC) can respond. However given the ramp rate of the traditional generator units and the limited spare capacity of the power system, it is necessary to smooth the component of the low frequency zone.

The smooth control strategy used by the energy storage system is the wind−PV power smooth control method based on the first-order filter and SOC feedback control technology. Its principle is similar to the one used in the introduction at Section 6.2.3. Meanwhile, in addition to using the first-order smooth method on the wind/PV output power, it also introduces the feedback control technology of the battery residual capacity SOC. When the SOC exceeds the set range, the storage control strategy will adjust the charge/discharge power of the energy storage battery in real time according to the current feedback value of the SOC. This method is used to modify the target power value of energy storage battery on line, so the energy storage system can work in the set of SOC range while smoothing the power fluctuation of wind−PV. The smooth control strategy based on first-order filter and SOC feedback control is shown in Fig. 6.5. If the active power reference value is gained through the SOC set range and actual external characteristics of the energy storage system, the service life of energy storage system and the power smooth of the grid-connected generation can be improved.

Figs. 6.6 and 6.7 are the hybrid wind−PV energy storage operation and output curve of a national hybrid wind−PV energy storage demonstration project in the mode of smooth power fluctuation. When the wind power generation is under the maximum power mode, the charge/discharge of the storage energy can smooth the wind power fluctuation.

From Figs. 6.8 and 6.9, the wind−PV power fluctuations are intense, and the fluctuation rates are generally ranging from 10% to 35%, having a larger impact on the voltage of the power grid in actual operation. The configuration of the energy storage system effectively controls the fluctuations and reduces the minimum fluctuation rate to 5%, improving the power quality of the power grid.

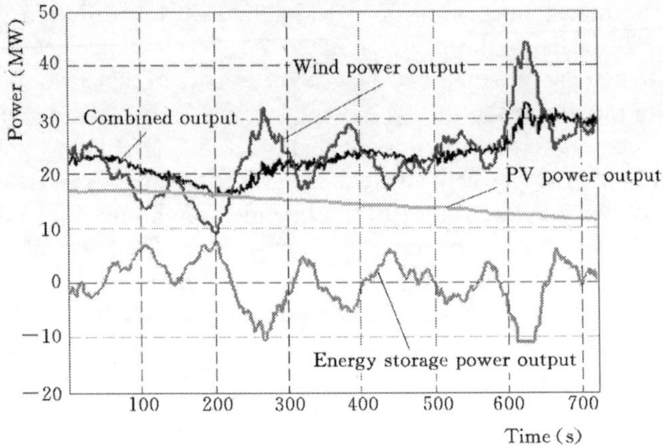

FIGURE 6.6 Hybrid wind–PV energy storage operation.

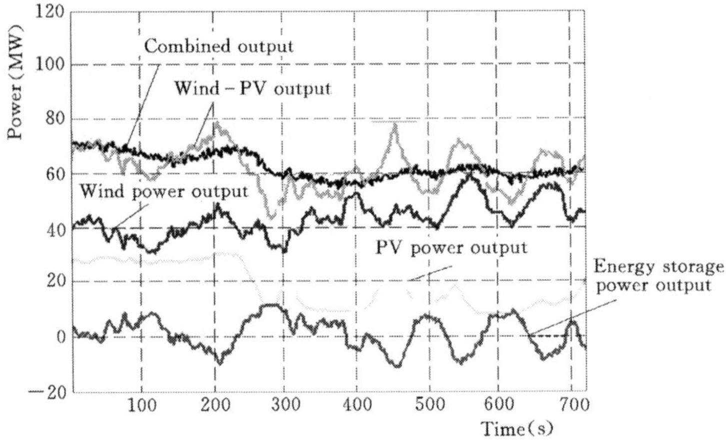

FIGURE 6.7 Hybrid wind–PV energy storage operation.

6.3.2 Trace schedule output

6.3.2.1 Operation mode

The operation mode of tracing schedule output is with the dispatch end's schedule curve as AGC. According to the schedule curve, it traces errors and the state of energy storage capacity, as well as dispatch the wind and PV power generation output in real time. At the same time, it makes the actual output of the hybrid generation system mesh with the generation schedule curve as much as possible, keeping the output power stable. When the generation schedule curve conforms to the requirements of actual operation, it means this mode agrees with the requirements of power grid dispatch and is widely used.

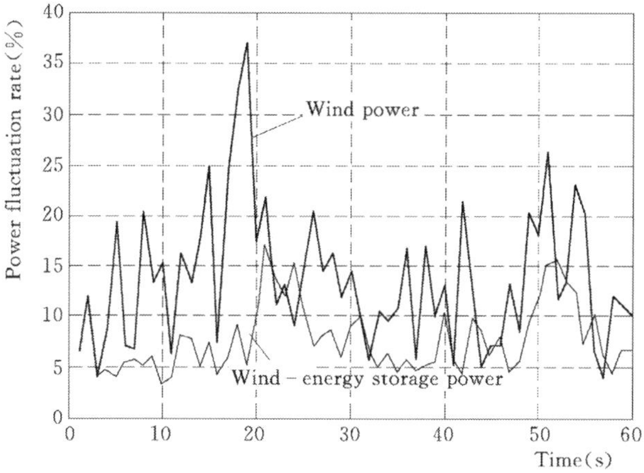

FIGURE 6.8 Fluctuation comparison of wind power and wind energy storage power.

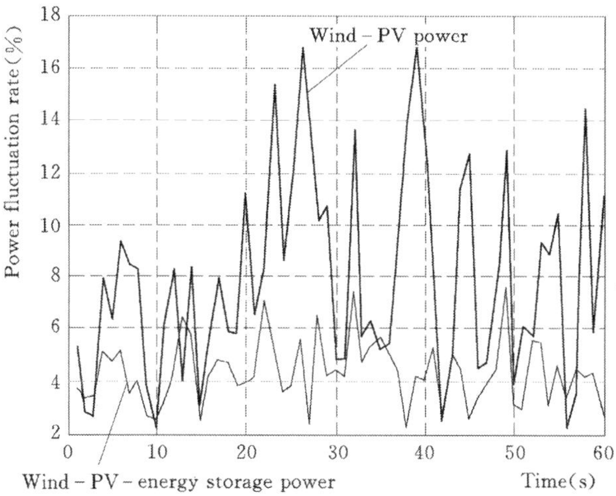

FIGURE 6.9 Fluctuation comparison of wind–PV power and wind–PV energy storage power.

It, therefore, just needs the dispatch end to select the appropriate control strategy based on the generation schedule curve, which is convenient and practical. When wind power resource, energy storage capacity, and dispatch schedule are in conflict, the hybrid control system will follow the different priorities set by users to automatically change the control strategy. In the national hybrid wind–PV energy storage demonstration project, the trace accuracy of the power generation instructions issued by the hybrid generation system is over 95%, with more than 70% at over 98%. The operation data confirm that the output power of the hybrid generation system can reach the output characteristics of the conventional power based on the cooperated operation of the control system.

In addition, according to the real-time output of the wind farm/PV power station as well as the capacity state of the energy storage battery, the dispatch end sets the output limit of the hybrid wind–PV energy storage power generation system. Then the schedule curve makes real-time adjustments, and the substation in the hybrid power plant receives the output limit. If the output value of the new energy is higher than the output limit of the hybrid generation system, the energy storage system stores the residual capacity. The mode is suitable for situations with strong wind and bright light, and it is based on controlling the hybrid wind–PV energy storage power generation system output to ensure the power grid is stable.

The generation system can be controlled at a certain time. That is to say, as the mode controls the power of the system, it also controls the start/end time while automatically switching to the schedule trace after the end time. This mode is geared to power control within a short period of time.

6.3.2.2 Control strategy

The output power of the wind turbine relies on wind speed. Featuring intermittent and random fluctuation, the wind speed causes the wind turbine output power to fluctuate rapidly. This affects the stability of the power grid, making it difficult for operators and dispatchers to dispatch an accurate schedule of each wind farm in the generation system. Presently, the predictive accuracy of wind speed and the wind farm's output power is not high, so it is difficult to gain the dispatch schedule of wind power grid-connected generation. Therefore, a dispatch mode suitable for grid-connected generation should be adopted to make a more accurate dispatch schedule and take full advantage of the wind power resource.

To date, domestic and foreign experts have put forward many control strategies to solve the wind farm dispatch. The grid dispatch decision-making model with acceptance of wind power is constructed based on the concept of economizing coal consumption, and then the allowable fluctuation range of wind power is determined to reduce the influence of wind power fluctuation on power grid. However, the thermal power unit cannot make real-time adjustment based on the change of the wind power. After the energy storage system is configured at the exit of the grid-connected generation system, we should then calculate the active power needed by the grid using predictable wind speed value and peak power schedule curve, and then use the energy storage system to reduce the difference between the wind farm's actual output power and dispatch power, making the generated power to follow the dispatch schedule and meet the grid's requirement for dispatch.

As shown in Fig. 6.10, P_w and Q_w are the output active power and reactive power of the wind farm. P_g and Q_g are the active and reactive power of a grid.

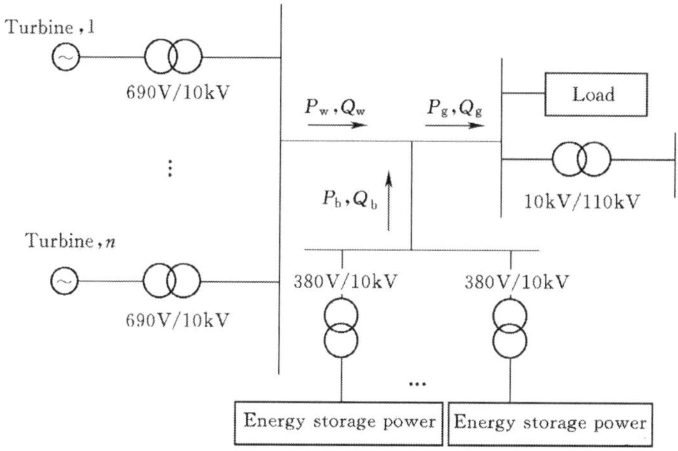

FIGURE 6.10 Wind farm system with energy storage devices.

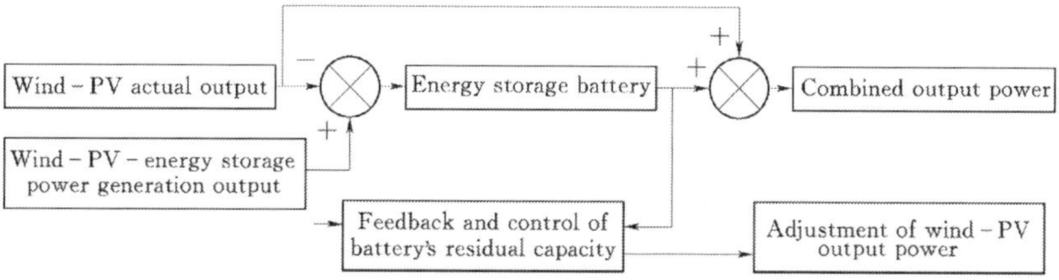

FIGURE 6.11 Control strategy of tracing schedule output.

P_b and Q_b are the active power and reactive power delivered/absorbed by the energy storage system.

As shown in Fig. 6.11, the DC/AC inverter is adapted to regulate and control the change/discharge of the energy storage system. When the voltage of the device increases through the inverter, the device is connected to the grid bus of the wind farm's exit and then to the grid.

The control strategy of tracing schedule output is to compensate the difference between actual and schedule power in real time. At the same time, according to the feedback value of current battery's power and residual capacity, the maximum work ability of the energy storage system is determined. After the dispatch end receives information such as current allowable capacity and maximum charge/discharge of the energy storage system, it requests the hybrid dispatch system to adjust related control variables. As the residual capacity of the energy storage battery comes close to its overcharge, the hybrid output is lower than the generation schedule value, so the dispatch end requests the hybrid dispatch system to limit generation. When the residual capacity of the energy storage battery approaches over discharge and hybrid output is

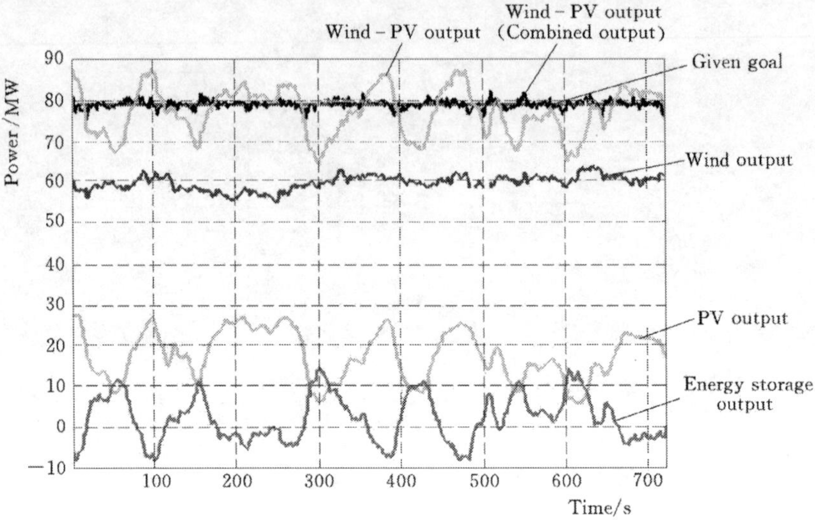

FIGURE 6.12 Hybrid wind–PV energy storage operation under schedule output mode.

still lower than the scheduled generation value, the dispatch end sends the maximum work ability of the energy storage system to the hybrid dispatch system and requests to adjust the control strategy.

Under the tracing schedule output mode in the national hybrid wind–PV energy storage generation demonstration project, the typical output curves of wind power generation, PV power generation, hybrid wind–PV generation, and hybrid wind–PV energy storage generation are shown in Fig. 6.12. From the figure, it can be seen that the charge/discharge of the energy storage system can make the hybrid wind–PV energy storage system operate as the schedule output curves.

From Table 6.4, it can be seen that trace accuracy of generation instructions in the hybrid generation system is over 95%, with more than 70% of the trace accuracy over 98%. This data fully confirms that through the coordinated operation of the wind–PV energy storage control system, the output power of the hybrid generation system can be in agreement with the output characteristics of the conventional power.

TABLE 6.4 Control accuracy in trace mode.

Trace accuracy	≥90	≥95	≥98
Percentage	100%	99%	73%

6.3.3 Frequency regulation of the system

6.3.3.1 Operation mode

For wind power grid-connected operation, an upsurge causing power fluctuation will affect the stability of the grid frequency because the power grid is weak in structure and is actively regulating power. The frequency response of an ordinary asynchronous generator based on constant speed wind turbines is similar to the conventional generator. However, the control system based on variable speed of the wind turbines decouples the rotary speed and grid frequency. In this case, the wind turbines cannot provide corresponding active power when grid frequency varies, reducing the inertia of the whole power grid. So when there is power shortage, the grid frequency rapidly varies and falls sharply, affecting the stability of the power grid.

The hybrid new energy storage generation makes full use of the energy storage system's ability to deliver/absorb power. In addition, the energy storage system with adequate regulation capacity can efficiently trace the grid's frequency fluctuation, helping regulate frequency. What is more, the regulation speed of the energy storage system is superior to the conventional thermal power unit. Presently, the battery energy storage of the demonstration project is limited in frequency regulation. With fast development in energy storage technologies, the system will have huge potential to be used for grid frequency regulation. For example, it can be used in the case of grid frequency shortage and tie-line power fluctuation.

Based on the regulation error of a local power grid, the actual energy output, and the capacity state of the energy storage battery, the dispatch end can give real-time output instruction to the hybrid new generation integrated energy storage system. When the hybrid generation system or energy storage system boasts of adequate regulation capacity and can quickly regulate active power, the dispatch end can regulate system frequency to ensure the control error is within specified range and stabilize the system frequency.

6.3.3.2 Control strategy

The hybrid new generation integrated energy storage system boasts of AGC. The unit participates in the system frequency regulation, meaning that based on the fast charge/discharge of the storage system, it receives instructions from the superior dispatch system in real time to meet the system power demand and support the corresponding power to the AGC plan. The energy storage station participating in system frequency regulation is required to respond to the power demand given by the superior dispatch system within 4 seconds. Fig. 6.13 is the control strategy of energy storage system participating in system frequency regulation. In regulating the frequency, based on the power demand of the energy storage station given by the superior dispatch system and the SOC of each energy storage unit, the converter coordinates control for

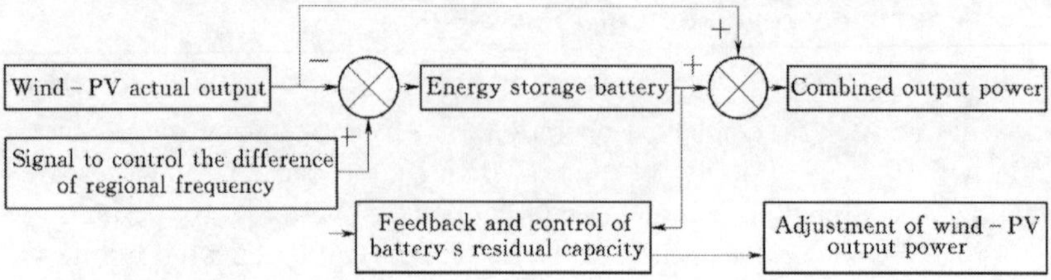

FIGURE 6.13 Control strategy of the system frequency regulation.

each energy storage battery's power and its distribution capacity to meet the system's demand and ensure the SOC of each energy storage unit is controlled within the expected range. When the system is at the over-frequency mode, it is necessary to judge the future output trend of the hybrid generation system by scheduling output. If the output is expected to reduce faster than the required speed, the hybrid generation system will remain at normal operation; if the output is expected to increase or have a slow fall speed, it is then necessary to reduce the output of the hybrid generation system by the required speed and to monitor the residual capacity of the energy storage battery. Based on the urgency of system frequency regulation, the charge/discharge strategy of the energy storage battery is considered adopted. Likewise, when the system is at the under-frequency mode, the hybrid generation system maximizes its ability to generate power and monitor the residual capacity of the energy storage battery in real time. In addition, based on the urgency of the system frequency regulation, the charge/discharge strategy of the storage device is considered used.

In the national wind—PV energy storage power demonstration project, the operation curve of an energy storage system under frequency regulation control strategy is shown in Fig. 6.14. It can be seen that the storage device can efficiently trace the grid frequency fluctuation, assisting it to regulate frequency. Due to limited storage capacity, the energy storage system is also restricted in frequency regulation. But related technology can ensure that the device has the ability to regulate the grid frequency.

6.4 Typical application cases

6.4.1 Tehachapi Wind Farm lithium-ion battery energy storage system in California, USA

As one of the smart grid and energy storage demonstration projects in the American Recovery and Reinvestment Plan (ARRAFP), Tehachapi Wind Farm, with a total investment of about $55 million, is backed by International Standards Organization in California, California State Polytechnic University, and Pomona and Pandora Consult Firm. Presently, Tehachapi Wind Farm has about 5000 wind turbines, known as the second biggest installation site in the

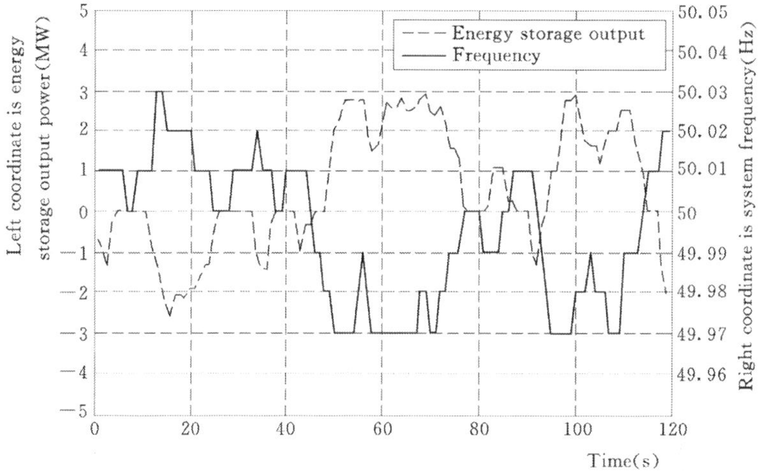

FIGURE 6.14 System frequency regulation control strategy.

world, as shown in Fig. 6.15. Since 1999, the power generation capacity of the wind farm has been the top in the world, and it is expected to reach 4500 MV in 2015. The annual power generation capacity is about 800 million kWh, satisfying the demands of 350,000 households.

The peak power generation capacity of the Tehachapi Wind Farm is about 310 MW, and the connected-line short-circuit capacity is about 560 MW, so it is a typical weak power grid-connected structure. If the long-distance transmission line is under a full load, the grid will be hit, resulting in collapsed voltage. LG Chemical Company provided the project with a set of 8 MW/32 MW·h lithium-ion battery energy storage systems in May 2013, which

FIGURE 6.15 Tehachapi Wind Farm and design of lithium-ion batteries energy storage system.

continues to be the biggest battery energy storage system in the United States. It, installed in Monolith substation, improves the stability of power grid.

6.4.2 Sodium–sulfur battery energy storage system of Rokkasho Village Wind Farm, Japan

The power grid located at Rokkasho Village, Aomori-ken, Japan, with small capacity has limited wind power generation capacity, and its power load difference between day and night is huge. As a result, to fully exploit the abundance of wind resources, meet the maximum storage capacity of wind power generation, and vie for government subsidy policy, it is necessary for the energy storage system capacity to be configured to achieve the expected 70% of wind power generation. Therefore, the Rokkasho Village Wind Farm with a capacity of 51 MW needed a 34-MW sodium–sulfur battery energy storage system. On August 2008, Tokyo Electric Power Company and Japan Electric Power Exchange started receiving power from the hybrid wind energy storage system.

It can be seen from Fig. 6.16, that the 34-MV sodium–sulfur battery energy storage system consists of 17 sets of 2-MV sodium–sulfur batteries, 17 sets of 2-MV PCS, 17 sets of battery management systems (BMS), a general control system, and a general monitor protection system. The PCS is connected to the 6.6-kV bus through the step-up transformer. The 2-MW sodium sulfide batteries are composed of 40 battery modules with the capacity of 50 kW; the voltage, current, temperature, and SOC that are monitored and collected by each BMS are transmitted to the general control system and the general monitor protection system; the PCS receives dispatch from the general monitor protection system every 0.2 s when the power grid is in normal

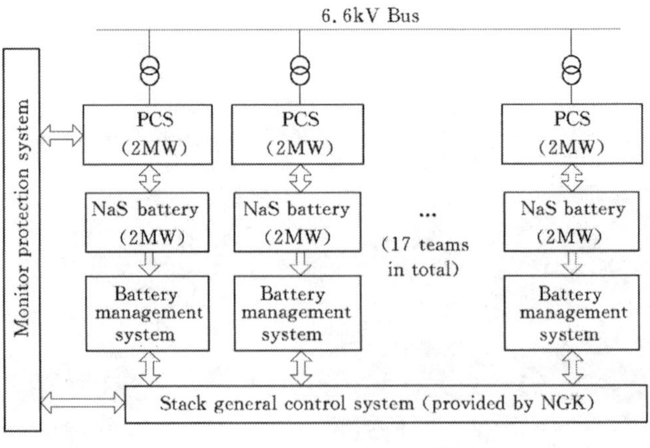

FIGURE 6.16 Topology system of a 34-MW sodium–sulfur battery energy storage system.

voltage to realize the charge/discharge control for the batteries. When the grid's voltage fluctuates, the PCS will be shut down.

To guarantee the 15-year service life of a sodium–sulfur battery energy storage system, NGK INSULATORS, LTD. (NGK) configured an additional set of 2-MW energy storage units that are alternately used with other sets of batteries for every 2 years to boost continuity when one set of energy storage unit fails. In addition, NKG can monitor the operation status of the energy storage system in real time through a remote monitor system, and regularly modify the parameters of the energy storage battery based on user's needs. In case of system failure, users receive a timely notice.

Following long-term operation, collected data indicates that the sodium–sulfur battery energy storage system can meet application requirements of smoothing a wind farm's output and peak load shifting. In peak load shifting mode, the wind farm predicts the next day's output schedule curve according to the power generation data from the previous day. The sodium–sulfur battery provides power output according to the schedule curve and remains operational under charge mode at night and discharge mode in the daytime. The deviation of the schedule curve and the actual operation curve is limited within $\pm 2\% * 40$ MW, and the fluctuation time of the instantaneous power for the energy storage battery is within 1–2 s.

6.4.3 National wind–PV energy storage power generation demonstration project

The national wind–PV energy storage power generation demonstration project is a large-scale, new energy comprehensive utilization platform that integrates wind power generation, PV power generation, energy storage, and smart transmission. The project, located in Heshangyi Town, Zhangbei County, Zhangjiakou, is expected to build a 500-MW wind farm, a 100-MW PV power station, and a 70-MW energy storage power station. In the Phase I project, a 100-MW wind power farm has been built, 40-MW PV power station, and 20-MW energy storage station, as well as the hybrid wind–PV energy storage control center and a 220-kV smart substation. The construction of the Phase II project was started in July 2014.

The overall framework of the wind–PV energy storage power generation demonstration project is shown in Fig. 6.17. When the voltage of each wind turbine and PV array is up to 35 kV through the inverter, all the wind turbines or all PV arrays are connected to the 35-kV bus of the substation through an integrated circuit. When their voltages reach 220 kV through the main transformer, they transmit power. When the voltage of the energy storage device is up to 35 kV through the DC/AC convertor, it is connected to the 35-kV bus of the substation, and it transmits power once their voltage is up to 220 kV through the main transformer.

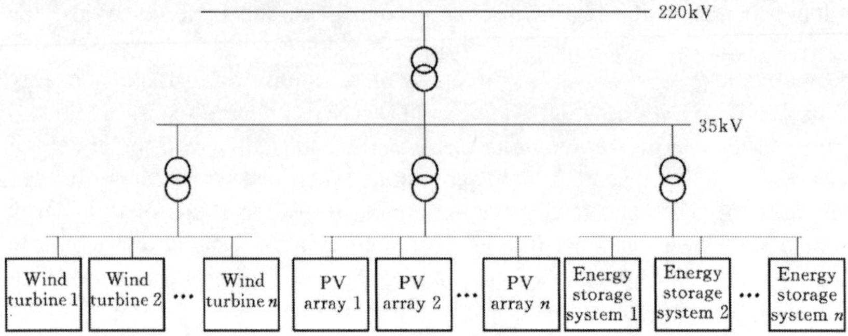

FIGURE 6.17 Overall framework of the wind–PV energy storage system.

The energy storage station consists of a 4-MW/63-MWh iron–lithium battery, a 2-MW/8-MWh all vanadium liquid battery, and a 2-MW/8-MWh sodium–sulfur battery energy storage device. Presently, the 14-MW iron–lithium battery has been put into operation, which is located in three workshops with an area of 8869 square meters and divided into nine energy storage units with a total of 27.4568 million monomer batteries.

According to the different types of lithium–iron–phosphate batteries, the energy storage system can be divided into "energy application" system and "power application" system. The "energy application" energy storage battery featuring high energy density is mainly used for the input/output of high energy. "Energy application" system with a total capacity of 52 MWh boasts of five units. The rated power of each unit is 2 MW, and its maximum charge/discharge power is 3 MW. Each unit consists of six sets of 500 kVA bidirectional converters, and its batteries include two types, namely 60-Ah battery and 200-Ah battery. Among them, the 4 MW * 4 h energy storage unit can be operated in the island. In addition, when the power station is in the total loss of power, the 4 MW * 4 h energy storage unit provides the emergency back-up power to ensure that the power station supplies power while driving the wind power, PV power, storage unit to generate power, therefore making the wind–PV energy storage demonstration station a stable and reliable black-start power supply. "Power application" energy storage battery with high cycle times advantages, high power density, and high discharge ratio is mainly used for the input/output of the instantaneous high power. "Power application" energy storage device with a total capacity of 11 MWh consists of four energy storage units. The rated power of each unit is 1 MW, and its maximum charge/discharge power is 2 MW. Each unit is composed of four sets of 500-kVA bidirectional converters, and its batteries include two types, namely 20-Ah monomer battery and 40-Ah monomer battery. The "power application" energy storage device with a relatively small charge/discharge capacity and large power is suitable for system frequency modulation or the high-power absorption/delivery.

The hybrid wind–PV energy storage power generation system, based on a scheduled plan and prediction of wind and illumination power, conducts the panoramic surveillance and smart optimization for the wind farm, PV power plant, energy storage system, and substation, realizing seamlessly switching among the six operation modes. These includes wind power system output; PV system output; hybrid force of wind power and PV system; the hybrid output of wind power and energy storage system; the hybrid output of PV and energy storage system; and the hybrid output of wind power, PV, and energy storage systems. Under a certain condition, the wind–PV energy storage system can switch seamlessly among six different operation modes.

The energy storage power station mainly plays the following functions in the wind–PV energy storage system; first, it is used to smooth the fluctuation of the hybrid wind–PV energy storage power generation and enhance its controllability; second, it traces the scheduled power generation to improve predictability of new energy generation; third, it participates in peak load shifting to improve the dispatch of the system; fourth, it participates in system frequency regulation to provide quality frequency regulation service for the power grid. From Figs. 6.18–6.21, these are fluctuations of the new

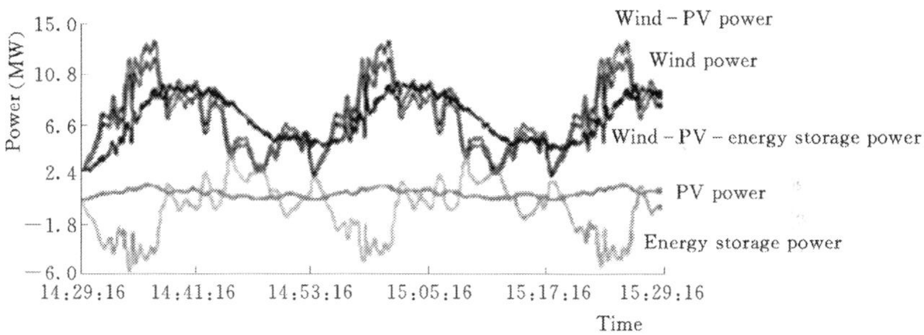

FIGURE 6.18 Fluctuation of the new energy generation.

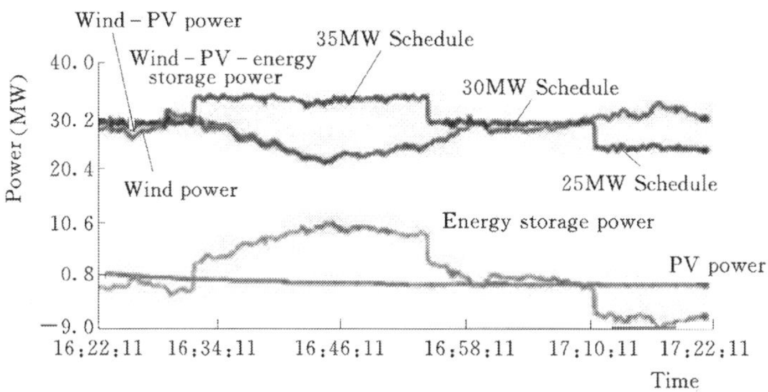

FIGURE 6.19 Dispatch schedule output.

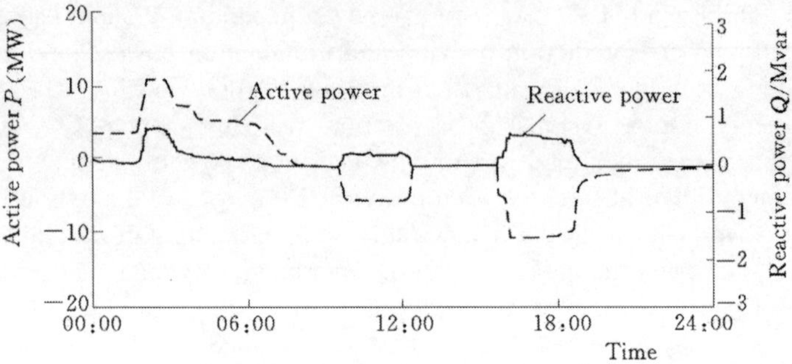

FIGURE 6.20 Peak load shifting.

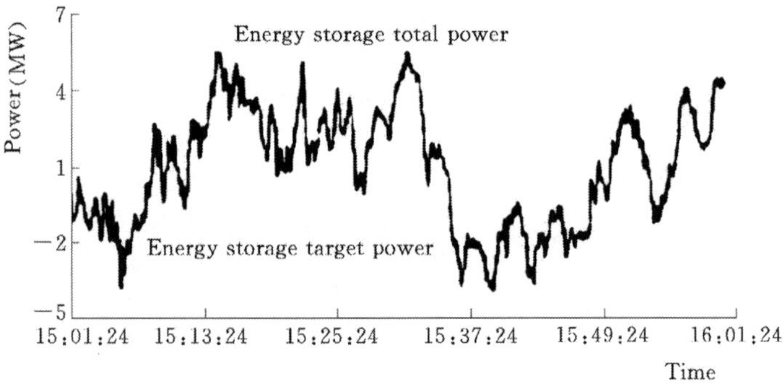

FIGURE 6.21 System frequency regulation.

energy generation, the dispatch schedule output, the peak load shifting, and the system frequency regulation.

6.4.4 The all vanadium redox flow battery energy storage system of the Woniushi Wind Farm, Liaoning

Located in Faku County, Shenyang city, Liaoning province, the Woniushi Wind Farm covers an area of about 16 km^2 with a total investment of about RMB 450 million. The project has 33 sets of 1500-kW wind turbines installed with a total capacity of 49.5 MW. The wind farm is connected to the Liaoning power grid through a 66-kV bus. The Woniushi Storage Station, connected to the wind farm's booster station, configures 10% of its energy storage system, that is to say the energy storage battery capacity is 5 MW * 2 h.

The project is the second largest energy storage power station among the national power grid firms, more than the Hokkaido 4 MW * 1.5 h energy storage power station of Sumitomo Electric Industries. It is known as the largest all vanadium redox flow battery energy storage power station in the world.

The energy storage power station is used to trace scheduled generation, smooth wind power fluctuation, and improve the ability of wind power to connect to the power grid. In addition, it plays an important role in emergency response to the transient active output and in providing emergency support to transient voltage when the wind grid is under operation, to ensure that the overall operation of the power grid is safe and reliable. The 5-MW/10-MWh all vanadium redox flow battery system adopts a modular design of 350 kW, and the rated output power of each power reactor is 22 kW.

The connection of the all vanadium redox flow battery energy storage power station to the wind farm is shown in Fig. 6.22. The station mainly consists of all vanadium redox flow unit battery system, BMS, power conversion system (energy storage inverter PCS), transformer, and energy management system (EMS). The energy storage system unit of 5-MW/10-MWh all vanadium redox flow battery storage power station has a rated output power of 352 kW, and the whole power station consists of 15 sets of 352 kW energy storage systems that are independently regulated. Each set of 352-kW all vanadium redox flow unit batteries is used together with a 352-kW energy storage inverter to form an independent 352-kW/700-kWh unit energy storage system. The AC output side of the energy storage inverter with 352 kW is connected to a double split transformer with the capacity of 1650 kVA in primary winding, and the secondary side of the transformer is connected to the 35-kV power grid in the wind farm via the composite switch. The 15 sets of energy storage system are dispatched by the EMS to realize the start−stop operation of the single unit energy storage system and the whole energy storage system. Based on the dispatch by the EMS, the energy storage system absorbs or releases power through the 35-kV bus of the wind farm to realize charge and discharge. After the superimposition between the charge/discharge of the energy storage system and the power generation of the wind turbines in the wind farm, the energy storage device is connected to the booster transformer with 35 kV, and when its voltage is up to 66 kV, it is connected to the power grid.

Under the unified dispatch of the EMS, the 5-MW/10-MWh all vanadium redox flow battery energy storage system is used to mainly realize the following effects: smoothing wind power output; improving the ability of the wind farm to trace the schedule output; improving the utilization of wind resources; emergency response to transient active power, and providing emergency support for transient voltage. Figs. 6.23 and 6.24 are the smooth output power curve of the wind farm and the operation curve of the wind farm planned output.

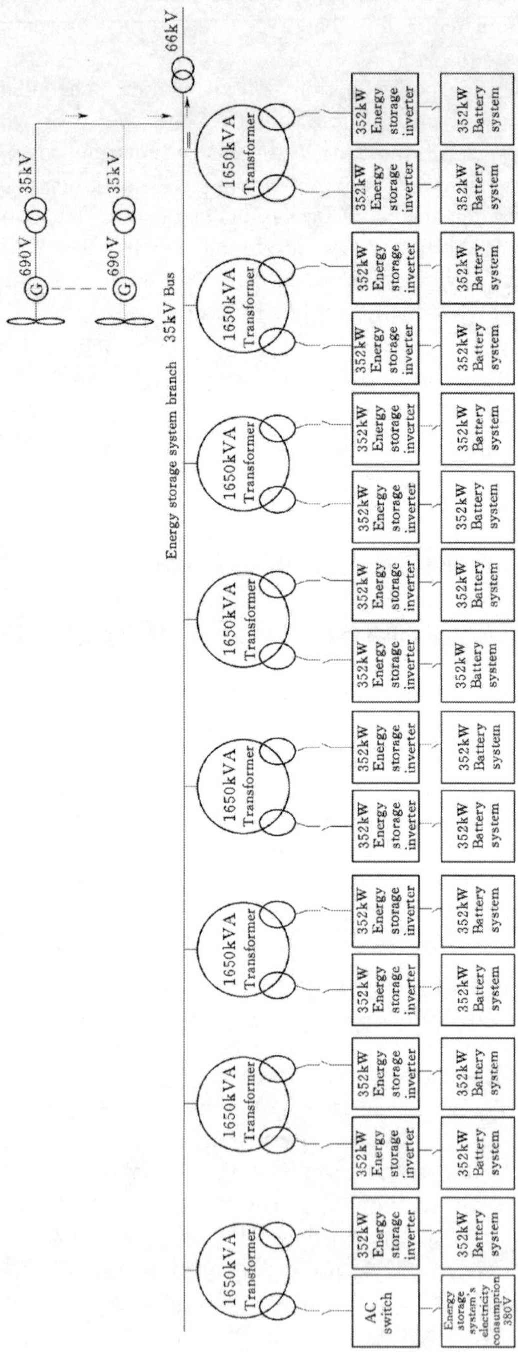

FIGURE 6.22 5-MW/10-MWh all vanadium redox flow battery energy storage system.

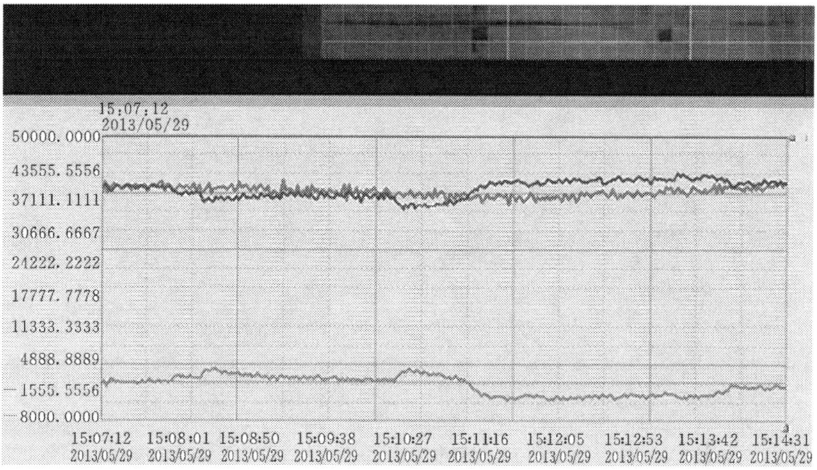

FIGURE 6.23 Smooth output power curve of wind farm.

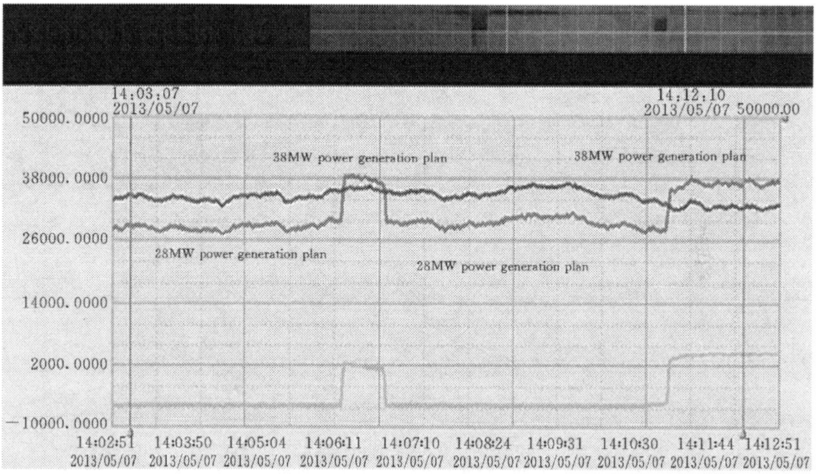

FIGURE 6.24 Wind farm power generation output curve.

References

[1] Sun T, Wang W, Dai H. Voltage fluctuation and flicker caused by wind power generation. Power System Technology 2003;27(12):62–6.

[2] Wu S. Application of energy storage system in wind farm. Jinan: Shandong University; 2012.

[3] Mear S LD, Gotscall HL, Key T, et al. Energy storage for grid connected wind generation applications. American; Electric Power Research Institute and Department of Energy; 2004.

[4] Zhang F. Research on correlation problem of stored energy wind farm integration into power network. Beijng: North China Electric Power University; 2010.

[5] Sun F. Wind farm battery energy storage technology based on power dispatching. Chinese Journal of Power Sources 2012;36:912–4.

[6] Li W. Study on smooth up wind farm output fluctuation by energy storage system. Chongqing: Chongqing University; 2012.

[7] Peng S. Battery energy storage system and its operation and control in the isolated grid based on wind - battery. Shanghai: Shanghai Jiao Tong University; 2013.

[8] Feng X. Study on capacity configuration of large scale energy storage system aim to improve wind power's integrate capacity. Jilin; Northeast: Dianli University; 2013.

[9] Wang Z, Su A, Lu S. Analysis on maximum wind power penetration into liaoning power grid based on power balance. Automation of Electric Power Systems 2010;34(3):86–90.

[10] Ding M, Xu N, Rui B. Modeling and evaluation of energy storage station in smoothing the new energy's power fluctuation. Automation of Electric Power Systems 2011;35(2):66–72.

[11] Bao G, Lu C, Yuan Z. Dynamic programming based on rReal-time optimization of peak shaving and valley filling in battery energy storagesystem. Automation of Electric Power Systems 2012;36(12):11–6.

[12] Xie S, Yang L, Li L. A chance constrained programming based on optimal configuration method of hybrid energy storage system. Power System Technology 2012;36(5):79–84.

[13] Wang C, Yu bo, Xiao J. Sizing of energy storage system for output smoothing of new energy systems. Proceedings of the CSEE 2012;32(16):1–8.

[14] Bingyu S, Yibin T, Gao Z. Research on topology and control strategy of the super - capacitor and battery hybrid energy storage. Power system Protection and Control 2014;42(2):1–6.

[15] Bingyu S, Deshun W, Yang B. Sizing of energy storage system for output smoothing of new energy systems. Proceedings of the CSEE 2014;34(22):3700–6.

[16] Peng Y, Zhao Y, Zhou W. Research on the method based on hybrid energy storage for balancing fluctuant wind power. Power system Protection and Control 2011;39(24):35–40.

[17] Zhang G, Tang X, Qi Z. Application of hybrid energy storage system of supercapacitor and battery in microgrid. Automation of Electric Power Systems 2010;34(12):85–9.

[18] Xie Y, Jiang X. Analysis of influence of energy storage system on unit commitment with wind farm. Automation of Electric Power Systems 2011;35(5):19–24.

[19] Zhang X, Liu B, Ye J. Application of energy storage device in wind – PV - ES hybrid system. East China Electric Power 2010;38(12):1895–7.

[20] Wang W, Qin M, Liu W. Research on economic development mode's transformation and independent innovation of large - scale energy storage technology in wind power generatio (2). In: 12th annual meeting China association for science and technology; 2010. p. 1–6.

[21] Liang L, Li J, Hui D. Optimization configuration for capacity of energy storage system in large - scale wind farm. High Voltage Engineering 2011;37(4):930–6.

[22] Zeng J. Study aon applying battery energy storage system to wind power generation. Guangdong Electric Power 2010;23(11):1–5.

[23] Tang Z. Modeling and controlling of vanadium redox flow battery to smooth wind power fluctuations. Northeast Electric Power University; 2011.

[24] Li B, Shen H, Tang Y. Impacts of energy storage capacity configuration of HPWS to active power characteristics and its relevant indices. Power System Technology 2011;35(4):123–8.

[25] Xin G, Liu P, Wang J. Analysis on wind - PV - ES hybrid technology. North China Electric Power 2012;1:64–7.

[26] Xie S. Research of control strategy and capacity configuration method about hybrid energy storage system. DH Hangzhou: Zhejiang University; 2012.

[27] Zhang Y, Guo L, Jia H. Age control method based on state of charge and variable filter time constant. Automation of Electric Power Systems 2012;36(6):34–62.

[28] Sandia National Laboratories. Energy Storage systems Program (ESS). http://www.sandia.gov/ess/.

[29] Starts MegawattPower generation/storage system demonstration project in Yokohama factory. Business Wire: Sumitomo Electric Industries, Ltd.; 2012. http://industry.ppsj.com.cn/2012-7-24/3980174330.html.

[30] Shibata T, Kumamoto T, Nagaoka Y, et al. Redox flow batteries for the stable supply of new energy. SEI Technical Review 2013;76:14–21.

[31] Liu Z, Zhang H, Gao S. The world's largest all - vanadium redox flow battery energy storage system for a wind farm. Energy Storage Science and Technology 2014;3(1):71–7.

[32] Word's First Wind–PV energy Storage - Smart Transmission Built in China. http://www.windosi.com/news/201201/362187.html.

[33] VRB Power System Inc. The multiple benefits of integrating the VRB - ESS with wind energy case studies in MWH applications. 2007. http://wenku.baidu.com//link?url=oPKG69vMF17W5Vni5DvfmgRUJJ92vBq82mXQjKc91_wzDPT6oF23ACw61T0iIopNlC5f0eH7O4ewc_WqGRkWeMHIPQPu5LEkz_ZwjgF193m.

[34] Gao M, Hui D, Gao Z. Introduction of national wind–PV energy storage demonstration project and analysis of typical operation mode. Automation of Electric Power Systems 2013;37(1):59–64.

Chapter 7

Application of energy storage technology in the microgrid

Chapter outline

Grid-scale Energy Storage Systems and Applications. https://doi.org/10.1016/B978-0-12-815292-8.00007-1

A microgrid is a small, low-voltage system consisting of distributed generation, energy storage, and load. A microgrid can operate under the off-grid mode or on-grid mode and realize seamless transition of the two modes. For a major power grid, a microgrid can be regarded as a "controllable unit" with certain predictability and schedulability programmed to quickly respond to the system needs; for users, a microgrid can be customized to meet varied demands, for example, enhancing regional power supply reliability, lowering feeder loss, improving energy efficiency, and achieving calibration of voltage lower limits or uninterrupted power supply. Presently, microgrids have become critical in enhancing safety and stability of electric power systems while realizing diversification and high-efficiency utilization of energy.

Unlike traditional grids, microgrids are powered by inverters or small-capacity asynchronous generators that have small system inertia, insufficient damping, and lack the antidisturbance capability of traditional grids. In a microgrid, the intermittency and randomness of power generation such as PV, wind, and other renewable energy power generation, the random switching of load, and the switching between on-grid and off-grid eventually influences the system stability and power quality, causing internal voltage and frequency fluctuation and making the microgrid unstable.

The energy storage system can realize flexible, four-quadrant operation through the power conversion device, and it boosts instantaneous rebalancing of active and reactive power of the microgrid, which equates to enhancing system inertia and damping while improving system stability. The energy storage system enhances the ability of the microgrid to balance the power supply-demand relationship between distributed generation and load, effectively reducing adverse impact of wind generation, PV generation, and other intermittent power supplies, while scaling up grid connection capacity of renewable energy. Moreover, the energy storage can boost performance in energy quality, supply reliability, and safety to meet varied requirements of different users.

7.1 Functions

7.1.1 Improving the distributed generation utilization

Presently, distributed generation is the commonly used form of renewable energy in most areas except for a few areas where wind power, solar energy, and biomass can be used reliably at a large scale. Because the distributed generations show obvious randomness, intermittency, and scattered distribution, it will have adverse impact on operation and control of the electric power system when connected to large grids.

Scattered distributed generations can converge at a microgrid and connect to the grid. However, because of unreliability of generations such as solar and wind power due to weather and other natural factors, disruptions of established power generation plans can occur such as wind and solar power curtailment, thus wasting a significant volume of new energy resources. The use of energy storage technology on the microgrid will guarantee the supply of electric power at the designated time, effectively balancing power deviation of distributed generations and ensuring optimum power generation according to the preset generation plan. Fig. 7.1 shows how the energy storage system improves utilization rate of distributed generations.

A microgrid with energy storage systems can offer a controllable and predictable power source or load reliability. Because the power supply and demand of distributed generation and load in the microgrid are highly volatile, the deployment of energy storage systems may realize power balance between them and precise control of system power at a variety of time scales, while providing guarantee for appropriate predictability and programming of the microgrid.

When the distributed generations are connected to the microgrid, the energy storage systems can manage its volatility, and the grid can control the hybrid output characteristics of the energy storage systems and distributed generations with the power dispatching, so the microgrid can participate in the peak-load shaving of the large grid to alleviate the system upgrade pressure and improve the load rate.

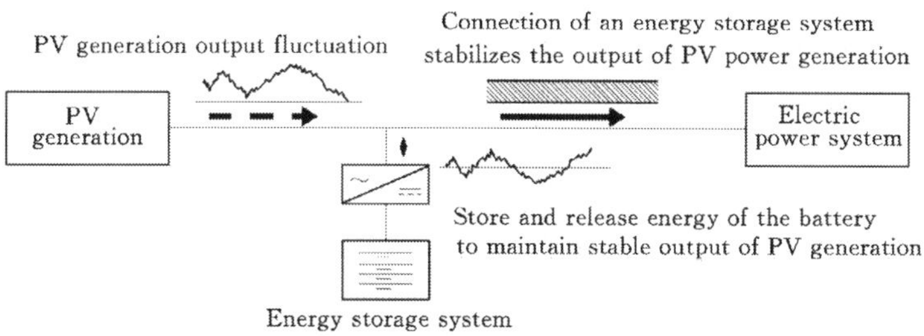

FIGURE 7.1 An energy storage system improves utilization rate of distributed generations.

7.1.2 Improving the microgrid off-grid operating stability

A microgrid has two typical operating modes: normally, it is connected to the conventional distribution network and operates in the grid-connected mode; when detecting grid failure or low energy outputs against high requirements, it disconnects from the grid and operates independently in the off-grid mode. To sustain smooth operation of distributed generation and load during the transition, it is critical to maintain the active and reactive power balance of the microgrid. Installation of energy storage devices will buoy power shortfall and realize smooth transition between the two modes.

The power supply of a microgrid is mainly inverter-based, and it lacks the system inertia and effective antidisturbance capability found in traditional grids. The intermittence of distributed generations and load random switching will cause instantaneous unbalance of active and reactive power of the microgrid, consequently causing system voltage and frequency fluctuation that affects the operation stability of the system. In addition, because the microgrid circuit has a relatively high impedance parameter, the system active and reactive power cannot fully decouple, making the traditional stability control unable to operate effectively.

To start the microgrid in the off-grid operating mode, it needs a stable generation to provide voltage and frequency support. The energy storage system can realize stable, controllable AC voltage output through the inverter, and it has the technical advantages of a stable power source. It is also helpful for rapid black start of the microgrid and maintaining the power supply to loads. Fig. 7.2 shows how the energy storage system improves the microgrid off-grid operating stability.

The energy storage system has a rapid uptake and produces active and reactive power using the conversion device, controls the node voltage and current distribution of the microgrid, and realizes voltage and frequency modulation. It is similar to primary frequency modulation on a traditional electric power system. Moreover, with the energy support of the energy storage system, it can realize smooth transition between on-grid and off-grid in case of system failure and improve the operating stability of wind and PV power generation and other intermittent power supplies.

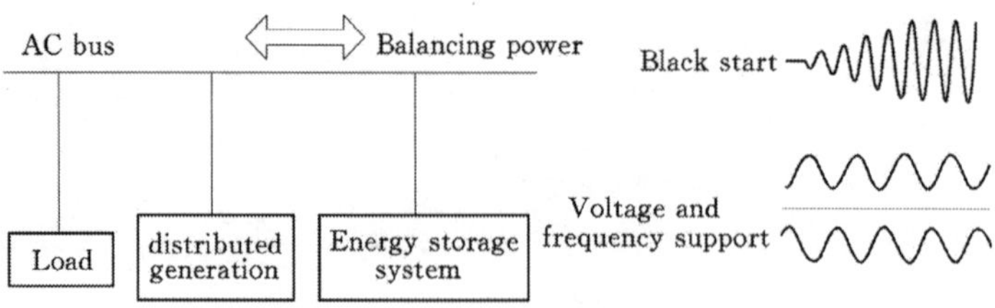

FIGURE 7.2 Am energy storage system improves the stability of off-grid operation.

7.1.3 Improving microgrid energy quality

The operating mechanism of a microgrid and characteristics of distributed generations increases its vulnerability to energy quality problems during operation. During the start and shutdown of the distributed generations on a microgrid, the switching process of distribution network, microsource, and load random power changes can result in energy quality problems such as voltage waveform deformation, DC drift, frequency fluctuation, power factor decline, or unbalancing of the three phases. This particularly hits microgrids using renewable energy power generation, where microgrid output power intermittency, randomness, and electric device–based generation mode will further exacerbate the energy quality problems. The energy storage system can rapidly adjust its power output according to the microgrid operating status, curb the system voltage and frequency fluctuation, reduce the main harmonic components of the system, realize balanced operation of the three phases, and improve energy quality of the microgrid.

Moreover, the energy storage system can avoid microgrid power supply interruption through smooth transition and rapid disconnection when the distribution network fails. In case of voltage drop and flicker, the energy storage system can rapidly provide reactive power support to improve regional voltage stability and the energy quality of the microgrid.

7.2 Capacity optimization

Compared with distributed generation, the generation capacity of a microgrid is determined by the load according to the principle of local consumption; the proportion between wind and solar energy should reflect the availability of the natural resources and thus stabilize the volatility of the total power output. While taking economic performance into account, the energy storage system can guarantee sustainable power supply over a certain period of time to important load components of the microgrid during extreme conditions. Since the distributed generation in the microgrid has a relatively low capacity, the fluctuation has little effect on the main grid. Therefore, the configuration of the energy storage system is mainly subject to the load demand.

According to the microgrid's different application modes, energy storage systems can be configured in the on-grid or off-grid mode. When operating on-grid, the storage system can be discharging during the daytime and charging at night according to the peak-valley price differences; when it is on the off-grid mode, the system charges in the daytime and discharges at night.

7.2.1 Power constraints of an energy storage system

While a microgrid is in the on-grid mode, it can receive energy from the main grid, and the energy storage system should make the longest cycle life as its optimal goal, and choose the appropriate type of energy storage system according to the maximum power and fluctuation of PV/wind power. Take the battery energy storage system as an example, the system operating power should fall in the allowable charge/discharge rate range, and when exceeding allowable scope, the energy storage battery should not be allowed to run.

In the off-grid microgrid, the energy storage system can solely meet the electric power demand and does not obtain energy from the main grid.

Take the wind/solar/energy storage microgrid, for example. If the microgrid system can operate in both on-grid and off-grid modes, the power of the energy storage battery must meet the following equation to meet the energy storage system state of charge (SOC) requirements in any period of time t in a year:

$$P_{\mathrm{Es},t} \geq \max\left|P_{\mathrm{L},t} - \left(P_{\mathrm{WG},t} + P_{PV,t}\right)\right| \tag{7.1}$$

Where $P_{\mathrm{Es},t}$ is the rated power of the energy storage battery; $P_{\mathrm{L},t}$ is the load power requirements; $P_{\mathrm{WG},t}$ is the instantaneous power of wind generation; $P_{\mathrm{PV},t}$ is the instantaneous power of PV generation.

7.2.2 Capacity optimization of an energy storage system in on-grid microgrid

The energy storage battery charges at night, the energy initially is from the wind power, and the remaining charge quantity is supplemented by the main grid. When the energy storage battery SOC reaches $SOC_{\max}$, it stops charging. The energy storage battery charge quantity is this:

$$E_{\mathrm{Es,ch}} = \max\left[E_{\mathrm{L,N}} - (E_{\mathrm{WG}} + E_{\mathrm{G}})\right] \tag{7.2}$$

Where $E_{\mathrm{Es,ch}}$ is the charge quantity of the energy storage battery (negative value); $E_{\mathrm{L,N}}$ is the electric quantity needed by load at night (positive value); E_{WG} is the electric quantity provided by wind power (positive value); E_{G} is the electric quantity provided by grid (zero or positive value).

In daytime operation, a PV and wind generation system supplies the load, and the shortfall will initially be supplemented by the energy storage battery. When the energy storage battery SOC reaches $SOC_{\min}$, it stops discharging. The discharge quantity of energy storage battery is this:

$$E_{\mathrm{Es,dis}} = \max\left[E_{\mathrm{L,D}} - (E_{\mathrm{WG}} + E_{\mathrm{PV}} + E_{\mathrm{G}})\right] \tag{7.3}$$

Where $E_{\mathrm{Es,dis}}$ is the discharge quantity of energy storage system (positive value); $E_{\mathrm{L,D}}$ is the electric quantity needed by load in daytime (positive value); E_{WG} is the electric quantity provided by wind generation (positive value); E_{PV} is the electric quantity provided by PV generation (positive value); E_{G} is the electric quantity provided by grid (zero or positive value).

In summary, the rated energy value E_{E_S} of the energy storage battery refers to Eq. (7.4).

$$E_{E_S} = \max\left(\frac{|P_{E_S,ch,max}|}{C_{E_S,ch,max}(SOC_{E_S,max} - SOC_{E_S,min})\eta_{E_S,ch}}, \frac{P_{E_S,dis,max}}{C_{E_S,dis,max}(SOC_{E_S,max} - SOC_{E_S,min})\eta_{E_S,dis}}, \frac{E_{E_S,max}}{(SOC_{E_S,max} - SOC_{E_S,min})\eta_{E_S,ch}}, \frac{|E_{E_S,min}|}{(SOC_{E_S,max} - SOC_{E_S,min})\eta_{E_S,dis}}\right) \tag{7.4}$$

Where $P_{E_S,ch,max}$ is the maximum charge power of the energy storage battery, kW; $C_{E_S,ch,max}$ is the maximum charge rate allowed by the energy storage battery, h^{-1}; $SOC_{E_S,max}$ is the allowable maximum SOC value, %; $SOC_{E_S,min}$ is the allowable minimum SOC value, %; $\eta_{E_S,ch}$ is the charge efficiency, %; $P_{E_S,dis,max}$ is the maximum discharge power, kW; $C_{E_S,dis,max}$ is the allowable maximum discharge rate, h^{-1}; $\eta_{E_S,dis}$ is the discharge efficiency, %.

Allowable charge/discharge rate, efficiency, and allowable SOC scope are determined by the battery characteristics parameters. Suppose the rated voltage of the energy storage battery is U_B (V), then its rated capacity C_B (Ah) is this:

$$C_B = \frac{1000E_B}{U_B} \tag{7.5}$$

It is noteworthy that in the actual simulation process, the battery SOC value and temperature shall be based on adjusting the smoothing constant r in real time, correct the real-time output power of the energy storage battery, avoid impact of the energy storage system caused by excessive charge and discharge, and prolong the service life of the energy storage battery.

7.2.3 Capacity optimization of energy storage system in off-grid microgrid

In the daytime, PV and wind generations supply load, and the surplus energy charges the energy storage battery. When the battery SOC reaches SOC_{max}, it stops charging. The energy storage battery charge quantity is this:

$$E_{E_S,ch} = \max\left[E_{L,D} - (E_{WG} + E_{PV})\right] \tag{7.6}$$

Where $E_{L,D}$ is the electric quantity needed by load in daytime (positive value); $E_{E_S,ch}$ is the charge quantity of the energy storage system (negative value).

The power supply of load at night jointly comes from wind turbines and the energy storage battery. When the battery SOC reaches $SOC_{\min}$, it stops discharging. The discharge quantity of the energy storage battery is this:

$$E_{\mathrm{Es,dis}} = \max\left(E_{\mathrm{L,N}} - E_{\mathrm{WG}}\right) \tag{7.7}$$

Where $E_{\mathrm{Es,dis}}$ is the discharge quantity of energy storage system (positive value); $E_{\mathrm{L,N}}$ is the electric quantity needed by load at night (positive value).

Moreover, extreme conditions are taken into account, for example, when there is lack of enough sunshine or the wind speed is unable to meet the expected power generation conditions, the battery maximum power supply time t, allowable SOC scope ($SOC_{\max}$, $SOC_{\min}$), system conversion efficiency η, system average capacity $\overline{E_{\mathrm{L}}}$, etc. Therefore the rated capacity of the energy storage battery should also meet the following:

$$E_{\mathrm{Es,L}} \geq \frac{\max\left|\overline{E_{\mathrm{L}}}t\right|}{(SOC_{\max} - SOC_{\min})\eta} \tag{7.8}$$

In conclusion, the value of the energy storage battery rated energy E_{Es} is shown in Eq. (7.9):

$$\begin{aligned} E_{\mathrm{Es}} = \max\Bigg(& \frac{\left|P_{\mathrm{Es,ch,max}}\right|\eta_{\mathrm{Es,ch}}}{C_{\mathrm{Es,ch,max}}\left(SOC_{\mathrm{Es,max}} - SOC_{\mathrm{Es,min}}\right)}, \\ & \frac{P_{\mathrm{Es,dis,max}}}{C_{\mathrm{Es,dis,max}}\left(SOC_{\mathrm{Es,max}} - SOC_{\mathrm{Es,min}}\right)\eta_{\mathrm{Es,dis}}}, \\ & \frac{E_{\mathrm{Es,max}}\eta_{\mathrm{Es,ch}}}{\left(SOC_{\mathrm{Es,max}} - SOC_{\mathrm{Es,min}}\right)}, \\ & \frac{\left|E_{\mathrm{Es,min}}\right|}{\left(SOC_{\mathrm{Es,max}} - SOC_{\mathrm{Es,min}}\right)\eta_{\mathrm{Es,dis}}}, E_{\mathrm{Es,L}}\Bigg) \end{aligned} \tag{7.9}$$

7.3 Hybrid energy storage system

7.3.1 Characteristics of hybrid energy storage

The uncertainties and fluctuation of the power output by distributed generations pose a huge challenge to the power quality during off-grid operation of the microgrid and to the power programming during on-grid operation. For stability, microgrid energy quality improvement, and uninterrupted power supply for critical loads, among other functions, the energy storage system should have the short-term, high-power support capability and long-time energy support, imposing high requirements on the technology and economic performance of the energy storage system.

Given today's energy storage technologies, whether electrochemical energy storage or the latest energy storage technology, it is difficult for a single energy storage technology to meet energy and power density, energy storage efficiency, service life, environmental characteristics, and cost effectiveness all at once. Moreover, the use of a single energy storage technology is not viable to prolonging the life of energy storage systems. Optimum technical and economic performance can be achieved by combining two or more mutually supplementary energy storage technologies to create a hybrid storage system.

For the energy storage equipment itself, supercapacity is characterized by a high power density and rapid charging and discharging qualities that can contain the instantaneous power fluctuation from the distributed generations; the battery system is characterized by high energy density and stable voltage to maintain the periodic power fluctuation; the hybrid energy storage system based on the combination of supercapacity and battery can effectively improve the microgrid energy quality and on-grid operation programming. It can use the supercapacitor to smooth the high-frequency power fluctuation during off-grid operation, and jointly use the supercapacitor and battery to maintain the low power fluctuation frequency during on-grid operation. With joint efforts by the supercapacitor and battery, the system can effectively avoid unprogrammed charging and discharging while improving the energy storage system power response performance.

A hybrid energy storage system can be a device-level combination within the energy storage system or a system-level integration of energy storage systems. For example the asymmetric supercapacitor is a device-level hybrid energy storage system. A system-level hybrid storage system features flexible application. Its performance is related to the combination mode and its control strategy, in addition to the performance of individual energy storage devices.

7.3.2 Topological structure of a hybrid energy storage system

7.3.2.1 AC-side connection in parallel

The supercapacitor and battery are normally connected in parallel at the AC side through a DC/AC converter. Each energy storage device, in the topological structure, needs an independent two-way DC/AC converter, as shown in Fig. 7.3, and it has rapid response to the grid-side voltage and frequency changes. It realizes rapid and precise tracking of reference power through the DC/AC converter, and it enables the energy storage systems to control and modulate the output power of the microgrid centralized and stabilizes voltage at the connection point of microgrid and main grid. The DC/AC converter of the supercapacitor and battery can be deployed separately to meet their respective power demand; the high-capacity energy storage system at or above MW level can be parallel to achieve capacity expansion, and it does not have high requirements on PCS power. But the topological structure has high requirements on controlling the grid-side converter. In case of on-grid operation, the energy

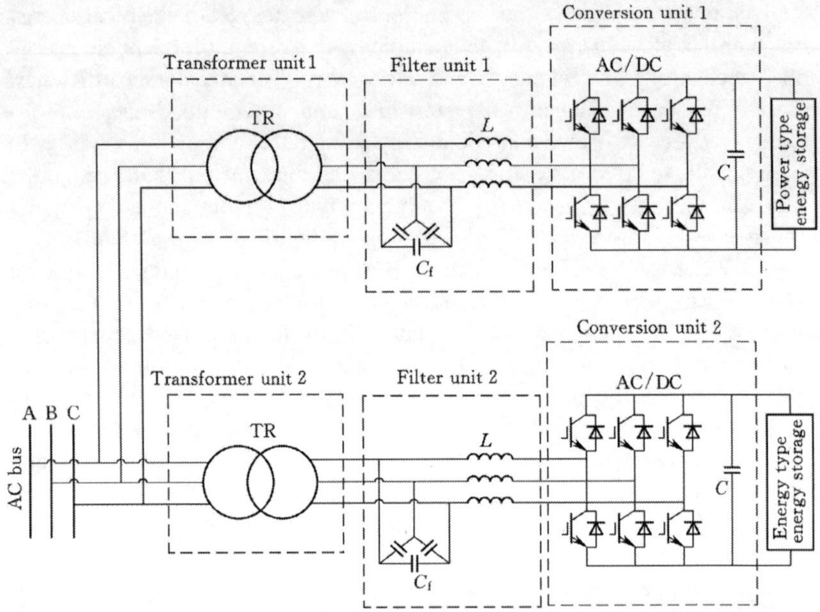

FIGURE 7.3 Topological diagram of AC-side connection in parallel.

management unit distributes power to energy storage devices; in the off-grid operation, the energy storage system is divided into main and auxiliary units; the power distribution strategy of the auxiliary energy storage unit and the coordinated control of main unit are complicated, and the cost of the DC/AC converter is high.

7.3.2.2 DC-side connection in parallel

Supercapacitors and batteries can be connected in parallel at the DC side and share a DC/AC converter to connect with the grid. It modulates power through the control of DC bus voltage. Its control is simple, but it needs to increase the power level of the DC/AC converter to meet the power requirement of the supercapacitor. Therefore, the system capacity is limited, and it is not suitable for the large-capacity energy storage system. The DC-side connection in parallel has different topological structures according to the applications.

1. Direct connection in parallel. The supercapacitor and battery are directly connected in parallel at the DC side, as shown in Fig. 7.4. This topological structure is simple, and it features low cost, high system efficiency, and rapid response speed. But the capacity of the energy storage system cannot be fully utilized. Current is automatically distributed between these two energy storage devices. The distributed quantity of current is subject to their respective internal resistance. Therefore, the power cannot be controlled.

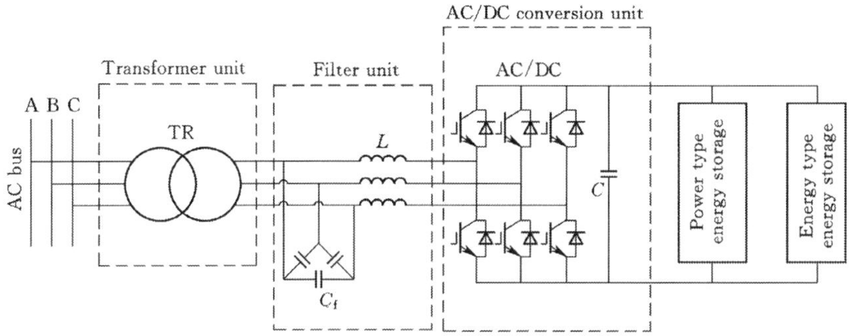

FIGURE 7.4 Topological diagram of DC-side direct connection in parallel.

Moreover, because the battery and supercapacitor have the same voltage, the supercapacitor's voltage is constrained, limiting the optimal use of the supercapacitor and selection of supercapacitor unit sequence. To reach the same voltage, it needs to connect more capacitor cells in series. More importantly, the voltage is uncontrolled and its change is subject to the battery SOC.

2. Connection in parallel through their respective DC/DC converter. The supercapacitor and battery are connected in parallel through the DC/DC power converter at the DC side, as shown in Fig. 7.5. The parallel connection through DC/DC converter has high flexibility in control. The DC/DC converter functions as a current converter and voltage modulator.

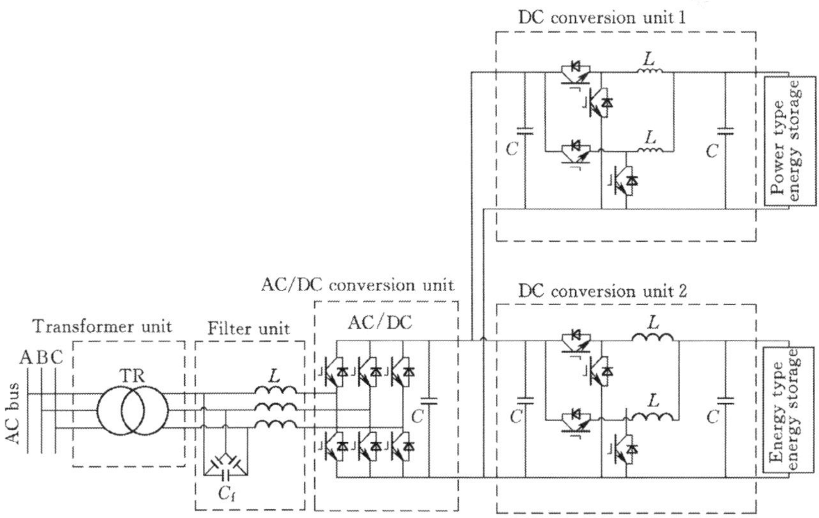

FIGURE 7.5 Topological diagram of connection in parallel through their respective DC/DC converter.

Therefore, it can be used to connect different energy storage components with different voltage and directly control each energy storage equipment. Meanwhile, it maintains the DC bus voltage at constant, optimizing the battery's discharge current and prolonging its service life. As the battery and supercapacitor have the capability of deep discharge, the stored energy can be fully used, optimizing the rated capacity of energy storage system accordingly. The DC/DC converter could make the energy management system configuration more flexible. But compared to parallel direct, the cost of a DC/DC converter increases together with the two-stage conversion power loss, thus pushing down system efficiency.

3. Energy type system in parallel connection through DC/DC converter. The battery is connected in parallel to the DC bus through a DC/DC converter, while the supercapacitor is directly connected to the DC bus, as shown in Fig. 7.6. This topological structure can control the battery's charge and discharge current, prolonging the battery's service life accordingly; the supercapacitor's direct output according to changes of the DC bus voltage elicits rapid response. The disadvantage is low voltage of the supercapacitor unit, so it needs more units to connect in series to achieve a high bus voltage. Moreover, in the pulse load current range, the terminal voltage and load voltage of the supercapacitor decline. If load or connected inverter needs a stable or normal working minimum voltage, the decline will affect stability of the system operation.
4. Power type system connection in parallel through DC/DC converter. The supercapacitor is connected to the DC bus through a DC/DC converter, and the battery is directly connected to the DC bus, as shown in Fig. 7.7. The advantage of this topology is that it reduces the battery DC/DC converter, lowering costs and power losses; the energy stored by the supercapacitor can be fully utilized, and its rated capacity can be optimized. However, the battery power is uncontrolled, and its charge and discharge currents cannot be optimized. The DC bus voltage is subject to the battery SOC changes. This is a variable that cannot be directly

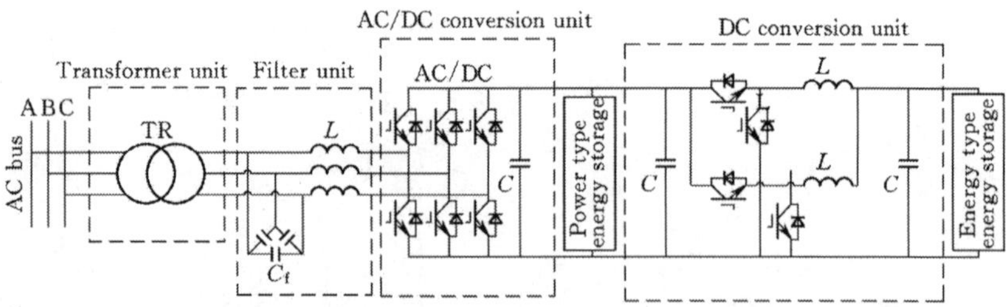

FIGURE 7.6 Topological diagram of energy type system connection in parallel through DC/DC converter.

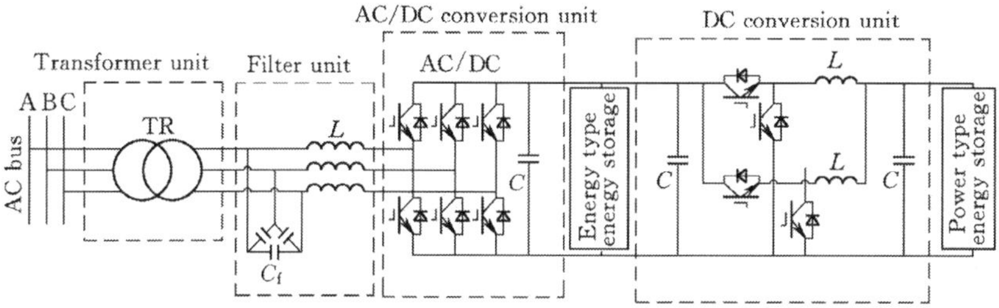

FIGURE 7.7 Topological diagram of power type system connection in parallel through DC/DC converter.

controlled, and it must be maintained within a given scope. Therefore the system operation is constrained.

7.3.3 Capacity optimization of a hybrid energy storage system

By making battery energy storage and supercapacitors as the research object of a hybrid energy storage system, the storage system is therefore used to make up and absorb the difference between the distributed generation P_{DG} and the target grid-injected power P_G^*; that is, the active power target value compensated by the hybrid energy storage system is this:

$$P_{MES,k}^* = P_{G,k}^* - P_{DG,k} = \frac{\tau_{MES}}{\tau_{MES} + T_c}\left(P_{G,k-1}^* - P_{DG,\,k}\right) \tag{7.10}$$

Where τ_{MES} is the smoothing time constant of hybrid energy storage system suppressing the power fluctuation of the distributed generation.

The supercapacitor is a power-type energy storage system, and it is suitable for frequent charging and discharging. It can be used for bearing the high-frequency fluctuation component of P_{MES}^*; the battery is an energy-type storage system, and it is suitable for long-term stable output. It can be used for the low-frequency component of P_{MES}^*. Normally in a hybrid energy storage system, the distribution of power is achieved by the high-pass filter, moving average filtering, or other means, and the basic principle is similar. For example, in the high-pass filter, the reference value of active power of hybrid energy storage system after being broken down by the high-pass filter is as follows:

$$P_{U,k} = P_{MES,k}^* \frac{\tau_{uc} s}{1 + \tau_{uc} s} \tag{7.11}$$

$$P_{B,k} = P_{MES,k}^* - P_{U,k} \tag{7.12}$$

Where $P_{U,k}$, $P_{B,k}$ is the power reference values of supercapacitor and energy storage battery, respectively; τ_{uc} is the smoothing time constant of high-pass filter.

The higher the τ_{uc} is, the higher the power compensation of supercapacitor is, while the smaller the power compensation of energy storage battery is. On the contrary, the smaller τ_{uc} is, the smaller the power compensation of the supercapacitor is, while the power compensation of the storage battery is relatively higher. Therefore, τ_{uc} value should be determined on the basis of optimizing the target in specific application. Meanwhile, on the basis of unified assessment standards (for example, power density distribution, power fluctuation reduction), the optimized power of a supercapacitor and energy storage battery is determined respectively, marked as $P_{U,k}$ and $P_{B,k}$.

Similarly, the energy of supercapacitor and energy storage battery is respectively expressed as shown:

$$E_{U,k} = E_{U,0} - \sum_{m=1}^{k} P^*_{U,k} T_c, k = 1, 2, \cdots, n \tag{7.13}$$

$$E_{B,k} = E_{B,0} - \sum_{m=1}^{k} P^*_{B,k} T_c, k = 1, 2, \cdots, n \tag{7.14}$$

Where $E_{U,0}$, $E_{B,0}$ is the is the initial energy of storage battery and supercapacitor. To facilitate calculation, the value is supposed to be zero.

Therefore, the energy configuration of supercapacitor and energy storage battery can respectively be expressed as follows:

$$E_{U,W} = E_{U,k,\max} - E_{U,k,\min} \tag{7.15}$$

$$E_{B,W} = E_{B,k,\max} - E_{B,k,\min} \tag{7.16}$$

The initial SOC value of a supercapacitor energy storage system is normally set up at 50%–60% in practice. The charge and discharge rate and efficiency of supercapacitors have relatively small effect on the system characteristics and are seldom taken into account. Compared with an energy storage battery, the supercapacitor rated energy is expressed differently, and it meets the following requirements:

1. In the charge status, it should ensure the highest charge rate input while meeting the SOC scope requirements.
2. In the discharge status, it should ensure the maximum discharge power output while meeting the SOC scope requirements.
3. It should ensure that the supercapacitor can discharge maximum real-time remaining energy determined by Eq. (7.13).
4. It should ensure that the supercapacitor can absorb the maximum charge quantity determined by Eq. (7.13).
5. It should ensure that the supercapacitor has rated energy determined through Eq. (7.15).

In summary, the supercapacitor's rated energy E_U (kWh) can be expressed as this:

$$E_U = \max\left(\frac{\left|P^*_{U,ch,max}\right| \times t_{ch}}{SOC_{U,max} - SOC_{U,min}}, \frac{P^*_{U,dis,max} \times t_{dis}}{SOC_{U,max} - SOC_{U,min}}, \frac{E_{U,max}}{SOC_{U,max} - SOC_{U,min}}, \frac{\left|E_{U,min}\right|}{SOC_{U,max} - SOC_{U,min}}, \frac{E_{U,w}}{SOC_{U,max} - SOC_{U,min}}\right) \quad (7.17)$$

Where $P^*_{U,ch,max}$ is the maximum charge power of the supercapacitor, kW; t_{ch} is the accumulated charging time of the supercapacitor, s; $SOC_{U,max}$ is the allowable maximum SOC value, %; $SOC_{U,min}$ is the allowable minimum SOC value, %; $P^*_{U,dis,max}$ is the supercapacitor's maximum discharge power, kW; t_{dis} is the accumulated discharging time of the supercapacitor, s.

Suppose the maximum voltage of the supercapacitor energy storage system is U_{max}(V) and minimum voltage is U_{min}(V), then the rated capacity of the supercapacitor energy storage system C_U(F) can be expressed as Eq. (7.18):

$$C_U = \frac{2 \times 1000 \times (E_U - E_\Omega)}{U^2_{max} - U^2_{min}} \quad (7.18)$$

Where E_Ω is the system internal resistance heat loss.

The instantaneous current of supercapacitor shall be lower than the allowable maximum current in operation. Similarly, the supercapacitors' SOC value should be based on adjusting the smoothing time constant τ_{uc} in actual simulation process and optimize the real-time output power of the energy storage system.

7.3.4 Coordinated control of hybrid energy storage system

7.3.4.1 Operation control strategy of smooth fluctuations

See Fig. 7.8 for the microgrid system composed of the hybrid energy storage and distributed power source. Setting the energy storage system discharge power as positive and charge power is negative, the relationship between variables in the diagram is shown in Eq. (7.19).

$$P_E = P_W + P_V - P_{line} \quad (7.19)$$

Where P_E is the energy storage output power; P_W is the wind turbine output power; P_V is the PV output power; P_{line} is the grid connection line power; it is positive when discharging to grid and negative when obtaining power from grid.

In Fig. 7.8, when the microgrid is operating on-grid, the power fluctuation of grid connection line should be smoothed to minimize effect of wind, PV, and other intermittent distributed generations to the power quality of the

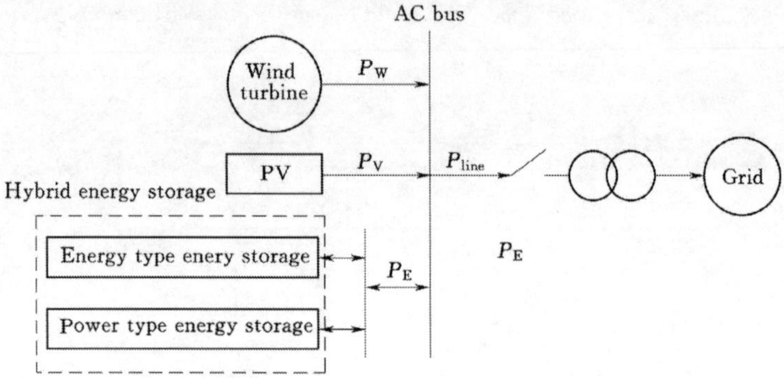

FIGURE 7.8 Wind/PV energy storage microgrid.

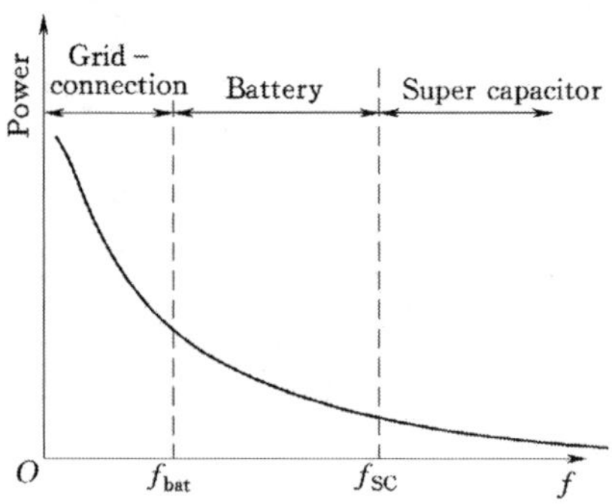

FIGURE 7.9 Distributed generation output fluctuation spectrogram.

distribution network. First, use Fourier transform to the distributed generation output power to get the power fluctuation spectrogram. The distributed generation output fluctuation distribution is shown in Fig. 7.9. The supercapacitor features a high power density, low energy density, and a long service life to compensate components with high power fluctuation frequency but small amplitude. Batteries feature a low power density and a high energy density, and they are used to compensate components with small fluctuation frequency but high amplitude.

Based on the power fluctuation spectrogram, the low-pass filtering method is used to determine the supercapacitor's and battery's power. The second-order low-pass filter shown in Fig. 7.10 is used. Where, P_T is the output of distributed generation, time constant $T_{sc} < T_{bat}$ correspond to the cut-off frequency $f_{sc} > f_{bat}$ in Fig. 7.7.

Based on the spectrum analysis of the distributed generation output power, and the distributed generation on-grid power fluctuation requirements and capacity of the hybrid energy storage system, the fluctuation frequency bands that should be stabled with the battery and supercapacitor can be determined to

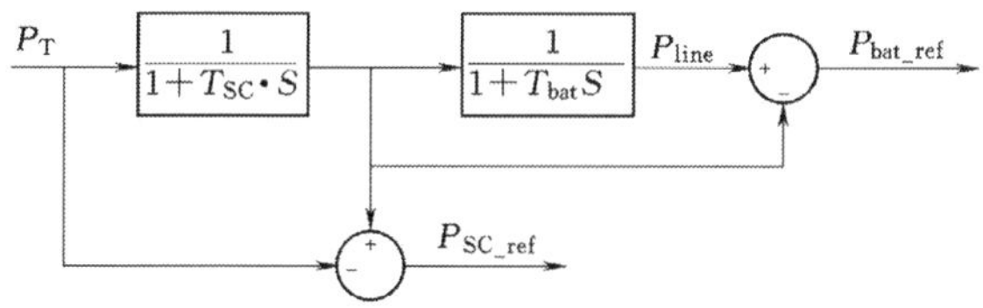

FIGURE 7.10 Second-order low-pass filter control.

establish the time constant corresponding to the low-pass filter and get the output power reference values, respectively. The reference values of power of supercapacitor and battery are as follows, respectively:

$$P_{SC\_ref} = -\frac{T_{SC}S}{1 + T_{SC}S}P_T \tag{7.20}$$

$$P_{bat\_ref} = -\frac{1}{(1 + T_{SC}S)}\frac{T_{bat}S}{(1 + T_{bat}S)}P_T \tag{7.21}$$

Making the reference power value obtained through the low-pass filter as the set value of the energy storage system, constant power operation can generate the control effect of smooth fluctuations.

7.3.4.2 Control strategy considering battery power limits

From Eqs. (7.19) and (7.20), we know that after determining the filter time constant, the energy storage system reference power is directly related to the distributed generation output. If the distributed generation output power has high fluctuation, the reference power compensated by the energy storage system fluctuates too. The requirements on the battery charge and discharge current are strict. For the battery to be economical, the charging and discharging state and power limits should be taken into account. For systems using battery energy storage only, it can smooth PV output power fluctuations and effectively manage the battery charging and discharging. In practice, it is prohibited to charge or discharge when the battery SOC reaches its limits. When it fails to reach the limit, the charge and discharge power rate should be constrained in the specified scope according to the rated capacity of the battery.

Therefore, to meet on-grid power fluctuation requirements, the reference power of second-order low-pass filter is corrected. The power limits are set up according to the battery charging and discharging requirements, and the supercapacitor compensates the part exceeding the battery charge and discharge power limits. See Fig. 7.11 for the correction block diagram.

In the diagram, P_{bat_set} stands for the battery limit power; P_{SC_ref}, P_{bat_ref} are reference power values of supercapacitor and battery before correction; P'_{SC_ref}, P'_{bat_ref} are the reference power values of supercapacitor and battery

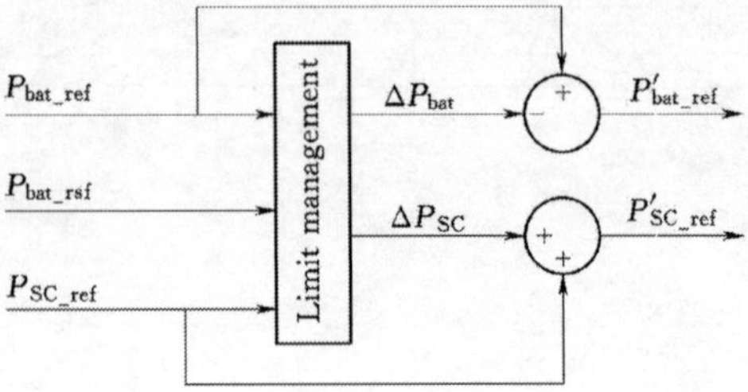

FIGURE 7.11 Battery power limit correction method.

after correction; ΔP_{SC}, ΔP_{bat} are corrected power, and $\Delta P_{SC} = -\Delta P_{bat}$. The specific method of limit management is as follows:

1. Normal working range: if $|P_{bat\_ref}| \leq P_{bat\_set}$, $\Delta P_{bat} = 0$, accordingly $\Delta P_{SC} = 0$.
2. Discharge power out of limit: if $P_{bat\_ref} > P_{bat\_set}$, then $\Delta P_{bat} = P_{bat\_ref} - P_{bat\_set}$, accordingly $\Delta P_{SC} = -\Delta P_{bat}$.
3. Charge power out of limit: if $P_{bat\_ref} < -P_{bat\_set}$, then $\Delta P_{bat} = P_{bat\_ref} + P_{bat\_set}$, accordingly $\Delta P_{SC} = -\Delta P_{bat}$.

The aforementioned limit correction method can keep the charge and discharge power within the limits and make good use of the high-power charge and discharge characteristics of the supercapacitor to contain peak power fluctuation while meeting the grid connection line power control requirements.

7.4 Operation control technology

7.4.1 Overview

When the microgrid runs normally, the distributed generation can solely maintain the basic operation of the microgrid. But the system has a relatively weak capability to withstand disturbance, especially in the off-grid operation mode, the system may face higher security risks given the unpredictability of wind power and solar power resources. Therefore, it is important to study how to make efficient operations of the system. Microgrid system with wind, PV, and energy storage in Fig. 7.12 is used to analyze the microgrid's operational requirements.

The main technical requirements of the operation control are as follows.

1. Microgrid automatic control structure and system

The distributed power sources of the microgrid have different controllable degrees. For example, the active power of the renewable energy power generation is subject to weather and other elements and can hardly be controlled nor adjusted; however, some energy sources are controlled by users and cannot be included in the unified automatic control system; moreover, the dynamic

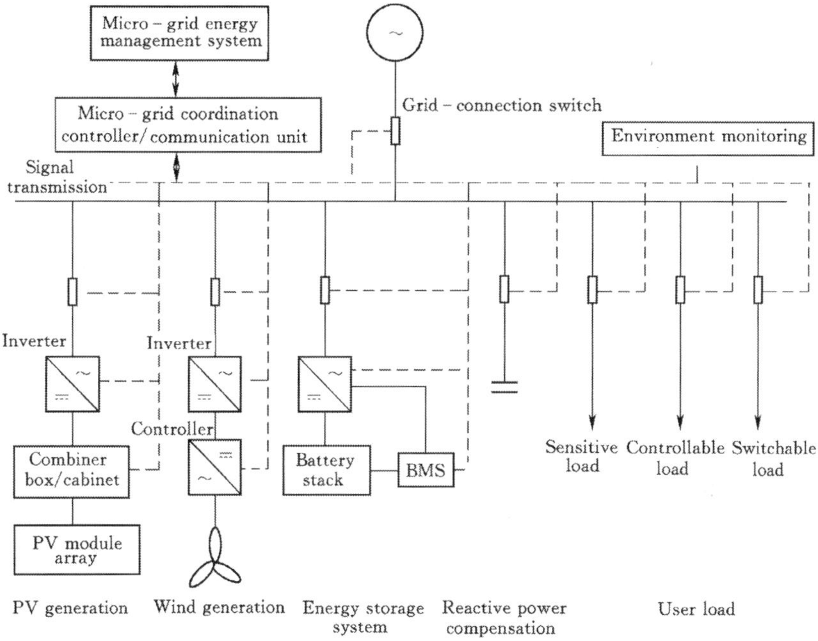

FIGURE 7.12 Microgrid system structure with wind, PV, and energy storage.

response characteristics of distributed generation highly varies. Therefore, there is need to determine proper automatic control structure and system of the microgrid according to the actual situation. It mainly adopts the microgrid energy management center for the concentrated monitoring and control. See Fig. 7.13 for the system structure of the typical microgrid energy management center.

2. Seamless switching

A microgrid has two operation modes, namely on-grid and off-grid operation. When a microgrid is detected to be islanding, or it needs to operate independently according to prevailing situation, it should rapidly disconnect from the public grid to switch into the off-grid operation mode. When the public grid recovers, or the microgrid needs to be on-grid, it should reconnect to the public grid again. During the switching between the two modes, operation control strategy should be taken into consideration to ensure smooth switching and transition.

3. Automatic generation/frequency control

The microgrid's frequency has little influence in on-grid operation mode because of the function of the main grid, but in some instances of off-grid operations, the frequency changes rapidly during the disturbance period due to the small system inertia. Automatic frequency controls must be put in place to maintain the system frequency within the allowable scope.

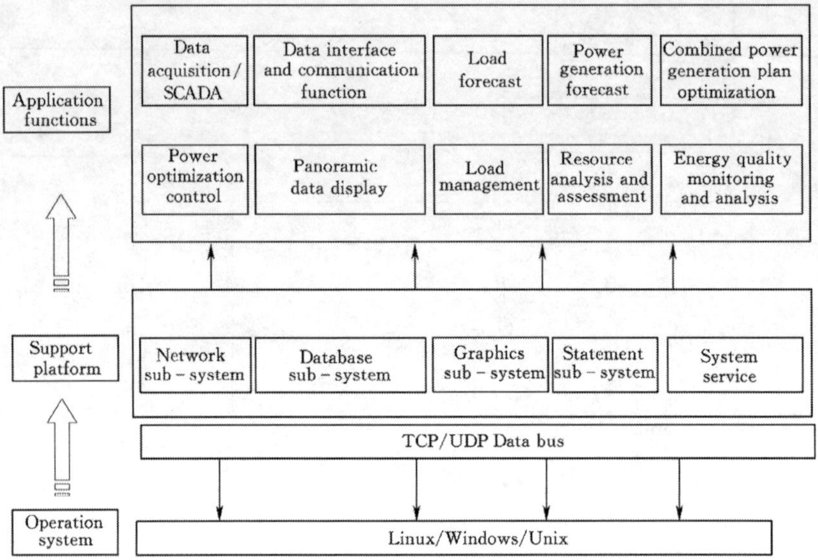

FIGURE 7.13 Structure of typical microgrid energy management system.

When the quantity and capacity of distributed power sources that participate in the main control frequency are relatively small, the microgrid's frequency control is more difficult.

4. Automatic voltage control

In microgrids, the renewable energy resource fluctuation and asynchronous wind turbine grid connection, among others, may cause voltage fluctuation; different loads (including the induction machine) are very close to the distributed power sources, making the voltage fluctuation and other issues complex. Automatic voltage control systems should be put in place to ensure the system's voltage is within allowable scope.

5. Rapid stabilizing system

Shutdown, failures, and great load fluctuation of key electric equipment may cause an upsurge in system frequency and voltage beyond allowable scope within the distributed power source and could cause other components to exceed the rated value, circulation, power oscillation, and other phenomena in the distributed power sources. In such cases, a corresponding stabilizing control strategy should be deployed quickly to stabilize the system and to maintain the frequency and voltage stability by cutting off the distributed power source or load, or by other means.

6. Black start

In certain cases, for example, when islanding transition failures or complete outage happens because of instability, a microgrid can black start by distributed power source with auto-start and isolated power supply characteristics and ensure power supply to important loads.

7. Function and control

Function and control of a distributed energy storage system on the safe and stable operation of a microgrid: This includes dynamic characteristics of different energy storage devices and improvement to the microgrid's dynamic operation characteristics; improvement to the microgrid's voltage/frequency modulation, disturbance, security and stability; and comprehensive control strategy of distributed energy storage systems in the microgrid.

The control system must ensure that the microgrid can safely and stably operate both in on-grid and off-grid modes; especially in the off-grid mode, the control system must be able to control local voltage and frequency and provide or contain instantaneous power fluctuations between power sources and load.

The energy storage system can serve both as a load and distributed power source in the microgrid, but its status is uncertain. Moreover, the energy storage system is characterized by its unpredictability; that is, its maximum and remaining capacity under different powers are uncertain. Therefore this device should be specialized and take a different control method. What is more, efforts should be made to analyze whether this device's capacity and time have any impact on the microgrid's operation when connected to the grid.

This book adopts the master—slave control strategy for the overall control of the microgrid with energy storage units, that is, adopting a conventional energy storage device (battery system) as the master power source to provide voltage and frequency support when the microgrid is in the off-grid operation; the other distributed power sources serve as the slave power sources that use the output voltage of the master power source as the reference voltage, and they are connected to give the rated power. The master power source of the master—slave control strategy adopts the current control method in the on-grid operation and the voltage control method in the off-grid operation. Therefore, it needs seamless switching between the two control modes. The other slave power sources adopt the current control method and transmit power to the microgrid as current. See Fig. 7.14 for the typical topological structure of the energy storage system.

When the energy storage system is operating in the microgrid, it could directly power the loads through the inverter output. When operating as distributed power source, it could change to the current control mode and

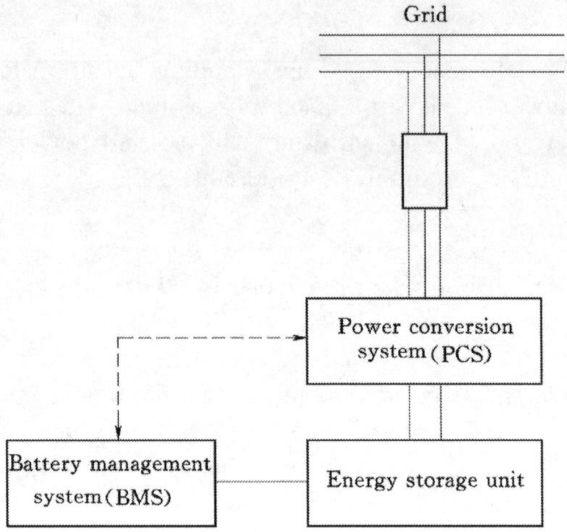

FIGURE 7.14 Typical topological structure of an energy storage system.

transmit power to the microgrid. Moreover, when the remaining capacity of the energy storage system approaches the critical value, it needs to undertake reasonable charging and discharging strategy of energy storage system and load and power source switching strategy according to demands of the current load and distributed power source.

7.4.2 Coordination control technology of a microgrid with an energy storage system

The microgrid system normally includes PV generation units, wind turbine units, energy storage units and loads, among others. Output of PV, wind, and other renewable energy generation is uncertain according to weather conditions, while loads change unpredictably and tend to have high fluctuations. Therefore, a control strategy to coordinate and control the power sources, energy storage units, and load units in the microgrid system needs to be developed, and active and reactive power output of equipment should be coordinated to keep the microgrid running stably for longer.

7.4.2.1 Microgrid operating status

A microgrid is a single controllable unit composed of distributed generation, energy storage, and load from an aspect of a system. The normal microgrid operation has on-grid and off-grid modes and on/off-grid and off/on-grid switching status. The following paragraphs will analyze the different operation modes from its operating status:

1. On-grid operating status: when the microgrid's grid connection switch is turned on, the distributed generation and energy storage units

operate on-grid. In such cases, the energy storage system charges, discharges, or is on standby operation according to its own capacity and dispatching requirements.

2. Off-grid operating status: when the microgrid's grid connection switch is turned off, the master power source is in the V/*f* control mode, provides voltage/frequency parameters to the system, and dominates the primary frequency modulation; only important loads are on operation while the rest loads stops operation. Normally the energy storage system, as the master power source of the microgrid, works at the V/*f* status and balances the power of distributed power sources and loads.
3. Stopping status: when the grid connection switch is switched off, all distributed power sources and loads are disconnected from the grid and stop operation; the energy storage system enters the shutdown status.
4. On/off-grid switching status: when the main grid becomes irregular or gives the off-grid order, the grid connection will be switched off, and the energy storage system, as the master power source, switches the control mode to provide voltage frequency support for distributed power sources and load and ensures continuous operation of the distributed power source and power supply to loads.
5. Off/on-grid switching status: when detecting that the main grid has recovered its power supply and given permission of grid connection from the main grid dispatcher, the energy storage system regulates the microgrid's voltage and frequency; when the synchronization switch-on conditions are met judging by the grid connection synchronization device, the microgrid transfers to the on-grid mode; then the energy storage system enters the on-grid operation status, and the distributed power source and loads operate in coordination to ensure successful and safe grid connection.

Fig. 7.15 shows the microgrid's operating process state machine, indicating that the energy storage system could provide the voltage frequency support during the whole switch process to improve operating stability of the microgrid as the energy buffer system. In the on-grid operation, the microgrid is equivalent to a controllable load or power source. In this state, the distributed power source, energy storage units, and loads operate according to the optimization goals allowed by the grid security and stability. The capacity of the energy storage system should be considered to formulate an active and reactive power dispatch plan and realize optimal operation of the system. Meanwhile, the energy management system keeps communication with the on-grid point protection device and calculates the operation strategy of the main power source, distributed power sources, energy storage unit, and loads and formulates appropriate operation control strategies at the moment of grid disconnection.

The important load is powered, the grid connection is switched off, and the bus voltage/frequency is stable.

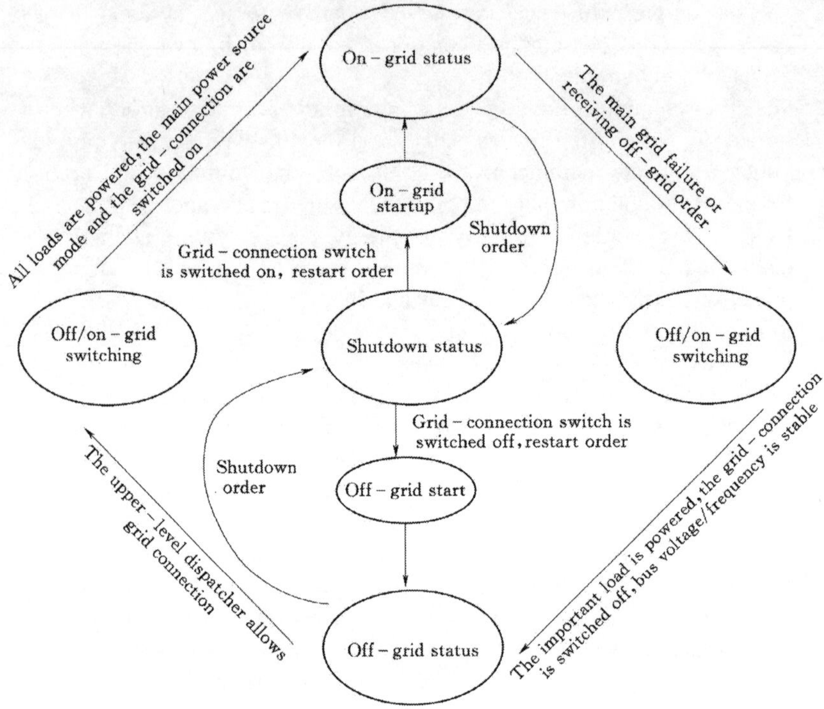

FIGURE 7.15 Microgrid operating process state machine.

7.4.2.2 Active and reactive power coordination control of the microgrid

1. Active power coordination control

When the microgrid is off-grid, the primary goal is that power is supplied to important loads. On this basis, it can selectively guarantee power supply to critical loads and ignore noncritical loads. The core of the off-grid active power coordination control strategy is to determine the adjustment methods of the microgrid power generation units and load units according to the energy storage capacity remaining in the energy storage unit of the microgrid. The microgrid's active coordination control procedure is shown in Fig. 7.16.

The following paragraphs explains (1)–(6) in the diagram.

1. In practice, the power difference value P_{ref} is compared with the preset power deviation dead band value, avoiding direct comparison with zero, lowering the given power volatility.
2. When the result of total intermittent power source subtracting the load is less than zero, or if the intermittent power source with the controllable active power does not operate at the maximum power, or the branch switch

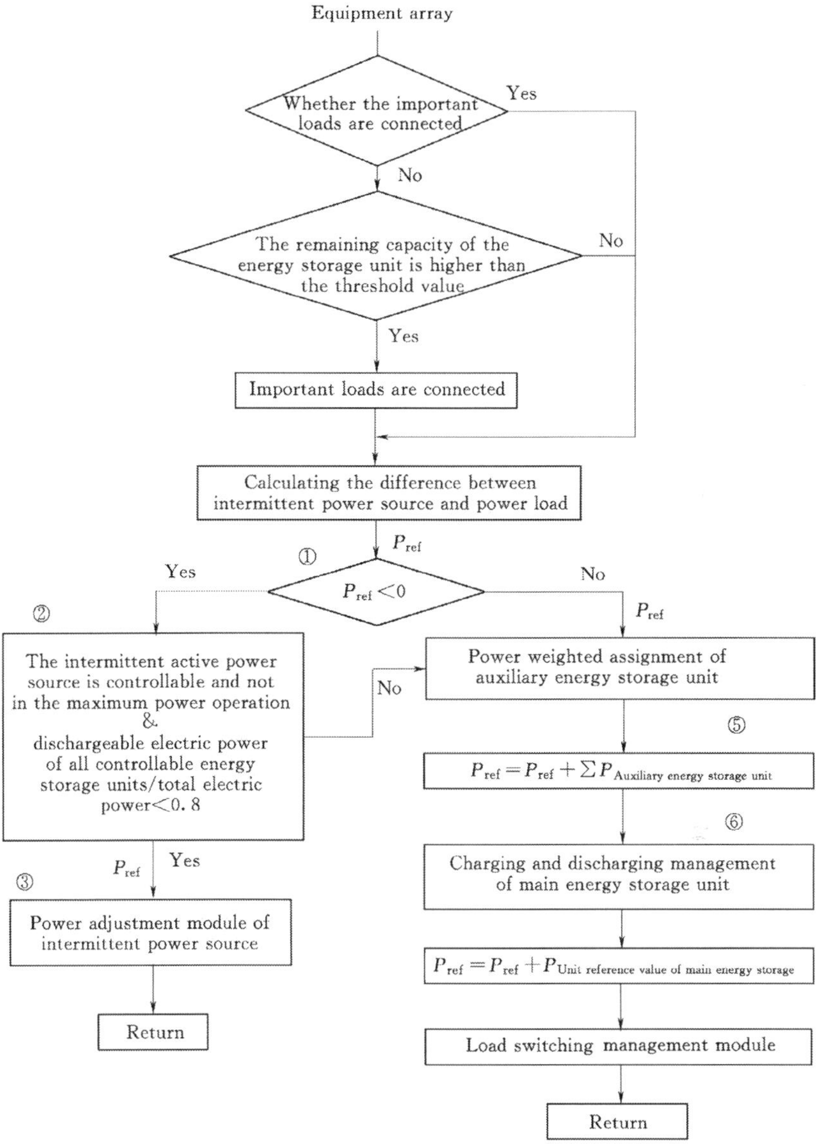

FIGURE 7.16 Procedure of complicated model of active power coordination control strategy.

that is corresponding to the intermittent power source with uncontrollable active power is switched off, the intermittent power source can be adjusted to avoid the total load power being higher than the total power of the intermittent power source. But if all energy storage units are fully charged, there may be a possibility that the total intermittent power sources are higher than the total load power after adjustment. Consequently the power of intermittent power source has to be reduced, which will result in "adjustment oscillation." Therefore a condition is introduced: that is, the proportion of dischargeable electric power from all energy storage units of the total electric power should be smaller than a given value (0.8 in the diagram); only when this condition is met can power of the intermittent power sources be adjusted.

3. The power adjustment method should adjust the intermittent power sources with controllable active power, and then those with uncontrollable active power. The adjustment method of intermittent power sources with controllable active power is based on the weighted assignment, assigning the increment value with the weighted method according to the increasable power when the power increases; when the power decreases, the reduced value is assigned according to the current generation power of the power source. The method of adjusting the intermittent power source with uncontrollable active power is to connect or disconnect the power source branch.
4. The auxiliary energy storage power weighted assignment method: during charge, assign according to the weight of the chargeable electric power of the auxiliary energy storage unit to the total chargeable electric power of all energy storage units; during discharge, assign according to the weight of the dischargeable electric power of the auxiliary energy storage units to the dischargeable electric power of all the energy storage units.
5. Uniform agreement: the power value is negative when the energy storage units are charging; when they discharge, the power value is positive, indicated by "+" here.
6. In the off-grid state, the main energy storage system is used to maintain the stability of voltage frequency and its power is controlled by the system itself and does not need to set up its power value. But to observe the power balance status of the microgrid, the main energy storage power setup reference value is formulated for the follow-up grid-connected power generation and load management. The charging and discharging management method of energy storage system is mainly to adjust the charging and discharging management module and then output the adjusted power value according to the energy storage charge and discharge curve shown in the diagram.

In Fig. 7.17, the power input value is P_f, E stands for the remaining capacity of the energy storage unit, E_{m1} is the low charge constant value, E_{m2} is the high charge constant value, E_{m3} stands for the maximum charge constant value, E_{max} is the maximum energy storage capacity in engineering, E_{n1} stands for the high discharge constant value; E_{n2} is the low discharge constant value, P_{n3} is the minimum discharge constant value, E_{min} is the minimum energy storage capacity in engineering cases. During charging, when E is smaller than E_{ml}, the power output value is P_f, when E value is between E_{ml} and E_{m2}, if $0.5P_f$ is higher than P_{m2} (a fixed minimum value), the output power value is

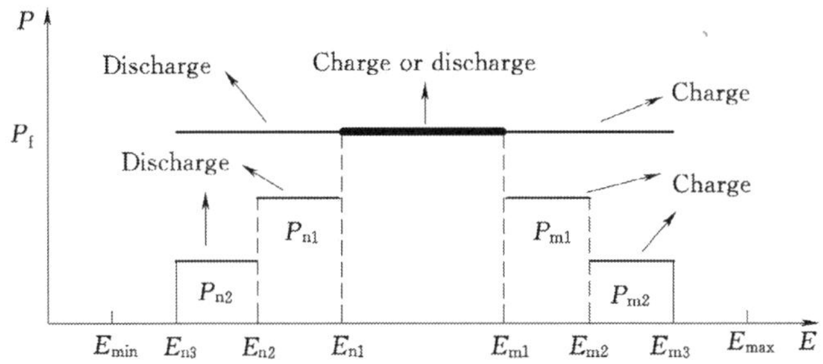

FIGURE 7.17 Charge and discharge curve of energy storage system.

changed to $P_{m1} = 0.5P_f$. Otherwise, it is still P_f; when E falls between E_{m2} and P_{m3}, if P_f is higher than P_{m2}, the output power value is changed to P_{m2}. Otherwise, it is still P_f. When E reaches P_{m3}, the output power value changes to 0. During discharging, when E is higher than E_{n1}, the output power value is still P_f. When E is between E_{n2} and E_{n1}, if $0.5P_f$ is higher than P_{n2} (a fixed minimum value), the output power changes to $P_{n1} = 0.5P_f$. Otherwise it is still P_f. When E is between P_{n3} and E_{n2}, if P_f is higher than P_{n2}, the output power value changes to P_{n2}. Otherwise it is still P_f. When E reaches P_{n3}, the output power value changes to zero. The purpose of designing the charge and discharge power setting curve this way is to reduce its charge power when the energy storage units' remaining capacity is high and reduce its discharge power when the remaining capacity of the energy storage unit is low.

2. Active power coordination control

In the off-grid operation, the main energy storage system of the microgrid operates at a constant voltage and frequency mode, and the reactive fluctuation of the microgrid is absorbed by the main energy storage system. When the reactive load value is low, it does not need to optimize the reactive power control. When the reactive load is high, the main energy storage system completely undertaking the reactive load may affect leveling of the main energy storage system to active fluctuation. Therefore it needs to optimize the control of reactive power.

Main functions of the off-grid reactive power coordination control include monitoring the main energy storage system reactive output value, so when it exceeds a given value, the surplus will be assigned to the other equipment that may provide reactive power. The reactive coordination control procedure is shown in Fig. 7.18.

In Fig. 7.18, the reactive management system of the main energy storage system gets the reactive reference value of the main energy storage system according to the measured value of active power and active and reactive allowance considerations. The auxiliary energy storage units and power source reactive weighted assignment algorithm is assigned according to weight of the current maximum allowable reactive value to the total allowable reactive value.

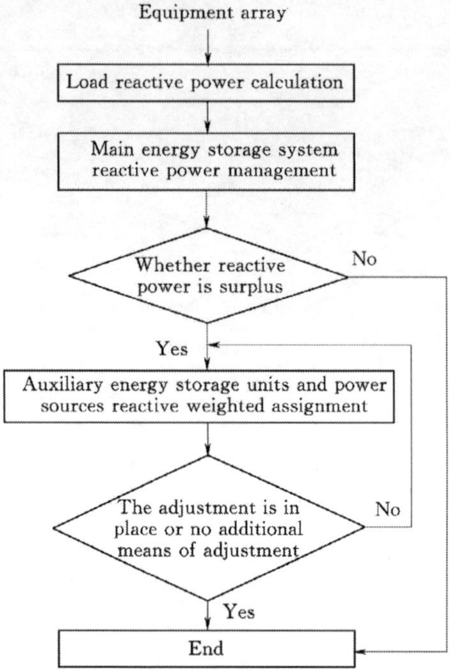

FIGURE 7.18 Reactive coordination control procedure.

7.4.2.3 Microgrid coordination control based on energy storage unit status assessment

The distributed system aims at simplifying and building a reliable operation while the PV and wind power sources of the microgrid operate at maximum power output. Therefore, the power source output should be adjusted to switch to the power source branches. Loads can simply be divided into two categories: noncritical and critical loads. Therefore, the load power choice should be capable of switching to the different load branches. As for the microgrid with an energy storage system, the energy storage units operate at constant voltage and frequency mode, and their output power is automatically controlled by the energy storage system and does not need the coordination controller. However, considering the capacity and service life of the energy storage system, an optimal procedure that meets the characteristics of energy storage can be formulated with automatic switching of the distributed power sources.

The optimized active control strategy has the following main functions: monitoring SOC value of the main energy storage battery; when SOC real-time value approaches its maximum, it needs to adjust the power source output; when SOC real-time value gradually approaches minimum SOC value, it needs to adjust the load power. The core of the control strategy is to exercise corresponding operations when the SOC value of the energy storage system falls in different value ranges. The specific process is as follows: energy storage unit overcharge protection, reconnection of the disconnected power sources, overdischarge protection, user load reconnection, etc.

1. Energy storage system overcharge protection
    1. Start conditions. When the SOC value is higher than the maximum capacity limit of the energy storage system, the power source output should be adjusted. The start conditions meet the following equation:

$$SOC_{\mathrm{now}} \geq SOC_{\mathrm{max}} - \frac{P_{\mathrm{power}} - P_{\mathrm{self}}}{W_{\mathrm{N}}} t_{\mathrm{inv}} - k_{\mathrm{r}} \tag{7.22}$$

    Where SOC_{now} is the current SOC value of the energy storage battery; SOC_{max} is the maximum SOC of the energy storage battery; P_{power} is the sum of maximum power of all intermittent power sources, kW; P_{self} is the maximum power of self-consumed load, kW; W_{N} is the rated energy capacity of the energy storage battery, kWh; t_{inv} is the interval of a strategy execution cycle, h; k_{r} is the confidence coefficient, percentage of W_{N}, recommended to be 0.01, expressed as 1% of W_{N}.

    The physical meaning of Eq. (7.22): when the SOC reaches a given value, and the intermittent power sources generate at full power, the SOC will reach its maximum after an execution cycle interval; to make the microgrid system more reliable, a confidence coefficient is introduced. In practice, the charge of an energy storage system should not reach its maximum SOC; therefore overcharge protection should be made, and the power source output should be adjusted accordingly.
    2. Operation process. The power source output adjustment method: when SOC_{now} exceeds the limits, SOC_{now} values in each calculation cycle should be maintained; when the value is at its highest, the power source should be switched off. Only one power source branch should be switched off in each calculation cycle. The switching-off procedure should be as follows: switching off the PV power branch and then the wind power branch during the daytime, and only switching off the wind power branch at night.
2. Reconnection of the power sources that are switched off
    1. Start conditions. When the current SOC value of the energy storage unit is lower than the minimum capacity limit, it should be reconnected to the power sources that were switched off. The start conditions should meet the following equation:

$$SOC_{\mathrm{now}} < SOC_{\mathrm{max}} - \frac{P_{\mathrm{power}} - P_{\mathrm{self}}}{W_{\mathrm{N}}} t_{\mathrm{inv}} - k_{\mathrm{r}} - \frac{P_{\mathrm{pow \cdot min}}}{W_{\mathrm{N}}} t_{\mathrm{char}} \tag{7.23}$$

    Where $P_{\mathrm{pow \cdot min}}$ is the minimum value of rated power of the intermittent power source branch; t_{char} is the charging time at $P_{\mathrm{pow \cdot min}}$ power value, h.

Compared with Eqs. (7.22) and (7.23) has an additional item at the right, which functions to form a hysteresis to avoid "switching oscillation" caused by frequent connection and disconnection of the power source branches.

2. Operation method. The reconnecting method of disconnected power sources: when SOC meets Eq. (7.23), if there are disconnected power source branches, the calculated values of SOC_{now} should be kept in each cycle, and if the calculated SOC_{now} value is the smallest, reconnection should be executed; only one power source branch should be reconnected each time. The PV power branch will be given priority, followed by the wind power branch.

3. Protecting energy storage system from overdischarge
   1. Start conditions. When the SOC current value of the energy storage system is smaller than the minimum capacity limit, the load power should be adjusted. The start conditions should meet the following equation:

$$SOC_{\text{now}} \leq SOC_{\text{min}} + \frac{P_{\text{load}}}{W_{\text{N}}}t_{\text{inv}} + \frac{P_{\text{self}}}{W_{\text{N}}}t_{\text{inv}} + \frac{P_{\text{self}}}{W_{\text{N}}}t_{\text{selfrun}} + k_{\text{r}} \tag{7.24}$$

Where SOC_{min} is the minimum SOC value of the energy storage battery; P_{load} is the maximum power of all loads (including self-consuming load), kW; t_{selfrun} is when there is no intermittent power source, the time that the self-consuming load can supply power uninterruptedly, h.

The physical meaning of Eq. (7.24): when SOC is lower than a given value, if the intermittent power sources do not generate at the moment but operate at full capacity, the energy storage unit SOC should reach its minimum after an execution cycle interval, and the confidence coefficient is introduced to make engineering more reliable. What is more, the energy storage system needs to reserve some electric power for self-consuming load for a period of time. The discharge of an energy storage system should not reach its minimum SOC value, so protection against over discharge should be made and the load power adjusted.

   2. Operation process. The load power adjustment method: when the SOC current value meets Eq. (7.24), all the calculated SOC_{now} values in each cycle should be kept, and when the calculated SOC_{now} value is the smallest, the user load branch is disconnected; otherwise the disconnection operation should not be made.

4. Reconnection of user load
   1. Start conditions. When the SOC current value of the energy storage system is higher than the maximum capacity limit of the energy storage system, it is necessary to reconnect the disconnected user loads. The start conditions meet the following equation:

$$SOC_{\text{now}} > SOC_{\text{min}} + \frac{P_{\text{load}}}{W_{\text{N}}}t_{\text{inv}} + \frac{P_{\text{self}}}{W_{\text{N}}}t_{\text{selfturn}} + k_{\text{r}} + \frac{P_{\text{load}}}{W_{\text{N}}}t_{\text{disc}} \tag{7.25}$$

Where t_{disc} is the discharging time at the power value P_{load}, h.

Comparing with Eqs. (7.24) and (7.25) has an addition item on the right, which functions to form a hysteresis to avoid "switching oscillation" caused by frequent connection and disconnection of the power source branches.

2. Operation process. The user load reconnection method; when SOC current value meets Eq. (7.25), all the calculated SOC_{now} values in each cycle should be kept, and when the calculated SOC_{now} value is the highest, the user load branches should be reconnected.

5. Microgrid system shutdown for power supply
   1. Start conditions. When SOC value is smaller than the minimum capacity limit of the energy storage system, it is necessary to shut down the microgrid system to partly reserve power of the energy storage battery for future normal start. Its start conditions shall meet the following equation:

$$SOC_{now} \leq SOC_{min} + k_r \tag{7.26}$$

   2. Operation process. When the SOC current value meets Eq. (7.26), the microgrid coordination controller switches from off-grid operation to shutdown; the microgrid system shuts down because of the energy storage system incapability to maintain the remaining capacity and meet the autorun requirements of the microgrid control system.

See Table 7.1 for the operation process of the off-grid control strategy of microgrid system.

7.4.3 Seamless switching technology with energy storage units

7.4.3.1 Switching technology from on-grid to off-grid

Normally, the microgrid is connected to the conventional grid to supply surplus electric power or obtain power supply to compensate for its shortage.

TABLE 7.1 Microgrid control strategy operation table.

Function	Start conditions
Overcharge protection of energy storage system	$SOC_{now} \geq SOC_{max} - \frac{P_{power} - P_{self}}{W_N} t_{inv} - k_r$
Reconnection of disconnected power sources	$SOC_{now} < SOC_{max} - \frac{P_{power} - P_{self}}{W_N} t_{inv} - k_r - \frac{P_{pow\,min}}{W_N} t_{char}$
Energy storage system protection against overdischarge	$SOC_{now} \leq SOC_{min} + \frac{P_{load}}{W_N} t_{inv} + \frac{P_{self}}{W_N} t_{selfrun} + k_r$
Reconnection of user load	$SOC_{now} > SOC_{min} + \frac{P_{load}}{W_N} t_{inv} + \frac{P_{self}}{W_N} t_{selfrun} + k_r + \frac{P_{load}}{W_N} t_{disc}$
Microgrid system shutdown for power supply	$SOC_{now} \leq SOC_{min} + k_r$

When detecting the grid's failure or the energy deficit, the microgrid is disconnected from the main grid and the distributed power source and energy storage device to power loads of the microgrid. In such cases, the microgrid switches to off-grid operation. During the switching process, the energy storage devices provide effective support for the voltage frequency, reduce fluctuation, and meet the technical requirements of seamless transition, so the power supply to loads is not interrupted, and the distributed power sources are not disconnected.

The smooth switching between on-grid and off-grid makes the microgrid operation more flexible, maintains uninterrupted power supply to loads, and improves the power supply reliability. Seamless switching refers to the state in which the microgrid voltage and frequency fluctuation is within allowable range. The general international standards for microgrid operation stipulates that to maintain safe and stable operation, after the operating mode switching, the microgrid's AC bus voltage deviation is $\Delta U \leq \pm 7\% U_N$ (U_N stands for rated voltage) and frequency deviation $\Delta f \leq 0.1$ Hz. The key of smooth switching is a proper control strategy between the microgrid and the energy storage system. The distributed power sources should sustain the on-grid control method and guarantee uninterrupted power supply to loads.

The microgrid switching process can be divided into planned and fault switching based on whether the command is predictable or not. The planned switching refers to when the system receives the switching command from the main grid in advance and adopts the seamless switching plan. The fault plan refers to the case when the main grid suffers from an unpredictable short circuit, voltage sag, or any other failure, and the microgrid energy management system cannot adjust the equipment control mode in advance, thus having to depend on the equipment's own devices to detect failure and switch the mode to guarantee stable operation of the microgrid. The following paragraphs will discuss the on-/off-grid switching control methods of the microgrid.

1. Planned off-grid switching

The microgrid energy management system (MEMS) receives a dispatch command from the main grid, controls the PCC power at zero to ensure the voltage frequency is stable during the off-grid switching, and prepares for off-grid operation. At the same time, the MEMS orders the main power source to switch the mode, transfer to V/f control, adjust the energy storage device output, and guarantee the connection line power to gradually decline. When the connection line power changes to zero and the main power source mode is successfully switched, the MEMS switches off the grid connection switch, and the microgrid runs off-grid. See Fig. 7.19 for the detailed control process.

The MEMS calculates the real-time power of the microgrid at the grid connection switch and chooses the control method according to the calculation result. If the microgrid transmits power to the main grid, the former's real-time

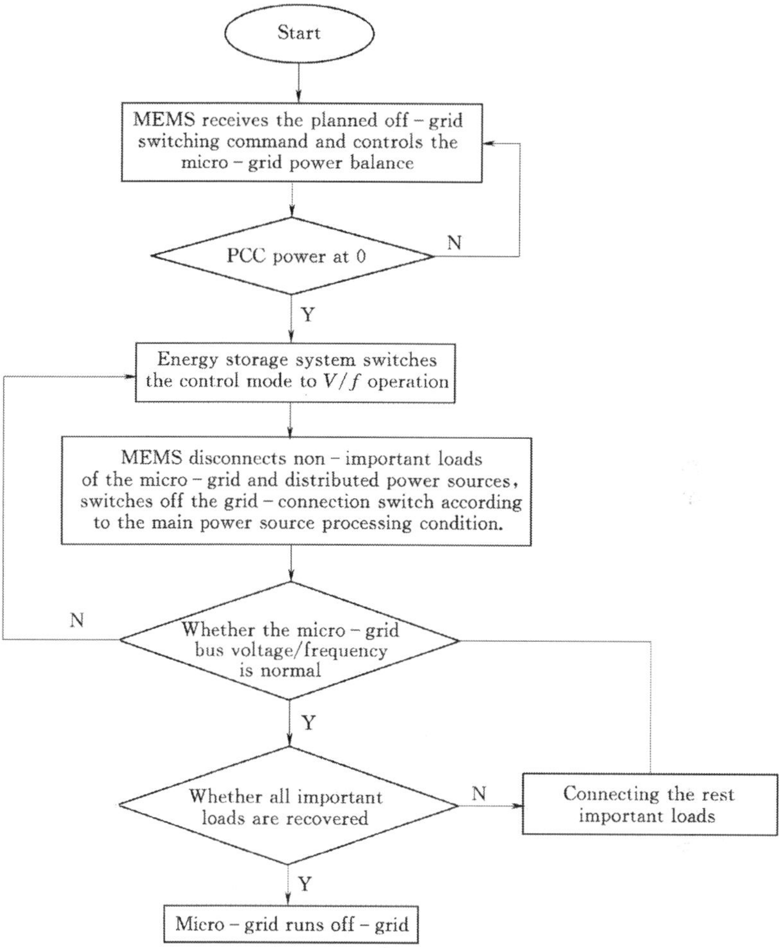

FIGURE 7.19 Planned off-grid switching control process.

output is lowered in the order of distributed power source output, energy storage devices charge, and disconnect part of the distributed power sources to make it meet the off-grid operating conditions. If the microgrid absorbs the main grid's power, the former's output is adjusted in the order of increasing distributed power source output, energy storage device discharge, and switching loads to increase it to zero. When the connection power line is adjusted to zero, the MEMS controls the main power source to switch the control mode and operates in the *V/f* control mode. At the same time, the microgrid's coordination controller further adjusts the equipment output to keep power of the connection line at zero. After the successful switching of the main power source, the MEMS commands the grid connection switch to switch off and run off-grid.

2. Off-grid switching in case of grid failure

The MEMS cannot predict the time that failures occur and cannot timely adjust the power of connection line or command the main power source to activate the control mode. It needs to formulate the off-grid switching control emergency plan according to its system requirements at certain intervals and make a hard link with the microgrid on-grid power protection device. Once the grid connection point protection device detects any grid failure, it will trigger the MEMS to switch from on-grid to off-grid, and transfer to the preset off-grid switching control plan. The protection device output triggers the signal to the microgrid's main power source to control mode switching. The microgrid coordination controller rapidly executes the command according to the preset on-/off-grid control plan. When the grid connection switch is turned on, the microgrid realizes seamless switching. During the process, the distributed power sources are still grid-connected and controllable, and switchable loads are disconnected. The main power source control mode is adjusted, and the MEMS coordinates energy storage devices according to its own output to meet the power supply of important loads. After the grid connection switch is turned off, the system enters the off-grid operation. See Fig. 7.20 for the switching process.

7.4.3.2 Synchronous grid connection technology

The microgrid's connection to the distribution network is a beat frequency connection, and the voltage phasor on both sides of the grid connection coordination point and the public grid has not only amplitude differences, but also frequency and changing phase angle differences. Therefore direct grid connection will produce a relative high impulse current. In the operating mode the microgrid's connection to the distribution network should meet the synchronous grid connection conditions. The specific determination conditions include the voltage difference, frequency difference, and phase angle difference. The microgrid's beat frequency grid connection is made by predicting the timing when the phase angle difference is zero and the voltage and frequency on both sides of the synchronization point are close.

The basic principles that the microgrid should follow when being connected to the public grid:

1. When the breaker is switched on, the impulse current to the microgrid should be small enough.
2. When the microgrid is connected to the public grid, it can rapidly enter the synchronous operation state.

The basic conditions that the microgrid voltage and the public grid voltage must meet as per aforementioned principles are that the voltage amplitude, frequency, and phase difference must be small enough. See Fig. 7.21 for the analysis of quasisynchronization parallel voltage vectors. The voltage at the

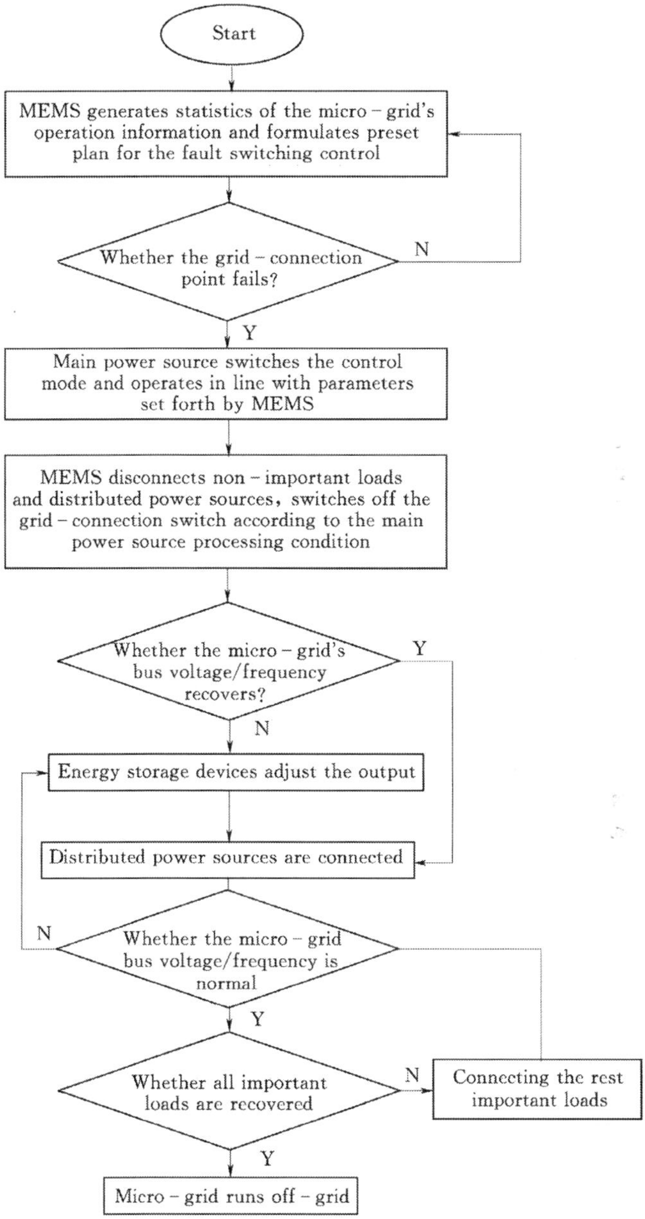

FIGURE 7.20 Off-grid switching control process in case of grid failure.

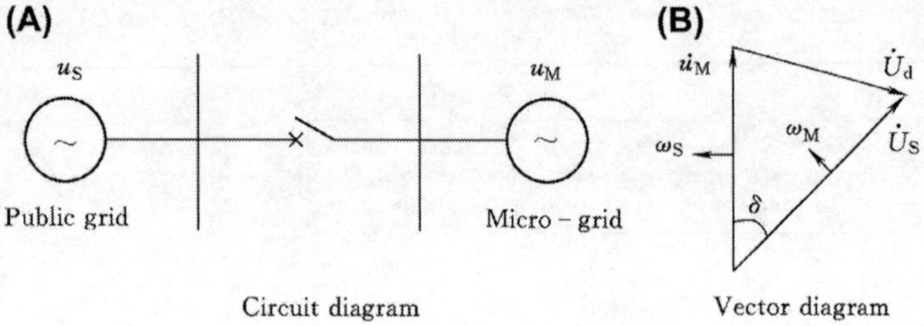

FIGURE 7.21 Vector diagram of synchronous grid connection conditions.

microgrid's side and the public grid side on both sides of the breaker before grid connection is as follows:

$$u_M = U_M \sin(\omega_M t + \delta_M) \tag{7.27}$$

$$u_S = U_S \sin(\omega_S t + \delta_S) \tag{7.28}$$

The voltage difference u_d between the microgrid and public grid is this:

$$u_d = u_M - u_s \tag{7.29}$$

Ideal conditions for circuit breaker closing are as follows: (1) The voltage amplitude of both the microgrid and the public grid is the same, i.e., $U_M = U_S$. (2) Voltage angle frequency is the same, i.e., $\omega_M = \omega_S$. (3) The phase angle difference at the moment of the switching on is zero, i.e., $\delta = 0°$.

If these three conditions are met concurrently, it means the voltage phasors on both sides of the breaker coincide and have no relative movement, and at the moment the voltage difference $u_d = 0$, impulse current is zero, the microgrid and the public grid immediately synchronize and have no disturbance. This is the ideal state of grid connection. In practice, if $\omega_M = \omega_S$, voltage on both sides is relative static; if there is phase angle difference, it is impossible to achieve $\delta = 0°$. Therefore, the equal angle frequency is expressed as close angle frequency.

The following paragraphs will analyze the effect of synchronous grid connection conditions.

1. Effect of voltage difference

Suppose the voltage of both parties share the same frequency and initial phase angle, and the initial phase angle on both sides is zero, but different in amplitude, then the voltage difference u_d is this:

$$u_d = (U_M - U_S)\sin \omega_S t = u_d \sin \omega_S t \tag{7.30}$$

From this equation, we can see the voltage difference is a sine wave that fluctuates at the system frequency. The peak impulse current passing the system at the moment of switching on $I_m = \frac{U_d}{X_c}$, X_c is the on-grid system impedance. The impulse current caused by the amplitude difference is the reactive current, and the electrodynamic force it generates has a relatively high effect on the synchronous generator winding.

2. Effect of frequency difference

Only when the voltage on both sides of the breaker is different, then the expression of the voltage difference u_d is this:

$$u_d = 2U_S \sin\left(\frac{\omega_d}{2}t\right)\cos\left(\frac{\omega_S + \omega_M}{2}t\right) \tag{7.31}$$

Where $\omega_d = \omega_M - \omega_S$ is known as the slip angle frequency. $f_d = f_M - f_S$ is known as slip frequency, referred to as slip.

The voltage difference amplitude envelope is called the ripple voltage, written as u_e. It is a sine wave with frequency at u_d and amplitude at $2U_S$.

$$U_e = 2U_S \sin\left(\frac{\omega_d}{2}t\right) \tag{7.32}$$

Then the phase angle difference between the two voltage vectors can be expressed as $\delta = \omega_d t$。

3. Effect of phase angle difference

If the microgrid voltage forward zero crossing point (ZCP) is the starting point, the sequential public grid voltage forward ZCP is the termination point, and the time interval between the two ZCPs is an impulse width and represents the phase angle difference between the generator voltage and system voltage, then the phase angle difference between the voltage on the two sides can be expressed with an impulse sequence that changes its width.

The change rule of phase angle difference is the same as the amplitude change rule of the ripple voltage u_e. The highest phase angle difference lies in the same place as the highest ripple voltage amplitude, and the smallest phase angle difference lies in the same place as the smallest ripple voltage amplitude. This shows that ripple voltage u_e contains all information needed for inspection for the quasisynchronization paralleling: rules of voltage amplitude difference, frequency difference, and phase angle difference to change with time.

From the analysis, we can see during the microgrid's quasiperiodicity, the frequency, voltage, and phase angle difference are factors directly affecting the grid connection. Of the three conditions for the quasisynchronization grid connection, only the phase angle difference can really affect the quality of grid connection.

Though the grid connection causes reactive power and active power impact when there are voltage and frequency differences between the two power sources, that is at the moment when the breaker is switched on, and the side with higher voltage will transmit a certain amount of reactive frequency to the lower side, and the side with high frequency will transmit a certain amount of active power. However, the consequence of grid connection is very serious when there is phase angle difference. Especially for the electrical devices in the microgrid, the grid connection phase angle difference will produce a relatively high impulse current, leading to power device damage and fast-acting fuses.

The phase angle difference between the microgrid and the public grid at the moment of ideal connection state is zero, i.e., $\delta = 0^\circ$. But in practice, with the measurement error and switching delay dispersion of the operating mechanism and breaker, normally, there is a certain phase angle difference at the moment of grid connection, and it produces some impact on the grid connection system. Referring to the synchronous generator grid connection conditions, the microgrid quasisynchronization grid connection conditions can be adjusted to the following: (1) The frequency difference between microgrid and public grid should not exceed 0.5 Hz. (2) The voltage difference between the microgrid and public grid should be less than 10%. (3) The moment the grid connection is switched on, the microgrid and public grid phase angle difference should not exceed 5 degrees.

The grid connection meeting the aforementioned conditions has a small impact on the microgrid and distribution network and can quickly enter the synchronous operating state.

The following procedure can be used to realize synchronization: the protection device at the grid connection point monitors the bus voltage and frequency in real time to determine when the grid power supply recovers, then it sends out the reconnection request to the microgrid coordination controller, which determines whether it should connect to the grid according to real-time system operation, and it applies to the main grid dispatcher, which gives out the order to the protection device upon receiving permission, and the microgrid coordination controller executes a seamless switching preset plan; the grid connection switch starts the synchronization inspection procedure; when it detects proper grid connection conditions, it automatically switches on; when the grid connection goes through, the microgrid's main power source switches the control mode, restarts the P/Q on-grid operation, annd eventually the microgrid's grid connection succeeds. See Fig. 7.22 for the specific control process.

7.5 Typical application cases

7.5.1 Chengde distributed generation/microgrid demonstration project

7.5.1.1 System overview

The demonstration project is located in Team 1, Yudaokou Village, Yudaokou Township, Weichang County, Chengde, a place identified by the township government for an "agritainment" community. It is at the tip of the 10-kV power supply line, and it features a single power supply, a long power supply radius, low-grade wire, low liability of local power supply, and serious low voltage. According to the new round of rural power grid upgrading,

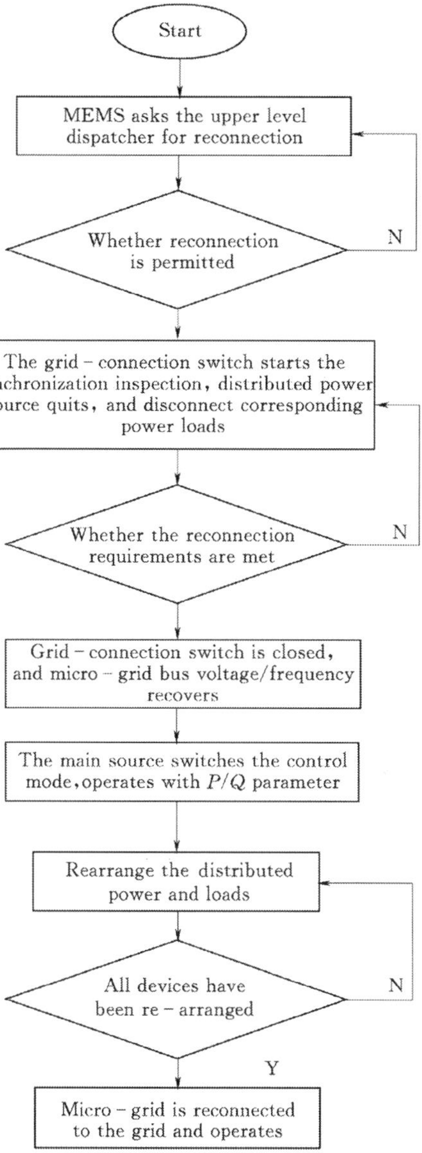

FIGURE 7.22 Synchronous grid connection control process.

the demonstration project can improve the power supply reliability and the low voltage issue of the pilot.

Loads of the pilots include the existing loads and future loads. Team 1 of Yudaokou Village consumed a total of 44,389 kWh in 2010. This project supplies 59 consumers, including 51 residential households and eight nonresidential households.

7.5.1.2 System structure

This project has combined the village mode and single-household mode, that is, using the village mode as the main mode while choosing two to five families to pilot the single-household mode.

1. System composition of the village mode

The microgrid in the village mode has an 80-kW PV generation subsystem, a 120-kW wind power subsystem, and an 80-kW/128-kWh energy storage subsystem as the whole village's power source, and the PV generation and wind power are connected to the 400-V low-voltage bus of the switching room microgrid, together with the energy storage subsystem, active filter, and reactive load compensation devices through the independent 400-V collector line. It supplies the village loads on one hand, and it connects to the public distribution network through the 400-V/10-kV distribution transformer on the other hand. It runs under the unified management of the MEMS.

The microgrid system in the village mode mainly includes a PV generation subsystem, wind power subsystem, low-voltage collector line, energy storage subsystem, energy calculation subsystem, energy quality monitoring and governance system, monitoring and energy management subsystem, civil engineering, distribution room, and other infrastructure.

In structure, the system adopts the distributed generation and unified grid connection mode. The PV generation subsystem, wind power subsystem, and energy storage subsystem are directly connected to the microgrid's 400-V low-voltage bus through the switches, the active filter and reactive compensation devices are directly connected to the 400-V low-voltage bus through the access switches. In this way, the connected subsystems are relatively independent, and the connection method is conducive to independent maintenance and system upgrade and renovation. See Fig. 7.23 for the general structure.

2. System composition and structure of single-household mode

The single-household mode includes a PV generation unit, wind power unit, battery energy storage unit, AC inverter unit, dual power source switching unit, and microgrid coordination control unit, among others.

The construction goals of the single-household mode are to implement the new national energy development strategy to vigorously develop low-carbon economy and adapt to the future need of large-scale application of distributed generation in rural areas, and the single-household microgrid construction mode study is implemented to make technical preparation for future large-scale application.

The microgrid in the single-household mode adopts a complementary system of wind power/PV generation/energy storage system to power loads of a single household. To study the proportion of these three different power sources, three different proportion modes are adopted according to load of

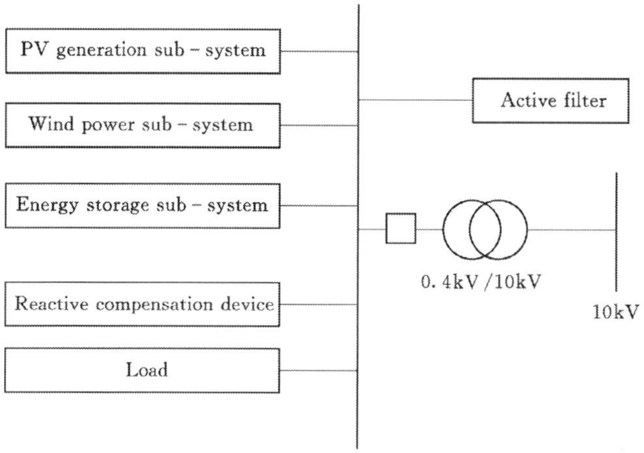

FIGURE 7.23 General structure of the microgrid in the village mode.

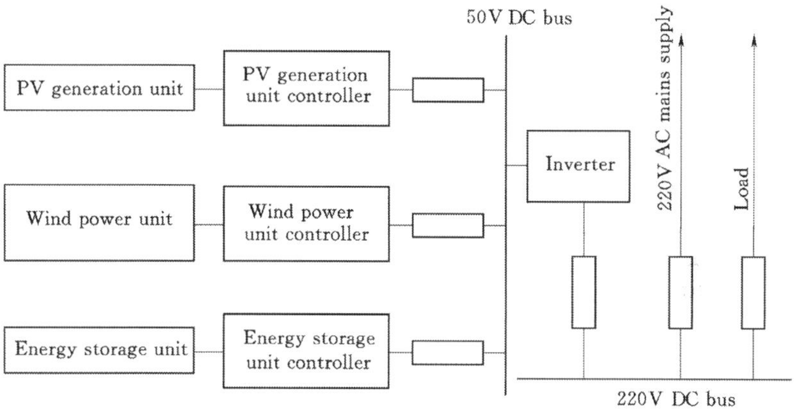

FIGURE 7.24 Structure of microgrid system in the single-household mode.

the users. The results have practical guiding significance for the future wind, PV, and energy storage proportion of a single household.

The microgrid in the single-household mode consists of a PV generation unit, wind power unit, battery energy storage unit, control unit, and AC inverter unit. See Fig. 7.24 for the main structure.

7.5.1.3 Operation summary

As this area enjoys annual sunshine time of 2899 h, 5000 h of effective average wind speed at 6.8 m/s, and given 80 kW rated power of PV generation, the annual power generation reaches 110 MWh; six 20-kW wind turbines generate 120 MWh, and the PV/wind power system can annually generate 230 MWh.

1. Analysis on economic benefits of renewable energy power generation

The connected wind power and PV generation can sufficiently supply power to local users and save conventional primary energy input while improving the electricity consumption structure of local users. The input and output are, however, not in proportion. But with technology advancement, the costs of wind and PV power generation equipment will decrease considerably. If there is no construction of microgrids but the main grid is used to meet existing costs issues, the input and output will quickly be balanced.

2. Improving power supply reliability for rural power grid users

The primary grid structure of initial power supply was not reliable, nor was the equipment configuration. Through upgrading of the grid's structure and primary equipment, a structure with "multiple sections and multiple power sources" has been formed, and branch lines have been equipped with automatic circuit breakers and protection functions. In case of failure of peripheral circuits, the outlet switch will not directly trip, sharply narrowing the outage scope. Meanwhile, the renovation of multisection and interconnection switches increased the line's ability to transfer load, thus bringing great benefits to the power supply reliability of the rural power grid.

3. Analysis of demonstration effect of smart grid technology in rural power grid

Following the project implementation, the North China Branch of State Grid has accumulated vast technological experiences in the smart grid field, especially the connection of distributed generation and operation of a microgrid to lay a technological foundation for comprehensive implementation of a smart grid.

7.5.2 Microgrid demonstration project in agricultural and pasturing area without electricity in Qinghai

7.5.2.1 Technical characteristics

The demonstration project contributes invaluable information for the implementation of a reliable power supply project in a remote area of Qinghai without electricity, and it provides technical support and experience accumulation for solving an electricity access issue to the population in the 12th Five-Year Plan period. It can be scaled up in Qinghai and the whole country as well. The wind/PV/energy storage complementary microgrid system has the following technical characteristics:

1. The wind/PV/energy storage complementary microgrid system consists of a PV power generation subsystem, wind power generation subsystem, energy storage subsystem, electric power measurement subsystem, power quality monitoring subsystem, low-voltage power distribution subsystem,

meteorological monitoring subsystem, coordination controller, and other subsystems, and these subsystems ensure flexible and stable operation of power plants through the high-speed, two-way communication system. The wind/PV/energy storage complementary microgrid system is characterized by unitization, modularity, and standardization.

2. The wind/PV/energy storage complementary microgrid system runs off-grid at the low-voltage of a 400-V environment, independently manages power sources and loads in the township area, achieves balance between loads and power supply of the microgrid, and maintains the system voltage and frequency stability. The system has characteristics of low voltage.
3. The wind/PV/energy storage complementary microgrid system can grasp operating information of the PV power station in real time through the remote monitoring system to improve the equipment operating efficiency, timely discovery of PV power station equipment failure, and make remote diagnosis and rapid maintenance to improve efficiency. This system runs unsupervised.
4. Equipped with an energy storage subsystem and coordination control subsystem, the system stores the surplus PV and wind power generated in the daytime for power supply to loads at night and meets the 24-h electricity demand of residents, and it realizes staggered balance and consumption of PV and wind power and resident electric power consumption and effective utilization of renewable energy. The system is conducive to optimal and reasonable configuration of resources.

7.5.2.2 System structure

1. Master–slave control structure of wind/PV/energy complementary microgrid at agricultural and pasturing areas without power supply

See Fig. 7.25 for the typical master–slave control structure of the complementary microgrid system.

In the case of microgrid off-grid operation, the energy storage unit adopts the *V/f* control mode to provide voltage and frequency reference for other power sources. It serves as the master controller of the microgrid, and the other power sources adopt the *P/Q* control and serve as slave controller. Its power dynamic fluctuation is absorbed by the master power source. When the power generation is higher than the load's power, the master energy storage unit charges; when the power generation is lower than the load's power, the master energy storage unit discharges.

Without any big grid in the agricultural and pasturing area, the microgrid runs off-grid and on an energy storage system, and the master power source adopts the *V/f* control mode. If the big grid extends to the agricultural and pasturing area without a power supply, the microgrid runs on-grid. In such a case the master energy storage system needs to change from *V/f* control to

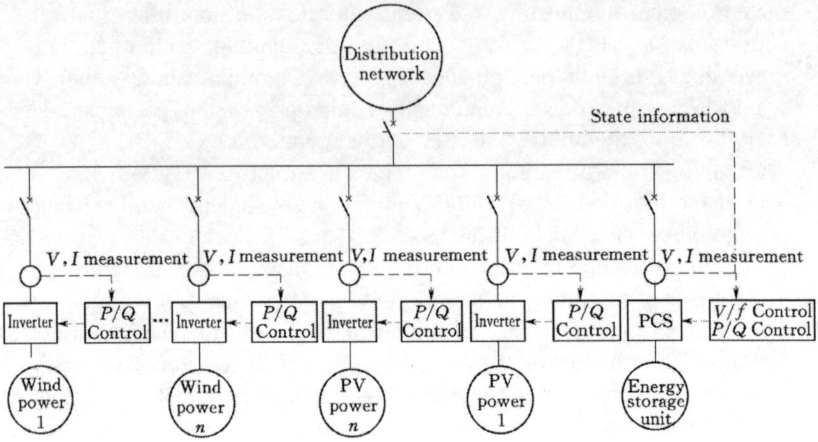

FIGURE 7.25 Master–slave structured microgrid in the agricultural and pasturing area without power supply.

P/Q control. Therefore PCS of the master energy storage needs the operation capability of on/off-grid dual mode for future development.

2. Two-level control system

Microgrids that use the master–slave control structure run steadily at constant voltage and constant frequency controlled by the energy storage system that serves as the master power source. There is no communication between the master energy storage system and the other power sources. The energy storage unit and the power sources run independently from each other. Consequently, there is a question about the power source *P/Q* value during the microgrid's off-grid operation.

Suppose the power sources run at the initial maximum power: when the master energy storage unit charges uninterruptedly to make the battery SOC approach the maximum limit, it is necessary to lower the set power source value of active power. When the main energy storage unit discharges uninterruptedly, which leads the battery SOC to approach the minimum value, it is necessary to increase the set power source value of active power. Therefore the microgrid needs an upper-level control system that observes the operating state and adjusts the power source and loads to keep the microgrid running steadily for a long time.

See Fig. 7.26 for the two-level microgrid control system. The coordination control system corresponds to the upper level control system, inverter, or PCS local control system corresponding to the lower level control system. The coordination control system receives equipment state information from the switch, then it forms and sends command to the equipment and switches after the strategic analysis.

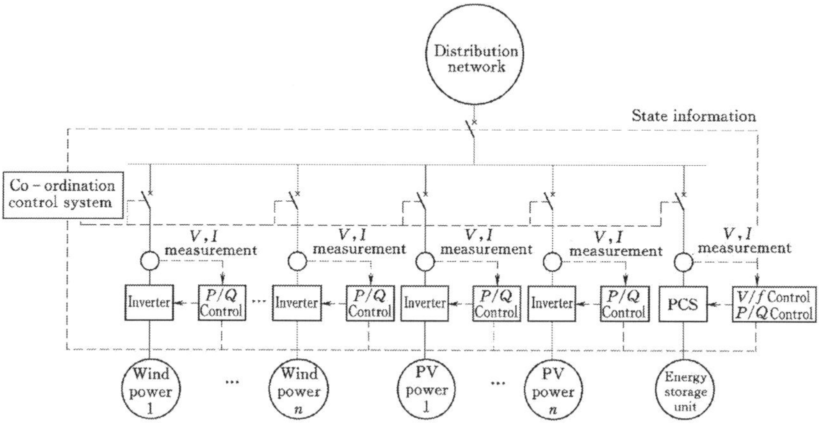

FIGURE 7.26 Structure of the two-level control system.

7.5.2.3 Operation analysis

1. Working characteristics of storage battery for wind/PV/energy storage complementary microgrid

PV power is characterized by high fluctuation, strong randomness, and unpredictability, and its output changes at any time. Equipping the PV power system with an energy storage battery improves the power supply quality and enables users to get uninterrupted and stable power supply. In Qinghai's off-grid solar energy storage system, the battery's operation characteristics are summarized as follows:

1. It charges in the daytime and discharges at night and during cloudy and rainy days, and it belongs to the cycling + floating charge hybrid working mode.
2. It has a low charge rate, normally 0.01–0.02C, rarely reaching 0.1–0.2C.
3. The one-time charging period is relatively short (the longest only 10 h during the daytime); consequently, the battery is often undercharged.
4. It has a low discharge rate, normally 0.004–0.05°C, but during long discharge time, as long as 7 days without sunshine, it tends to overdischarge.

Similar to the PV power generation system, the wind power system features strong randomness and intermittency. And the theoretic analysis and practical operation show that the wind power grid connection has some effect on the system voltage, frequency, and stability. Therefore the large-scale wind power grid connection should have proper configuration of energy storage devices to solve the large-scale energy storage and conversion.

The energy storage system used in the wind farm should have the following characteristics:

1. With a long discharge time, due to diurnal difference, there lacks continuous charging or discharging for more than 10 h. Therefore the battery is rarely overdischarged or undercharged.
2. The instantaneous wind power is high, and the battery charge rate is high accordingly.
3. The wind power features high power fluctuation frequency, showing obvious changes in the demand requirements of the energy storage system, and frequent charge and discharge transitions.

Moreover, in Qinghai and other areas with rich wind and solar resources, the geographic conditions are special. For example, there is a wide range of temperature changes, varying from 45°C to −30°C; and the altitude can be as high as 5000 m. The prevailing conditions demand the energy storage system to meet simple installation requirements, high system reliability, and be maintenance free, among others.

2. Lead−acid battery suitable for the wind/PV energy storage system

According to the energy storage operating conditions of the wind/PV power generation system, the lead battery should have the following strengths: (1) a long cycle life under the high discharge depth; (2) strong resistance to overcharge; (3) the long-time undercharge has little impact on the cycle life; (4) excellent small current discharge performance; (5) a wind operating temperature range, reducing the phenomena of shortened battery service life caused by temperature changes. The battery service life and advance failure are important elements hindering application of a lead−acid battery in the PV/wind power system. Considering the characteristics and working conditions of a lead−acid battery, a suitable battery should be chosen together with effective operation management to improve its viability.

Currently, the barren solution AGM technology has been used in the small PV power generation system. Due to its inability to meet the harsh climate of the plateau, the service life is not ideal. In recent years, the battery manufacturers have started research and development of a battery for high-altitude use. This kind of battery has a long cycle life, excellent deep discharge performance, a wide range of applicable temperature (−40−65°C), low self-discharge, and free of maintenance, meeting Qinghai's requirements of high altitude and low temperature.

Compared with the barren solution battery, the gel cell battery has better deep discharge recovery capability, small current discharge characteristics, and 20%−30% higher charge efficiency and longer service life in the extreme ambient temperature. Because of electrolyte characteristics, the gel cell battery can almost eliminate the stratification of the internal electrolyte and reduce the possibility of sulfation. The gel cell battery is divided into tubular and flat panel structure. The positive plate of the tubular structure has relatively high constraint forces on the active substance and can delay shedding of active substance while performing better than the flat panel structure. Presently, the tubular gel cell battery has not been promoted on a large scale in China because of its dependence on import for materials, and this structure complicates the manufacturing process, lowers productivity, and doubles the investment costs of the barren solution battery. However, considering the gel cell battery's design life is two times that of the barren solution batteries, the gel cell battery is comparable to the barren solution batteries judging from the economic performance over the entire service life. Therefore after future mass production and application, the tubular gel cell battery will become competitive in the PV/wind power system.

7.5.3 Dongfushan Island wind/PV/energy storage/diesel and seawater desalination integrated system

7.5.3.1 System overview

With continuous improvement of economic and living quality, the remote islands have increased their demand for reliable electricity supply. These islands have normally adopted autonomous systems, which mostly use diesel generators, and only a few are mixed with low-permeability wind power, PV power, and other renewable energy sources. It will be technologically and economically costly to connect with the main grid to improve supply reliability and quality considering the rising power generation costs caused by transportation difficulties and impacts of noises and emissions of diesel power generation on the island's vulnerable ecologic resources. It is therefore not desirable to continue with the diesel-based mode.

The islands normally have rich wind and solar resources. Reasonable use of such resources can significantly improve the power supply of the island's grid, reduce diesel use, and improve conservation of the ecologic environment. However, given the intermittency and fluctuation of wind and solar resources, and operating constraints of diesel generators, the wind/PV/diesel generation system normally makes the diesel generation the main source, supplemented by wind and PV power generation to maintain stability. To further improve penetration rate of renewable energy, the energy storage system can be adopted to effectively level intermittency and improve their utilization efficiency, which will sharply reduce use of diesel and effectively improve the system stability.

Dongfushan Island is an inhabited easternmost sea frontier of China, adjacent to the open sea on the east, and it has an area of 2.95 km^2. The island hosts a natural village and about 300 residents living on fishing and tourism. The island is far from the mainland and the garrison's diesel power generation costs are very high. Before implementation of the project, residents accessed power through the garrison's diesel generators, and their water supply is dependent on the existing reservoir to collect and purify rainwater and shipping water from Zhoushan Island.

To solve power and water supply challenges, the wind/PV/energy storage/diesel and seawater desalination integrated system of Dongfushan Island is a project designed to reasonably exploit the renewable energy resources of the island, with the total installed capacity of 510 kW and renewable energy installed capacity of 310 kW, which includes seven 30-kW wind turbines, a 100-kW PV power generation system, and a 200-kW diesel generator, a 2000-Ah energy storage battery, and a seawater desalination system that can process 50 t of seawater daily. Presently, the maximum load is about 120 kW and is predicted to reach about 200 kW in the future. The system is designed according to future maximum load. The seawater desalination system features an adjustable load for effectively increasing utilization rate of renewable energy while solving the water supply problem of the island.

7.5.3.2 System structure

See Fig. 7.27 for the system structure of the Dongfushan Island Microgrid. PCS in the diagram stands for the PV/energy storage integrated converter.

For an independent microgrid, the operating strategy is one of the deciding elements for optimal configuration and operating parameter settings of the energy storage system and critical for operation stability. Because either the diesel generator or energy storage system can serve as the main power source at Dongfushan Island, only one can be allowed to work as the main power source at a given time to prevent asynchronous parallel operation of the two main power sources and adopt *V/f* control to provide stable voltage and frequency support for the system. The characteristics of the energy storage

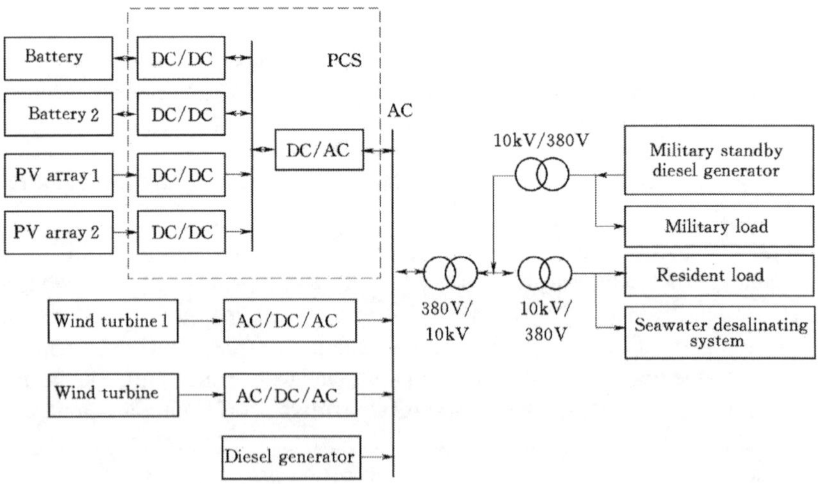

FIGURE 7.27 Microgrid system structure schematic diagram on Dongfushan Island.

system shall be taken into account when designing the operating strategy, and regular maintenance should be made to improve its cycle life.

Dongfushan Island microgrid mainly has the following three operating modes:

1. Diesel generator as the main power source

The diesel generator works as the main power source and produces constant voltage and frequency to support the system operation; the energy storage system works at a constant current charging or constant voltage charging state till the SOC or DC end voltage reaches the ceiling limit. Considering the limits such as minimum operating power and rated power of the diesel generator, the method of controlling PV power and wind turbines can be used to keep the diesel generator operating within a reasonable range. When the SOC reaches its limit, the system will transfer to energy storage system mode as the main power source.

2. The energy storage system as the main power source

When the energy storage system operates as the main power source, the diesel generator should be shut down to avoid asynchronous parallel operation, and the energy storage converter adopts *V/f* control to provide the voltage and frequency support for the AC bus. Its power output automatically makes up for the difference between wind power and loads. The optimized control of the energy storage system based on SOC or DC end voltage demands strong wind and solar resources to enable SOC or DC end voltage to reach the ceiling limit. The method of controlling PV power and the wind turbines ensures that the energy storage system works at the discharge state only and is within the preset range. When the SOC or DC end voltage drops to the minimum limit, the method of controlling PV power and wind turbines ensures that the energy storage system works at the charging state only and within the preset range. In normal operation, the optimized control strategy should keep the energy storage system working at the fixed state to avoid frequent switching between charge and discharge. When the SOC or DC end voltage is lower than the critical conversion value, the system should make the diesel generator serve as the main power source and change the PCS control strategy to charge the batteries, meaning the mode where the diesel generator is the main power source.

3. Full charge and discharge mode of the energy storage system

To reach the maximum service life of a battery, manual maintenance is required on a regular basis. Set the full charge and discharge mode at the control strategy, manually start the diesel generator to complete the four stage-standard-charge process "precharge—fast charge—equalized charge—floating charge" till charge is completed, and then control the battery to discharge fully.

Fig. 7.28 shows the system operating mode diagram of Dongfushan Island microgrid. During normal operation, the system mainly realizes operating mode switching manually or automatically. When the diesel generator or

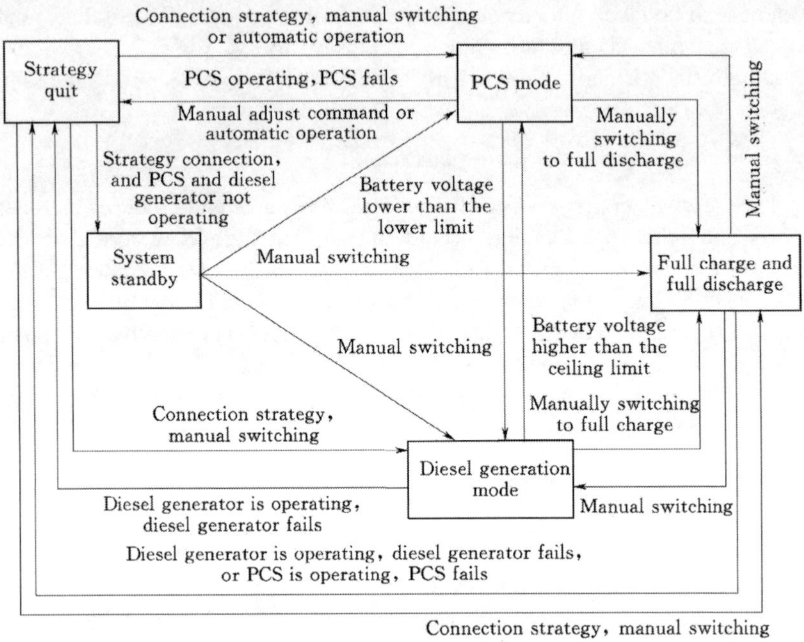

FIGURE 7.28 Microgrid system operating mode diagram on Dongfushan Island.

PCS fails, the corresponding strategy quit program and reconnection strategy should be offered according to the actual situation. When the system stops, it should first enter the standby state and then switch to the target operating mode manually or automatically.

7.5.3.3 Summary of operation

Dongfushan Island System was officially handed over to the owner in July 2011, and all equipment was operational and the energy storage system reached optimal performance as designed. According to the system design, the renewable energy electric power penetration rate is predicted to reach about 55% annually. The measured data analysis shows that from October to March, Dongfushan Island enjoys excellent wind power resources, and the diesel consumption will be reduced by 60% and the diesel generator's operating time by 70%; from April to June, the wind power resource is moderate, and the generator's diesel consumption and operating time is predicted to reduce by 40% and 50%, respectively; from July to September, the wind power resource is poor, and the generator diesel consumption and operating time is predicted to reduce by 35% and 40%, respectively.

References

[1] Huang W, Zhang J. Micro-grid operation control and protection technology. Beijing: China Electric Power Press; 2010.

[2] Nian H, Zeng R. Electric power quality of distributed generation in off- grid operating mode. Proceedings of the CESS 2011;12(31):22–8.

[3] Zhang B, Zeng J, Mao C, et al. Application of battery energy storage system in improving on-grid wind farm energy quality and stability. Power System Technology 2006;15(30):54–8.

[4] Zhang G, Tang X, et al. Application of super capacitor and battery hybrid energy storage system in micro-grid. Automation of Electric Power Systems 2010;12(34):85–9.

[5] Zhang Y, Guo L, Jia H, et al. An energy management method of hybrid energy storage system based on smoothing control. Automation of Electric Power Systems 2012;36(16):36–42.

[6] Sang B, Tao Y, Gao Z, et al. A study on topology and control strategy of hybrid energy storage system of super capacitor and battery. Power System Protection and Control 2014;42(2):1–6.

[7] Qiu P, Ge B, Bi D. A study on battery energy storage-based power stabilizing control for grid-connected PV power generation system. Power System Protection and Control 2011;39(3):29–33.

[8] Zhao D, Zhang N, et al. Synthetical control strategy for smooth switching between grid-connected and islanded operation modes of micro-grid based on energy storage system. Power System Technology 2013;02(37):301–6.

[9] Zhang T. An analysis of grid-connection control strategy and stability of micro-grid. Wuhan: Huazhong University of Science and Technology; 2008.

[10] Zhao B, Zhang X, et al. Optimal design and application of energy storage system in independent micro-grid of Dongfushan island. Automation of Electric Power Systems 2013;01(37):161–6.

Index

Note: 'Page numbers followed by "f" indicate figures, "t" indicate tables'.

C

F

I

N

O

T

U

V

Made in the USA
Monee, IL
13 September 2026

62977006R00177